Groundwater Residue Sampling Design

ACS SYMPOSIUM SERIES **465**

Groundwater Residue Sampling Design

Ralph G. Nash, EDITOR

EPL Bio-Analytical Services, Inc.

Anne R. Leslie, EDITOR

U.S. Environmental Protection Agency

Developed from a symposium sponsored
by the Divisions of Agrochemicals and
of Environmental Chemistry
at the 199th National Meeting
of the American Chemical Society,
Boston, Massachusetts,
April 22–27, 1990

American Chemical Society, Washington, DC 1991

Library of Congress Cataloging-in-Publication Data

Groundwater residue sampling design /   Ralph G. Nash, editor,
Anne R. Leslie, editor.

   p.   cm.—(ACS Symposium Series, ISSN 0097–6156; 465)

"Developed from a symposium sponsored by the Divisions of
Agrochemicals and of Environmental Chemistry at the National Meeting
of the American Chemical Society, Boston, Massachusetts, April 22–27,
1990."

Includes bibliographical references and indexes.

ISBN 0–8412–2091–3

1. Agricultural chemicals—Environmental aspects—Measurement—
Congresses.   2. Water,   Underground—Pollution—Measurement—
Congresses. 3. Soil pollution—Measurement—Congresses.

   I. Nash, Ralph G., 1930–   . II. Leslie, Anne R., 1931–   .
III. American   Chemical   Society.   Division   of   Agrochemicals.
IV. American Chemical Society. Division of Environmental Chemistry.
V. American Chemical Society. Meeting (199th : 1990 : Boston, Mass.)
VI. Series.

TD427.A35G78   1991
628.1′684—dc20   91–15752
                                   CIP

The paper used in this publication meets the minimum requirements of American National
Standard for Information Sciences—Permanence of Paper for Printed Library Materials, ANSI
Z39.48–1984.   ∞

PRINTED IN THE UNITED STATES OF AMERICA

ACS Symposium Series

M. Joan Comstock, *Series Editor*

1991 ACS Books Advisory Board

Foreword

THE ACS SYMPOSIUM SERIES was founded in 1974 to provide a medium for publishing symposia quickly in book form. The format of the Series parallels that of the continuing ADVANCES IN CHEMISTRY SERIES except that, in order to save time, the papers are not typeset, but are reproduced as they are submitted by the authors in camera-ready form. Papers are reviewed under the supervision of the editors with the assistance of the Advisory Board and are selected to maintain the integrity of the symposia. Both reviews and reports of research are acceptable, because symposia may embrace both types of presentation. However, verbatim reproductions of previously published papers are not accepted.

Contents

Preface

THE USE OF AGRICULTURAL PESTICIDES, once thought to be harmless to groundwater, has become a cause for concern over the past decade. The reason for the change is the discovery that under the right conditions, pesticides, like fertilizer nitrogen, can move through soil into groundwater, a phenomenon once thought improbable. Movement of agrochemicals in surface water flow was a recognized fact, but was not believed to be a major threat to drinking water, as is groundwater contamination.

The data collected over the past few years by the U.S. Environmental Protection Agency indicate that agrochemicals can contaminate groundwater in vulnerable areas.

The organizers of the symposium from which this book is derived shared the belief that an interdisciplinary discussion of groundwater research that focused on sampling methods and design would contribute to the planning, resource allocation, and implementation of future studies. The authors represent a number of disciplines that have been employed to evaluate groundwater and surface water quality, and their work should be a valuable contribution to the literature, offering insights to experienced researchers and those scientists just beginning groundwater studies.

The variety of investigative systems included here emphasizes the need for a multidisciplinary, team approach. Comprehensive coverage of **all** aspects of planning, designing, and conducting research on groundwater is beyond the scope of this book. Rather, the chosen focus is amply covered, and the case studies presented will serve as useful examples.

Because groundwater research is young, multidisciplinary, and site specific, few methodologies are available. The present state of the science is more a strategy than a method list. As stated in Chapter 1, "The initial design [should] be a pilot program that can be expanded as knowledge is gained from the study." Nevertheless, for given situations certain methods may be adaptable to other sites. Examples of proven methods include: the aseptic sampling in unconsolidated heaving soils described in Chapter 20; the techniques used by the Rhone Poulenc Ag Company for collecting data necessary for registration of pesticides (Chapters 9, 12, and 20); the U.S. Geological Survey Delmarva Peninsula designs for regional groundwater monitoring (Chapters 6 and 7); the U.S. Geological Survey site specific monitoring (Chapter 14); and the minimum cost sample allocation (Chapter 5).

Despite the increased allocation of research dollars toward groundwater research, wise husbandry of these resources is demanded to ensure the attainment of Congress's goal of protecting and improving our water supply.

We would like to thank all the authors, the two ACS divisions that sponsored the symposium, the numerous reviewers, our employers, and ACS Books for making this publication possible.

RALPH G. NASH
EPL Bio-Analytical Services, Inc.
Decatur, IL 62525

ANNE R. LESLIE
U.S. Environmental Protection Agency
Washington, DC 20460

Chapter 1

Groundwater Residue Sampling

Overview of the Approach Taken by Government Agencies

Ralph G. Nash[1], Charles S. Helling[2], Stephen E. Ragone[3], and Anne R. Leslie[4]

[1]EPL Bio-Analytical Services, Inc., Box 1708, Decatur, IL 62525
[2]Agricultural Research Service, U.S. Department of Agriculture, Beltsville, MD 20705
[3]Office of the Assistant Director for Research, U.S. Geological Survey, 104 National Center, Reston, VA 22092
[4]Office of Pesticide Programs, U.S. Environmental Protection Agency, Mail Stop H7506C, Washington, DC 20460

Recognition that nitrogen applied as fertilizer may reach groundwater has been known for two to three decades. It is only in the past decade that evidence has become available suggesting pesticides may leach to groundwater, also. The evidence, though mostly anecdotal, has raised the nation's awareness of the potential for contamination of our water resources, the need to ascertain the extent of the problem, and ways to prevent it. Because of the complexity of natural systems, an interdisciplinary study approach is needed to provide information for cost-effective solutions to the problem.

The Problem

Water resources are an aggregation of numerous dynamic, individual and interactive ground- and surface-water systems. These multiphase and multicomponent systems have their own hydrogeologic and mineralogic characteristics and are found in a variety of climatic settings. Therefore, the time it takes contaminants to reach and move throughout the system can vary from days to centuries. The pathways through which contaminants are transported also vary and, depending on conditions, may move between groundwater and surface water or from one aquifer to another. Contaminants may also transfer between the aqueous phase and the solid and gaseous phases. These factors may also affect the contaminant's ultimate fate by determining the reaction types or extent that can cause alteration (Ragone, S.E. Water-Quality Contamination: A Systems Approach Towards Its Protection and Remediation. In Proc. of the Conf. The Environment: Global Problem - Local Solutions Hofstra Univ. Long Island, New York, June 7-8, 1990).

0097–6156/91/0465–0001$06.00/0

Scope

This book emphasizes approaches needed to study the effect of agrochemicals on groundwater. Most of the principles that apply to agrochemicals apply to other anthropogenic substances. These are specifically related to two main sources of potential agrochemical contamination of groundwater: nitrogen and pesticides.

Nitrogen. Nitrogen, an essential element for plant growth, is cycled and recycled throughout the environment in a series of transport and transformation steps known as the nitrogen cycle. Atmospheric N_2 gas is incorporated into plant tissue and soil through symbiotic nitrogen fixation processes. Other inputs occur through plant residues, animal wastes, commercial fertilizer, atmospheric deposition, and lightening. Organic forms (plant residues and animal wastes) of nitrogen undergo mineralization, i.e., transformation to ammonium-, nitrite-, and nitrate-N, the three species assimilated by plants and soil microorganisms. The nitrogen source (natural or manmade), then, is immaterial. Nitrogen losses from soil, in addition to plant uptake, can include denitrification (conversion to gaseous nitrogen oxides and N_2) and ammonia volatilization. Groundwater contamination can result when an excess of nitrogen, primarily nitrate-N, is present and when percolation occurs.

Commercial nitrogen inputs to the United States cropland during 1985-87 averaged 9.6 million metric tons N per year. By comparison, inputs from other sources (manure, crop residue, rainfall, biological fixation) were estimated as 8.3 to 9.8 million metric tons (1). Significant influx of nitrate-N to groundwater may derive from geological deposits (2) or from forests, pastures, and human wastes (3).

Nitrogen from agricultural use may be a potential problem where N fertilizer consumption is greatest. On that basis the Corn Belt states of Iowa, Illinois, and Indiana are particularly large consumers, as well as Nebraska, Michigan, and Kansas (4). On a state-wide per-acre basis, N use in Iowa and Indiana greatly exceeds that in large states such as California and Texas. State use is consistent with cropping patterns: among the four major cash crops in the United States, N use is dominated by corn (4.6 million metric tons) and wheat (1.6 million metric tons), with cotton and soybean use of N far less (5). Except for cotton, nitrogen use has greatly increased when comparing 1965 and 1985 production years.

Tile drains provide an early warning of possible groundwater nitrate-N and pesticide contamination problems from agricultural use. At the least, these residues will be transported relatively quickly into surface waters. The four states with the most area under tile drainage are Illinois (ca. 3.1 million ha), followed by Iowa, Indiana, and Ohio, all in the Corn Belt (6). Nelson (7) evaluated N found in tile drainage water from the Eastern Corn Belt and concluded that, on average, 7 kg ha^{-1} y^{-1} derived from natural background and 20 kg ha^{-1} y^{-1}, from fertilizers. In addition to being a source of contamination, this clearly represents a significant economic loss.

The national picture of N in groundwater was summarized in 1985 by a major survey--124,000 wells over a 25 y period (8). Expressed as nitrate-N, the results showed that 80.4% of the wells contained <3 ppm, 13.2% had 3 to 10 ppm, and 6.4% had >10 ppm. The >10 ppm represents the health advisory level (HAL) for nitrate-N, reflecting concern about development of methemoglobinemia in infants drinking this water. There have been a number of reviews concerning nitrate-N and groundwater quality, especially as related to agricultural practices, and that by Hallberg (9) is among the most recent.

Pesticides. Pesticides in ground water have become a concern primarily in the past decade. There are two possible reasons for this, one scientific and one analytical. Much early research failed to demonstrate that pesticide leaching to groundwater was occurring, though this was partially because of limited investigations. First it was generally believed that pesticides would not leach because most degrade to innocuous compounds [within hours to (for some of the formerly used chlorinated hydrocarbon pesticides) years] and they also tend to adsorb tightly to soil. Second, our analytical capability has increased to the point that parts-per-billion (ppb) detection limits is often routine and parts-per-trillion (ppt) levels have been reported (10-11). It is now generally accepted that some pesticides will leach to groundwater: what is more surprising is how rapidly this may occur.

Leaching depends upon the chemical, soil, site, weather, and management. Agrochemicals whose adsorption is low, especially if coupled with an inherently slow tendency to degrade in soil, are more at risk to excessive leaching. Conversely, soils that are coarse textured and with low organic matter content usually increase the vulnerability of underlying groundwater, all other things being equal. Important site factors include shallow depth to groundwater, the presence of sinkholes, or improperly sealed abandoned wells. Weather may be the most dominant factor in certain cases (Table I) when high rainfall occurs shortly after application, producing high runoff

and fast, deep percolation in macropores. Within the
fifth category, management, the improper storage, use, and
disposal of pesticides, including their containers,
increases the probability of groundwater contamination.
High application rates also increase the risk of leaching.
On the positive side, as mentioned earlier (Figure 1),
recommended herbicide rates have tended to decline as
newer, more active and/or selective compounds are
developed.

Table I illustrates three cases of deep and/or very
rapid agrochemical movement. Bromacil [5-bromo-6-methyl-
3-(1-methylpropyl)-2,4(1$\underline{H}$,3$\underline{H}$)-pyrimidinedione] leached to
4.9 m within 4 months in a Florida sand (12). Alachlor
[2-chloro-$\underline{N}$-(2,6-diethylphenyl)-$\underline{N}$-
(methoxymethyl)acetamide], atrazine [6-chloro-$\underline{N}$-ethyl-
$\underline{N}$'-(1-methylethyl)-1,3,5-triazine-2,4-diamine],cyanazine
{2-[[4-chloro-6-(ethylamino-1,3,5-triazin-2-yl]amino]-2-
methylpropanenitrile, and carbofuran (2,3-dihydro-2,2-
dimethyl-7-benzofuranyl methylcarbamate) were all detected
in shallow (ca. 0.6 to 1.5 m) groundwater 6 d following
pesticide application to a Maryland silt loam (13). In
the latter case, leaching was attributed to macropore
flow in the no-till plots, because over 4 cm of rainfall
occurred beginning 12 h after application. In the third
case, nitrate and bromide ions were found in tile drainage
(1.1 m depth) beneath an Iowa loam within 1 h after a
precipitation event; the authors (14) stated that flow
predictions based on the usual convective-dispersion flow
equations would not have predicted such rapid movement,
hence macropore flow was suspected.

Table I.   Three Examples of Deep or Fast Pesticide
             Leaching

Pesticide	State	Soil	Time	Depth m	Reference
Bromacil	Florida	sand	4 mon	4.9	12
Atrazine	Maryland	siltloam	<6 d	0.6-1.5	13
Alachlor	"	"	"	"	"
Cyanazine	"	"	"	"	"
Carbofuran	"	"	"	"	"
NO_3^-, Br^-	Iowa	Loam	1 h	1.1	14

Therefore, the increased number of investigations
coupled with our improved analytical capability has
demonstrated that a very small quantity of some pesticides
can indeed move with percolate water. The movement of
small pesticide amounts has led to a review of our
predictive transport equations. Predictive equations are
sound for the greatest portion of applied pesticide. They

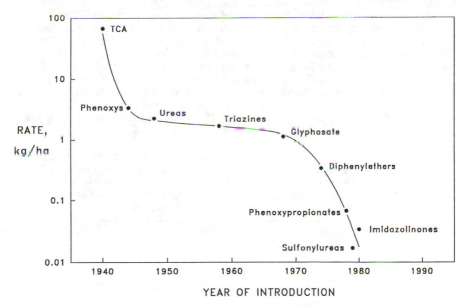

Figure 1. Historical trends in recommended herbicide application rates, by chemical class.

fail when rapid water movement occurs through macropores, i.e. pores so large, that even sorption to the pore walls fails to remove all pesticide from the percolating water.

Many new pesticides introduced are so pesticidal active, that the trend is toward lower application rates. Figure 1 shows how recommended herbicide rates have tended to fall since the introduction of synthetic organic herbicides in the early 1940's. Other things being equal, the trend toward reduced application rates will lessen the impact on groundwater.

The information summarized in Table II represent some milestones and trends related to finding pesticides in groundwater. Two key events occurred in 1979: the insecticide/nematicide aldicarb [2-methyl-2-(methylthio)propanal] was found in numerous wells in Long Island, New York (15) and the nematicide DBCP (dibromochloropropane) was detected in many California wells (16). Evidence was strong that contamination resulted from normal agricultural use, except for some of the latter. As a consequence of the aldicarb and DBCP cases, far more attention was paid nationally to the potential for pesticides to reach groundwater from agricultural practices.

Table II.   Pesticides in Groundwater from Normal Agricultural Use: Selected Key Findings

Year	No. of Pesticides Found	No. of States where Found	Reference
1979	(Aldicarb)	(New York)	15
1979	(DBCP)	(California)	16
1984	12	18	17
1985	17	23	18
1985	56	(California)	19
1988	67	33	20
1988	74 (46)[a]	38 (26)[a]	21

[a]The values 74 and 38 represent all data except known poor quality or known point source contamination. The values 46 and 26 are from confirmed studies where contamination is attributed solely to normal agricultural use.

The United States Environmental Protection Agency, Office of Pesticide Programs (USEPA/OPP) in 1984 (17) initially estimated that 12 pesticides were found in groundwater in 18 states as a consequence of normal agricultural use; in 1985 (18), this had increased to 17 pesticides in 23 states. In California, 56 pesticides were reported in groundwater (19), but originating through a variety of nonpoint and point sources, some industrial.

Two recent surveys deserve special attention. In the first (Table II), all 50 state agencies that have regulatory responsibility for pesticides in groundwater were contacted (20). Thirty-three of the 35 states reporting data had found pesticides, totaling 67 different compounds. One-hundred-two pesticides that were analyzed for were not detected. At about the same time, the U.S. EPA published (21) a survey based on monitoring studies conducted by federal and state agencies, universities, and pesticide registrants. Through screening procedures, they attempted to isolate reliable data and data clearly indicating the source of contamination. On this basis U.S. EPA reported that 74 pesticides had been detected in 38 states, when using a base that included all data except that of known poor quality and known point sources or misuse (Table II). When restricted to confirmed data attributed to normal agricultural use, the corresponding numbers were 46 pesticides in 26 states.

The pesticides that occur most frequently and at levels higher than the HAL (Table III) have been reported (20). For 1,3-dichloropropane, dinoseb [2-(1-methylpropyl)-4,6-dinitrophenol), and atrazine, only a small proportion of the positive findings were >HAL levels. Overall, 17 pesticides in 17 states were detected one or more times above the HAL.

Table III. Pesticides detected at concentrations greater than the health advisory level (HAL) and detected in (1) more than one state, or (2) more than four wells. Adapted from ref. 20.

Pesticide	Number of States		Number of Wells	
	Tested	Detected	Total	>HAL
1,2-Dichloropropane	7	3	7035	205
1,3-Dichloropropane	5	2	5517	5
Aldicarb	21	3	4004	175
Atrazine	28	5	5569	11
DBCP	4	1	7040	124
Dinoseb	10	2	1347	8
Ethylenedibromide	12	7	5133	520

The U.S. EPA completed (in 1990) sample collection for their National Pesticide Survey of the nation's community and domestic wells. They sampled 555 community system wells (from a population of 51,000) and 783 domestic wells

(from 13 million) (22). Samples from all 50 states were
analyzed for the same group of 127 pesticides, pesticide
metabolites, and nitrate-/nitrite-N. The wells were
selected to provide a statistically sound base (see
Chapter 5) for national projections of groundwater quality
trends. Preliminary results from 180 community system
wells found 6 with pesticides and 79 with nitrate-N (23).
From 159 domestic wells, 9 had measurable pesticides and
66, nitrate-N. None of the community wells contained
nitrogen >HAL of 10 ppm nitrate-N. Among domestic wells,
3 contained high pesticide levels and 8, nitrate-N. (23)

An additional, recent review (24) has reported that 39
pesticides have been detected in groundwater from 34 U.S.
states or Canadian provinces. Besides nonpoint sources,
these statistics include some commercial point sources
such as samples collected near pesticide supply and mixing
sites.

Study Approach

Any study of ground water quality must be cognizant of the
regulatory aspects, primarily those of the U.S.
Environmental Protection Agency (USEPA). Two approaches
have been described to investigate groundwater quality and
its amelioration [The U.S. Geological Survey, Water
Resources Division (USGS/WRD) (Ragone, S.E. Water-Quality
Contamination: A Systems Approach Towards Its Protection
and Remediation. In Proc. of the Conf. The Environment:
Global Problem - Local Solutions Hofstra Univ. Long
Island, New York, June 7-8, 1990) and the U.S. Department
of Agriculture, Agricultural Research Service (USDA/ARS)
(1)]. Both provide a framework to guide the allocation
of resources to study a wide-ranging and complex problem.
The individual researcher, or research team that actually
conducts the groundwater research or monitoring, needs
additional guidance over and above U.S. agency programs.
Most of the chapters in this book were written by
investigators that have conducted some aspect of
groundwater quality research, or at a minimum have
provided investigative tools that can be applied to best
design and execute research on agrochemical movement/loss
from the root or vadose zones and groundwater.

USEPA. Regulation is a tool (by requiring certain
preregistration information) to minimize the potential for
agrochemicals to reach groundwater. The USEPA-mandated
laboratory and field studies needed for pesticide
registration in the United States, require information on
the mobility and persistence in the environment. Certain
pesticides may be restricted based on various risk
factors, e.g., those showing actual leaching to
groundwater or movement deep into the root zone, those
showing potential to leach (through evidence of high

mobility and persistence), or those of special toxicological concern.

USGS/WRD. A hierarchial approach, that includes point, nonpoint, and continental studies has been described elsewhere (25). The approach for the three contamination type problems is similar in that it establishes the hydrogeologic framework in which the distribution of contaminant mass is identified. The approaches differ in detail because of scale.

For point-source studies, the focus is on the scale of lithologic variability. Groundwater contamination from a point source may have an overall areal dimension from several m^2 to >1 km^2 to as small as <1 m in the vertical direction. Recent observations of both the saturated and unsaturated zones suggest that interacting physical, chemical and microbiological reactions in this heterogeneous system may alter the specific chemical form of the contaminant and its distribution throughout the contaminant plume. (Ragone, S.E. Water-Quality Contamination: A Systems Approach Towards Its Protection and Remediation. In Proc. of the Conf. The Environment: Global Problem - Local Solutions Hofstra Univ. Long Island, New York, June 7-8, 1990.)

For nonpoint-sources, such as those introduced from crop or forest areas, the focus is on the scale at which major parts of the environment (ground water, surface water, the unsaturated and soil zones, biomass and the atmosphere) may store, transform or transport contaminants. The nonpoint scale represents an areal dimension of 10's to 100's of km^2 and generally represents concentrations barely above background levels. (Ragone, S.E. Water-Quality Contamination: A Systems Approach Towards Its Protection and Remediation. In Proc. of the Conf. The Environment: Global Problem - Local Solutions Hofstra Univ. Long Island, New York, June 7-8, 1990)

At the continental scale, major ground- and surface-water basins which cover 1,000's to 100,000's of km^2 are used to characterize the contaminant mass. Unlike the point and nonpoint scales, this scale should be considered "source-independent" in that it provides the broad perspective needed to understand the relative importance of different point- and nonpoint-source effects on water quality. At this scale, decisions to protect or enhance water quality can be evaluated as to their overall effectiveness with regard to their potential for transferring a contaminant from one environmental compartment to another, for instance. Hydrologic-unit maps have been published (25) that define national

surface-water units at several levels. "The largest
Hydrologic Units are called Regions and encompass the
drainage areas of the Nation's major river systems. The
21 Regions are divided into 222 Subregions, 352 Accounting
Units and 2150 Cataloging Units. These units create a
hierarchical structure that provides the flexibility to
address water-quality problems at the appropriate scale.
Point-source problems, singly or as aggregations in areas
where numerous point sources exist, may best be addressed
at the smallest, or cataloging unit scale, while nonpoint
source problems may best be addressed at the accounting
unit or subregional scale. Acid rain or global change
problems should be addressed at the regional scale."
(Ragone, S.E. Water-Quality Contamination: A Systems
Approach Towards Its Protection and Remediation. In Proc.
of the Conf. The Environment: Global Problem - Local
Solutions Hofstra Univ. Long Island, New York, June 7-8,
1990.)

USDA/ARS.   The USDA/ARS approach focuses research
primarily at a field level, but which would be applicable
to representative fields in a given area. The USDA/ARS
approach would fall primarily into the nonpoint source
category. Furthermore, the USDA/ARS narrows its studies
primarily to the root and vadose zones and attempts to
develop fundamental processes and mechanisms to explain
contaminant dissipation within, and movement from, one
zone to the other. The USDA/ARS has ongoing projects to
compile a reliable database of pesticide properties; study
pesticide adsorption, desorption, and dissipation; better
characterize surface and subsurface flow paths; and
investigate frost effects on nitrogen and pesticide fate.
Evaluating farming systems is a project receiving major
attention, one key component of which is to establish
Management System Evaluation Areas. This major effort
will involve cooperation with the USGS, with additional
cooperation expected from the involved States of Iowa,
Minnesota, Missouri, Nebraska, and Ohio and with the
USEPA.

For groundwater quality investigations, the soil
profile often is divided into three zones for study: 1)
root zone, 2) vadose zone (the unsaturated zone between
the root zone and the saturated zone), and 3) the
groundwater or saturated zone. What happens in the zone
above greatly influences the contaminant amount moving
into the zone below.

Design Strategy. A complete design strategy would involve
all of the following components:

- Statistical design   - Soil characterization
- Sampling methods   - Aquifer characterization
- Data collection   - Land management evaluation

    - Meteorological   - Data evaluation
       evaluation

This book focuses primarily on the first two components--
- those associated with design and sampling.

    Individuals who wish to initiate groundwater studies
should realize that no one person has the expertise to
address all the components of a complete groundwater
investigation. It is a multidisciplinary problem and is
best conducted by team research.

    The book contains statistical designs, one an actual
design   for   the   USEPA   National   Pesticide   Survey;
descriptions   of   regional,   field,   and   plot   designs
underway; methods for groundwater, vadose, and root zone
sampling; and methods for measuring agrochemical surface
loss.

Benefits

    Improved groundwater and soil sampling strategy will:
1) increase reliability of residue estimates; 2) identify
sources   of   contamination;   3)   identify   mechanisms;   4)
increase   sampling   efficiency;   and   5)   improve   model
validation. Furthermore, improved strategy may 6) assist
in the assessment of management alternatives, with respect
to   their   possible   impact   on   groundwater   quality;   7)
provide   a   basis   for   better   regulation,   8)   provide
examples   of   successful   interdisciplinary   research,   and
perhaps   most   important,   9)   provide   us   with   the
information necessary to protect our groundwater.

    This book brings together the ideas and approaches of
an interdisciplinary group of investigators. We believe
this volume will be useful and timely, not only to those
engaged   in   groundwater   research,   but   especially   to
researchers just beginning groundwater studies.
    Underlying the overview of agricultural chemicals in
groundwater is always the issue of choices and balance.
What is the best way to conduct a study that will yield
meaningful   results,   given   the   constraints   of   time,
funding, manpower, location, etc.? In a larger context,
what   is   the   cost   of   preserving   and   protecting   our
groundwater resource? How is that weighed against the
benefits of agricultural production as we know it, and
what are the risks associated with trace contamination?

Limitations

Perhaps, groundwater quality research presently is limited
less by funding than by having the necessary wide-range
of   disciplines   at   any   one   site.   The   assembling   or
cooperation of such a large group of scientists working

in a given study will be trying for both management and the scientists involved. Some aspects of a study will be short-range, while others long-range. Coordinating such an effort so that each scientist can get his of her deserved recognition from a large group will not be easy.

Breakthroughs

A common theme throughout the several chapters written by authors with practical experience in groundwater quality research was that the initial design be a pilot program that can be expanded as knowledge is gained from the study. All stressed that not enough information is available for below-the-horizon-studies to conduct a complete program from an initial design.

Advancement in determining groundwater quality likely will come from studies, such as several described in this book, that provide sound data from properly designed and conducted investigations. From these databases, mathematical models (many already in existence) to estimate agrochemical movement to groundwater can be tested. The tested models then can be used not only to estimate agrochemical movement under given conditions, but as an aid to further groundwater quality research by helping to define the most important parameters in which to direct the largest resource effort.

Literature Cited

1. U.S. Dept. Agric., Agric. Res. Service. ARS Strategic Groundwater Plan: 1. Pesticides, Feb. 1988; pp 8. 2. Nitrate. March; 1988; pp 9.
2. Viets, F.G., Jr.; Hagaman, R.H. Factors Affecting the Accumulation of Nitrate in Soil, Water, and Plants; U.S. Dept. Agric., Agric. Handbook 413, 1971; pp 63.
3. Keeney, D. Environ. Control. 1986; 16, 257-304.
4. Hauck, R.D. (Natl. Fert. Develop. Ctr., Tennessee Valley Authority). Cited by White, W.C. Sources of Nitrogen and Phosphorous; Fert. Resources Proj., Resource Washington, Inc., Washington, DC, 1989; pp 82.
5. Vroomen, H. Statistical Bull. No. 750; U.S. Dept. Agric., Econ. Res. Service, Washington, DC, 1987; pp 45.
6. Farm Drainage in the United states: History, Status, and Prospects; Pavelis, G. Ed.; Econ. Res. Serv. Misc. Pub. 1455; U.S. Dept. Agric.; Washington, DC, 1987; pp 170.
7. Nelson, D. In Plant Nutrient Use and Environment Symposium; The Fert. Inst.; Washington, DC 1985; pp 173-209.
8. Madison, R.J.; Brunett, J.O. In National Water

Summary 1984: Hydrologic Events, Selected Water-Quality Trends, and Ground Water Resources; U.S. Geol. Survey Water-Supply Paper 2275; Washington, DC, 1985; pp 93-105.

9. Hallberg, G.R. In Nitrogen Management and Ground Water Protection; Follett, R.F. Ed.; Elsevier Sci. Pub.: Amsterdam, 1989; pp 35-74.

10. Davoli, E.; Benferati, E.; Bagnati, R.; Fanelli, K. Chemosphere, 1987; 20, 1425-1430.

11. Nash, R.G. J. Assoc. Offic. Anal. Chem. 1990; 73, pp 438-442.

12. Hebb, E.A.; Wheeler, W.B. J. Environ. Qual. 1978; 7, pp 598-601.

13. Isensee, A.R.; Nash, R.G.; Helling, C.S. J. Environ. Qual. 1990; 19, pp 434-440.

14. Everts, C.J.; Kanwar, R.S.; Alexander, E.C., Jr.; Alexander, S.C. J. Environ. Qual. 1989; 18, pp 491-498.

15. Zaki, M.H.; Moran, D.; Harris, D. Am. J Public Health. 1982; 72, pp 1391-1395.

16. Peoples, S.A.; Maddy, K.T.; Cusik, W.; Jackson, T.; Cooper, C.; Frederickson, A.S. Bull. Environ. Contam. Toxicol. 1980; 24, 611-618.

17. Cohen, D.B. In Evaluation of Pesticides in Ground Water; Garner, W.Y; Honeycutt, R.C.; Nigg, H.N. Eds.; ACS Symp. Series No. 315, Am. Chem. Soc.: Washington, D.C., 1986; pp 499-529.

18. Cohen, S.Z; Eiden, C.; Lorber, M.N. In Evaluation of Pesticides in Ground Water; Garner, W.Y.; Honeycutt, R.C.; Nigg, H.N., Eds.; ACS Symp. Series No. 315, Am. Chem. Soc.: Washington, D.C., 1986; pp 170-196.

19. Cohen, S.Z. In Evaluation of Pesticides in Ground Water; Garner, W.Y.; Honeycutt, R.C.; Nigg, H.N., Eds.; ACS Symp. Series No. 315, Amer. Chem. Soc.: Washington, D.C., 1986; pp 499-529.

20. Parsons, D.W.; Witt, J.M. Pesticides in Groundwater in the United States of America: A report of a 1988 Survey of State Lead Agencies; Oregon State Univ. Exten. Serv. EM 8406, Corvallis, OR, 1989; pp 93.

21. Williams, W.M.; Holden, P.W.; Parsons, D.W.; Lorber, M.N. Pesticides in Ground Water Data Base: 1988 Interim Report. Office Pestic. Programs, U.S. Environ. Protection Agency: Washington, D.C. 1988; pp 141.

22. National Pesticide Survey: Project Update--March 1990. U.S. Environ. Protection Agency, Office of Drinking Water; Washington, D.C. 1990; pp 3.

23. National Pesticide Survey: Project Update--Fall 1989. U.S. Environ. Protection Agency, Office of Drinking Water; Washington, D.C. 1989; pp 2.

24. Hallberg, G.R. Agric. Ecosyst. Environ. 1989; 26, pp 299-367.

25. Seaber, P.R., Kapinos, P., Knapp, G.L. Hydrologic Unit Maps, U.S. Geological Survey Water-Supply Paper 2294, 1987; pp 63.

RECEIVED December 26, 1990

REGULATORY ASPECTS

Chapter 2

Health Advisories and Alternative Agricultural Practices

Regulatory Basis for Concern and Its Influence on Legislation

Anne R. Leslie and Michael Barrett

Office of Pesticide Programs, U.S. Environmental Protection Agency, Mail Stop H7506C, Washington, DC 20460

Farming practices that require intensive use of pesticides and fertilizer are believed to contribute to groundwater contamination in agricultural areas. Dependency on these practices is attributable in part to certain requirements of the federal commodity programs that result in continuous cropping patterns to retain government subsidies. There is increasing evidence of groundwater contamination; a number of studies have examined alternative approaches to prevent further contamination. The USEPA assigns risk-based advisories or regulatory standards to drinking water contaminants; this has had an important influence on legislative changes already made or being considered to provide incentives for farmers to reduce their dependence upon agricultural chemicals.

The editors have brought together in this book information on ways of sampling agrichemical contamination of groundwater. We believe that proper study design and technique are essential if researchers are to gather meaningful data. Although nitrate pollution of groundwater has been a demonstrated phenomenon for decades, groundwater pollution from leaching pesticides was first demonstrated in the late 1970s. Concern over the deterioration of this important resource is high because of the potential for long residence time of contaminants in groundwater and the difficulty in attaining successful remediation. It is important that the USEPA have meaningful data to determine how to approach the problem.

The number of investigations of the source and extent of pesticide contamination has increased rapidly in recent years. However, interpretation of the resulting data is often difficult because of variations in the way the studies were done. If researchers can choose the most appropriate study design and emphasize characterization of the sites where samples are taken, they will be able to identify more accurately the agricultural sources of groundwater contamination. We hope that this book will clarify the approach to the problem.

Early Assumptions about Pesticides in Groundwater

Although there has been interest in the impact of nutrients from fertilizers, sewage, and animal waste for several decades, interest in pesticide impact has developed relatively recently. Before significant efforts commenced in the 1970s to investigate the impact of pesticides on groundwater, it was assumed by most agricultural scientists that little, if any, of the applied pesticides would move below a meter or so in depth in the soil. Furthermore, it was assumed that if residues did somehow manage to reach groundwater, the concentrations likely to be found would be well below those of toxicological concern. This premise was so prevalent that there was generally very little interest in the scientific community to investigate pesticide leaching to the saturated zone until at least the early 1980s. The premise was based upon the knowledge from field dissipation studies that only a small percentage of applied pesticides moves much below the depth of incorporation and that the soil residue levels that were being found at a depth of 50, 100, or perhaps 150 cm were well below those known to demonstrate acute toxicity to target organisms.

A New View of Pesticide Contamination

The premise that pesticides could not pose a significant groundwater contamination hazard was challenged in the 1980s as studies directly investigating pesticide impact on groundwater proliferated, drinking water health advisories and standards were established, and public interest and concern emerged. A significant part of the current drive for alternatives to reliance on chemical inputs in agriculture (variously referred to as integrated pest management, low-input sustainable agriculture, alternative agriculture, organic farming, or best management practices) stems from the rise to prominence of the agrichemicals in groundwater issue.

The concerns of those responsible for assuring a clean supply of drinking water as well as those who drink it have led Congress to allocate funds for further studies and to call for increasing regulation. Congress commissioned a study of the problem, and this study (*1*) concluded that there must be significant changes in farm management to prevent further deterioration of groundwater. It also indicated that farmers are finding little information on integrated management practices and that public-sector sources of information such as the Cooperative Extension Service can play an important role, encouraging farming practice changes and assisting farmers in making management decisions.

In this chapter we will present our procedure for evaluating the risk to human health of the detected chemicals in drinking water. We will show how Health Advisories and Maximum Contaminant Level Goals (MCLGs) are derived, and how they are used to support drinking water standards (Maximum Contaminant Levels; MCLs).

In addition, we will review two analyses of the problem which describe possible ways of reducing the contamination and some of the arguments for and against these initiatives. These analyses recommended changes in the 1990 Farm Bill. Congress completed its revision and enacted the "Food, Agriculture, Conservation, and Trade Act of 1990" (2) on November 28, 1990. We will examine briefly the relevant sections of the Act to see whether the recommendations made were incorporated.

Rationale for Regulation of Pesticides That Contaminate Groundwater

Since the late 1970s the USEPA has been investigating groundwater contamination from agricultural applications of pesticides. Regulatory action has been based on the extent of contamination (many locations and soil types) and the level and kind of toxicity. For example, dibromochloropropane (DBCP) and ethylene dibromide (EDB) were found to be widespread contaminants which are cancer-causing; most uses were cancelled between 1979–1983. Aldicarb, on the other hand, was detected only in certain use areas and is a potent acute toxin, with no apparent carcinogenicity. In the 1980s, evidence of aldicarb residues in groundwater was gathered from numerous locations.

In general, the percentage of the applied pesticides found moving below the root zone (roughly the first meter of soil) is quite low. However, from the detection data the Agency has collected, almost any pesticide has a finite potential to reach groundwater—even immobile materials of low persistence. For example, in a recent survey of rural domestic well water in Iowa, either of the soil-applied herbicides trifluralin or pendimethalin were detected in 2% of the samples even though these compounds should be among the least likely of all soil-applied pesticides to migrate to any significant extent in soil (3).

A more significant conclusion, from a regulatory standpoint, is that some impact is inevitable as long as significant amounts of the chemical are applied to soil surfaces. The USEPA will have to make hard choices on what levels of pesticides are acceptable in groundwater.

Evaluation of Human Health Risk from Drinking Water Contaminants

The USEPA has been concerned with the effects contaminants may have on human health. In an effort to evaluate risks from drinking water contaminants while national regulations were being formulated, USEPA devised a yardstick for judging the significance of contaminant levels. This yardstick is known as a Health Advisory; it is calculated by a mathematical formula from toxicology data. The method of calculation and a list of Health Advisories established for some 50 pesticides as of 1989, were published by USEPA in a book, *Drinking Water Health Advisory: Pesticides* (4).

Health Advisory levels (HA's) are described as "nonregulatory concentrations of drinking water contaminants at which adverse health effects would not be anticipated to occur over specific exposure durations" (5). For example, a 10-day HA is set at a level such that human exposure not exceeding this level for a period of 10 days or longer should result in no manifestations of adverse health effects. Although HA's, in and of themselves, have no legal significance from the Federal perspective, they are widely used by OPP and by state agencies as guidelines to identify levels of concern for pesticides in groundwater.

Method of Calculation of Risks. HA's for pesticides are usually calculated from the No-Observed-Adverse-Effect Level (NOAEL) identified in the (usually) animal toxicity study deemed most appropriate by USEPA scientists. HA's are calculated for different durations of exposure. Data from animal studies usually must be relied upon in developing health standards—especially for the chronic toxicity data needed to calculate longer-term or lifetime HA's—since valid human data are seldom available.

The Agency assumes a standard human body weight, i.e., 70 kg for an adult, and a standard daily water consumption of two liters for that adult in the calculation. The goal is to determine the highest level of exposure that would not pose a risk.

Uncertainty Factors (UF). To derive a Health Advisory number, uncertainty ("safety") factors are applied to the appropriate NOAEL. A UF of 100 generally is applied when that NOAEL is identified in an animal study. This 100-fold reduction reflects the assumptions that humans are more sensitive than animals to the effects of the chemical and that within the human population, there also are varying degrees of sensitivity. Larger UFs (e.g., 1000 or 10000) may be used if no NOAEL is identified, but rather there only are data which identify the Lowest-Observed-Adverse-Effect-Level (LOAEL) and/or there are significant deficiencies in the toxicology data base for the substance. Smaller UFs (e.g., 10) may be used when the HA is derived from human data.

Example of Calculations: Atrazine. The following example illustrates the calculation of the lifetime HA for atrazine:

1. A NOAEL of 0.48 mg kg^{-1} d^{-1} is determined from the results of a 1-year dog study, then a Reference Dose (RfD) is calculated by using an uncertainty factor (UF) of 100.

$$RfD = (0.48 \text{ mg } kg^{-1} d^{-1})/100 = 0.005 \text{ mg } kg^{-1} d^{-1}$$

   The chosen RfD is conservative in the protective sense since the chosen NOAEL is from the study of the most sensitive test animal (dog) and the RfD is set at one hundredth of the NOAEL to ensure that no adverse effects are likely in humans.

2. The Drinking Water Equivalent Level (DWEL) is calculated by assuming an adult weighing 70 kg consuming 2 liters of water per day.

$$DWEL = (0.0048 \text{ mg } kg^{-1} d^{-1})*(70kg)/(2 \text{ L } d^{-1}) = 0.168 \text{ mg } L^{-1}$$

The DWEL represents "the concentration of a substance in drinking water that is not expected to cause any adverse noncarcinogenic health effects in humans over a lifetime of exposure" (5).

3. To account for possible exposure to the chemical from other media such as food or air, the DWEL is adjusted for the relative source contribution (RSC) of drinking water, generally assumed to be 20%, unless empirical data exist to indicate otherwise. In the calculation of the HA, an additional UF of 10 is included for group C chemicals (possible human carcinogens). This is the category for carcinogenic potential into which atrazine currently falls.

$$\text{Lifetime HA} = (DWEL)*(RSC)/UF = (0.168 \text{ mg } L^{-1})*(0.2)/10 =$$

$$0.003 \text{ mg } L^{-1}$$

Calculation of Health Advisories for Various Categories of Human Carcinogens. Lifetime HA's, per se, are not calculated for pesticides classified as known human (Group A) or probable human (Group B) carcinogens. In these cases, estimates are made which conservatively represent different risk levels (e.g., 1 in 10,000, 1 in 100,000, and 1 in 1,000,000 chance of developing cancer over a lifetime of exposure).

Legally Enforceable Drinking Water Standards. In response to the mandate of the Safe Drinking Water Act (as amended in 1986), Maximum Contaminant Levels (MCLs) are being established for "contaminants in drinking water which may cause any adverse effect on the health of persons and which are known or anticipated to occur in public water systems" (5). MCLs are legally enforceable standards for community water systems which serve more than 25 people or 15 connections.

To determine the MCL, USEPA first proposes a non-mandatory Maximum Contaminant Level Goal (MCLG), which for non-carcinogens is calculated in the same way as the HA. MCLGs are automatically set at zero for any contaminant that is classified as a known or probable human carcinogen. The MCL is established "as close to the MCLG as is feasible based on technological and economic considerations." As of November 1990, MCLs had been established or proposed for only 18 pesticides and three metabolites of these pesticides (5). On January 30, 1991, USEPA promulgated standards for an additional 12 pesticides (6) and proposed MCLGs/MCLs for two more (7).

USEPA's Proposed Strategy for Agricultural Chemicals in Groundwater

In 1988 the USEPA's Office of Pesticide Programs (OPP) released a proposed strategy concerning agricultural chemicals in groundwater (8). The goal of the strategy is to prevent contamination of groundwater. The strategy emphasizes protecting future and potential drinking water sources, as well as affected surface water. Under the proposed strategy, MCLs developed under the Safe Drinking Water Act will become the reference points for assessing groundwater contamination. Groundwater contamination as a result of use would be managed primarily through State Management Plans tailored to local conditions of pesticide use and groundwater vulnerability.

Contaminants Present in Greater Concentration Than the Regulatory Levels. HA's and MCLs promulgated to date are often equal to or lower than concentrations detected in groundwater. DBCP and EDB, for example, have often been found in groundwater at concentrations near their respective MCLs of 0.2 and 0.05 μg/L. Numerous detections of pesticides in groundwater in the range of 0.05 to 1 μg/L (and sometimes higher) have been attributed to agricultural use (*9, 10*). Low level concentrations on the order of analytical detection limits are often referred to as "trace" or some similar descriptor which is often interpreted as having little or no human health or environmental concern. HA's established for many pesticides indicate that we should be concerned about such "trace" levels in groundwater.

The Future of Pesticide Regulation Related to Groundwater Contamination. It is likely that pesticide use, even with the best pollution prevention practices, will result in at least some limited impact on groundwater. There are many uncertainties, of course, in determining the risk to human health of the levels of pesticides that might be found in drinking water, so the process for establishing levels of concern warranting regulatory action is likely to continue to undergo change for many years to come.

Expected regulatory actions include reauthorization of the Safe Drinking Water Act when it expires in 1991 and establishment of MCL's for an increasing number of pesticides and other chemicals.

Studies of Opportunities for Groundwater Protection

The widespread use of pesticides has, in many instances, resulted in clear evidence of diminishing returns over time. There is thus an economic incentive to change farming practices, but there are also a number of disincentives. Two significant documents were published in 1990 that address the problem of groundwater protection.

Report from the Office of Technology Assessment. Four Congressional committees and five subcommittees requested the Office of Technology Assessment (OTA) in 1988 to find out whether there were agricultural technologies that could reduce groundwater contamination by agricultural chemicals. OTA commissioned a number of authors to contribute to a comprehensive treatment of the subject, and the result is *Beneath the Bottom Line: Agricultural Approaches to Reduce Agrichemical Contamination of Groundwater* (*1*).

The book includes: a primer on contamination of the hydrogeological system; a review of technologies to improve nutrient and pest management; a chapter on decisions farmers can make on how agrichemicals are to be handled and applied, on whether to reduce their use or employ nonchemical practices, on the range of technical assistance available through Federal, State and local programs; and a review of the options Congress has available, with a discussion of the obstacles to solving the contamination problem and possible legislative solutions.

Though the problem is complex, there must be a major effort to reduce reliance on chemical controls. Existing research programs have focussed on new chemicals. Less than 1% of the 1985 Farm bill was for research on alternatives to chemically-intensive production methods (11). OTA recommends substantial increases in funding for research on biological controls.

The authors of the report propose several strategies such as improving decision-making, changing federal agricultural programs, and fostering a national effort to reduce agrichemical mismanagement and waste. They stated that the latter strategy could be accomplished if Congress authorized the USEPA to maintain an overview of State pesticide programs, including their certification and training requirements. OTA predicted that Congress' action would depend on whether it sees a need to integrate environmental protection into agricultural policy as a whole. The advantages and disadvantages of a number of policy decisions are discussed extensively in OTA's final chapter on findings, issues and options for Congress.

Report from the Environmental and Energy Study Institute. The second document is a policy report, "The 1990 Farm Bill: Opportunities for Groundwater Protection"(12), published by the Environmental and Energy Study Institute (EESI). This report focusses on the Farm Bill issues discussed in the OTA report. It shows how efforts to prevent groundwater contamination differ depending on whether the supporter of the effort believes that the major source is "point" (e.g., leaks and spills) or non-point (normal application). Point source contamination can be reduced by technology and by regulation, but non-point sources likely require an overall reduction of agricultural chemical use.

EESI recommends the adoption of "integrated farm management" systems that minimize all chemical applications, and that include cultural practices such as crop rotation. They present evidence that there has not been widespread adoption of such systems because the federal commodity program encourages farmers to grow the same crops on the same number of acres year after year in order to obtain the subsidies. Although the system may not be applicable to specific crops, the subsidy program may also not apply to these crops, according to the authors.

This report concluded that reduction of chemical inputs is necessary to reduce groundwater contamination, and it made the following recommendations for congressional action on the 1990 Farm Bill that could remove many of the disincentives for adoption of integrated farm management:

- Adoption of a change in base flexibility (allowing a number of acres of a base crop to be planted to a commodity other than the specific program crop)

- Adoption of a proposal for research on integrated farm management systems

- Approval of the House or Senate Agriculture Committee proposals for voluntary, multi-year programs as incentives for integrated farm management

- Funding to develop farmer manuals describing integrated farm management practices

- Passage of a proposal to train the U.S. Department of Agriculture's (USDA) Cooperative Extension Service (CES) and Soil Conservation Service (SCS) in integrated farm management practices to provide farmers with information and technical assistance.

The 1990 Farm Bill

The final 1990 Farm Bill enacted by Congress (2) addresses these issues:

- *Base flexibility:* Title XI—General Commodity Provision, Subtitle A, Section 1101. This section amends Title V of the Agricultural Act of 1949 to allow up to 25 percent of the crop acreage base to be planted to a commodity other than the specific program crop.

- *Research on integrated farm management systems:* Title XVI—Research. Subtitle B, Chapter 1,—Best Utilization of Biological Applications (BUBA), Section 1621. In addition to requiring research projects that will show how to reduce the use of chemical pesticides, fertilizers and toxic natural materials in agricultural production, this section supports improvements in low-input farm management and promotes crop, livestock and enterprise diversification. It proposes to study farms using these methods and to transfer information about the technology to farmers and ranchers.

- *Incentives for integrated farm management:* Chapter 2, Section 1627 aims to encourage producers to adopt and develop individual site-specific integrated crop management practices.

- *Information and technical assistance; development of farmer training manuals:* Chapter 3, Section 1628 requires the development within two years of books and technical guides describing sustainable agriculture production systems and practices, including integrated pest management practices.

- *Information and technical assistance; training the CES and SCS in integrated farm management practices:* Chapter 3, Section 1629 directs the Secretary of Agriculture to establish a National Training Program in Sustainable Agriculture to develop understanding and competence in Cooperative Extension Service agents and other professionals so that they can teach these concepts to farmers and urban residents who need the information. All agricultural agents of CES are to complete the training within a five year period.

Other sections of the 1990 Farm Bill address the issue of groundwater contamination, but the above provisions deal with the major issues in the reports. The USEPA is participating in review of FY91 research proposals solicited by the ARS/USDA. Funds were to be awarded in February, 1991, and they propose to support some much needed research in this area.

Problems in Interpreting Groundwater Data

As so much legislation is aimed at insuring that agrichemicals are used in a fashion which preserves environmental quality, researchers are trying to understand the enormously complex array of processes related to whether and how these chemicals get into groundwater. There are numerous difficulties with interpreting available data including:

1. a lack of appropriate design in the sampling program, so that the target population is never defined,

2. variations in quantitative detection limits and level of confirmation of residue identity,

3. defects in well construction allowing direct channeling of residues,

4. concentration of studies on sampling in atypical situations (in excessively well-drained soils, in highly- developed karst regions, from unconfined aquifers only a few feet in depth, near agricultural chemical storage facilities or other potentially potent point sources, etc.) which yields data which are difficult to extrapolate to more typical situations,

5. variations in sampling technique, and

6. lack of collection of site-specific information on agrichemical use, well characteristics, and hydrogeologic environment. The proportion of detections of a given agrichemical in most studies is probably more a function of how much sampling concentrated on extremely vulnerable groundwater than the extent of groundwater contamination in the study area.

Conclusion

The processes for determining whether or not agrichemicals impact groundwater are complex, and it is critical for researchers to develop techniques for sampling design, collection, preservation, analysis, and interpretation so that we can improve our understanding of these processes. We are likely to see more legislation designed to minimize the environmental impact of agriculture at both the federal and state level. This will continue whether or not a sufficient foundation is established for understanding what measures will be most effective in preserving the environment and protecting human health. The other chapters in this book provide an update on what is being done to close the gap between scientific understanding and legislation.

Literature Cited

1. U.S. Congress, Office of Technology Assessment, *Beneath the Bottom Line: Agricultural Approaches To Reduce Agrichemical Contamination of Groundwater,* OTA-F-418 (Washington, DC: U.S. Government Printing Office, November 1990).

2. "Food, Agriculture, Conservation, and Trade Act of 1990" (Public Law 101-624, November 28, 1990).

3. Kross, B. C. et al., *The Iowa State-Wide Rural Well-Water Survey, Water-Quality Data: Initial Analysis;* Technical Information Series 19; Iowa Department of Natural Resources, Larry J. Wilson, Director: Iowa City, IA, 1990; pp 53–3y.

4. United States Environmental Protection Agency Office of Drinking Water Health Advisories. *Drinking Water Health Advisory: Pesticides;* Lewis Publishers: Chelsea, MI, 1989; 819 p.

5. United States Environmental Protection Agency. *Drinking Water Regulations and Health Advisories;* Office of Drinking Water, USEPA: Washington, DC, 1990, 10 pp.

6. United States Environmental Protection Agency. National Primary Drinking Water Regulations; Final Rule. *Federal Register* **1991**, *56*(20), 3526–3597.

7. United States Environmental Protection Agency. National Primary Drinking Water Regulations; Proposed Rule. *Federal Register* **1991**, *56*(20), 3600–3614.

8. United States Environmental Protection Agency. 1987. *Agricultural Chemicals in Ground Water: Proposed Pesticides Strategy;* Office of Pesticides and Toxic Substances, Washington, DC.

9. Barrett, M. R.; Williams, W. M. In *Pesticides in Terrestrial and Aquatic Environments;* Weigmann, D. L., Ed.; Proceedings of a National Research Conference; Virginia Polytechnic Institute and State University: Blacksburg, VA, 1989, pp 39–61.

10. Williams, W. M.; Holden, P. W.; Parsons, D. W.; Lorber, M. N. *Pesticides in Ground Water Data Base: 1988 Interim Report.* U.S. Environmental Protection Agency, Office of Pesticide Programs: Washington, DC, 1988.

11. Cohen, S. Z. *Pesticides and Nitrates in Ground Water: An Introductory Overview,* contractor report prepared for the Office of Technology Assessment (Springfield, VA: National Technical Information Service, August 1989).

12. Bird, J. C.; Edmond, J. *The 1990 Farm Bill: Opportunities for Groundwater Protection;* Policy Report; Environmental and Energy Study Institute: Washington, DC, 1990.

RECEIVED March 20, 1991

Chapter 3

Field-Scale Monitoring Studies To Evaluate Mobility of Pesticides in Soils and Groundwater

Elizabeth Behl and Catherine A. Eiden[1]

Office of Pesticide Programs, U.S. Environmental Protection Agency, Mail Stop H7506C, Washington, DC 20460

The United States Environmental Protection Agency (EPA) may require data from ground-water monitoring studies to support the registration of pesticide products under the Federal Insecticide, Fungicide, and Rodenticide Act (FIFRA) sections 3(c)5 and 3(c)7. Data from ground-water monitoring studies are used both to determine the likelihood that a pesticide will leach and to detect the presence of a pesticide in ground water from years of use. Field-scale monitoring studies are necessary because patterns of pesticide degradation and movement in the field are influenced by a wide variety of natural environmental factors that cannot be duplicated in the laboratory. Monitoring studies have been required for 37 compounds when residues of the pesticide are reported in ground water or when the Agency has evaluated the pesticide as a potential "leacher," based on a review of it's persistence and mobility. This paper explains the history of ground-water monitoring requirements for the Office of Pesticide Programs of EPA and events and issues that led to the development of the monitoring guidance. New directions in small-scale monitoring studies are described. Large-scale ground-water monitoring is mentioned briefly.

In 1987, the Office of Pesticide Programs (OPP) of the United States Environmental Protection Agency (EPA) began requiring ground-water monitoring studies for pesticides that have the potential to contaminate ground water. Since then, over 37 monitoring studies have been required as part of the registration process for these products. Results of the studies are used to establish regulations to insure that pesticides are not applied in environments that are vulnerable to contamination.

---

[1]Current address: United Nations Development Program Office, Port Louis, Mauritius

The history of the monitoring study requirement is discussed below, followed by a description of the small-scale ground-water monitoring study designs. These designs incorporate revisions that correct deficiencies identified in recently completed ground-water monitoring studies.

Historical Development of Monitoring Study Requirement

The Role of Environmental Fate Data in Pesticide Registration. The registration process for pesticides under the Federal Insecticide, Fungicide, and Rodenticide Act (FIFRA) requires that the environmental fate characteristics of a pesticide be established by performing laboratory and field studies. Study requirements are described in more detail in 40 CFR 158 (*1*). The detail on how these studies are to be performed is in the Pesticide Assessment Guidelines (*2*). Environmental fate studies are listed in Table I.

Table I. Environmental Fate Studies Required for All Terrestrial Uses

Reference No.	Guideline Name
63-1	Physical and Chemical Characteristics: solubility, vapor pressure, Henry's Law constant, etc.
161-1	Hydrolysis
161-2	Photodegradation: water
161-3	Photodegradation: soil
162-1	Aerobic Soil Metabolism
162-2	Anaerobic Soil Metabolism
163-1	Mobility in Soil: column leaching and adsorption/desorption
164-1	Terrestrial Field Dissipation

These studies are reviewed and accepted or rejected by the Environmental Fate and Effects Division of OPP. Validated fate parameters are compiled in OPP's Environmental Fate One-Liner Database. Information on the pesticide's persistence and mobility is used to assess the potential exposure from pesticide residues, and to evaluate the fate of the pesticide and its degradates in the environment. Of the studies listed above, only the field dissipation study is performed in the field, all others are laboratory studies.

Prior to 1980 it was thought that pesticides would degrade in the biologically active root zone and in the vadose zone and therefore would not pose a threat to ground-water quality. In the 1970's, investigations indicated that pesticides were being detected in ground water nationwide (*3–4*). In the late 1970's and early 1980's, concerns about ground-water contamination resulting from the normal use of agricultural chemicals led

OPP to evaluate approximately 800 active ingredients of pesticides in an effort to screen out those that had the potential to leach to ground water. The pesticides were screened based on broad criteria related to the mobility and persistence of the compounds (4) listed in Table II.

Table II. Physical and Chemical Characteristics of Pesticides Found in Ground Water (Source: Adapted from ref. 4)

Characteristic	Leaching Criteria
Water solubility	>30 ppm
Henry's Law Constant	$<10^3$ Pa$(m^{-3})(mol^{-1})$
Hydrolysis half-life	>25 weeks
Photolysis half-life	>1 week
Soil adsorption: K_d	<5 (usually <1–2)
Soil adsorption: K_{oc}	<300–500
Aerobic soil metabolism half-life	>2–3 weeks
Field dissipation half-life	>2–3 weeks
Depth of leaching in field dissipation study	>75–90 cm

If a pesticide was found to be persistent and mobile relative to these criteria, it was selected for further evaluation. A total of 141 registered pesticides were identified in this initial screen. In 1984, data related to ground-water contamination potential was requested for the 141 pesticides by issuance of a Ground-Water-Data-Call-In (GWDCI) as authorized under Section 3(c)(2)(B) of FIFRA for currently registered pesticides. From the 141 compounds, four have been determined not to pose a ground-water problem, 19 have been canceled, and approximately 37 ground-water monitoring studies have been required. Studies for the remaining compounds are under review or awaiting submission of data required for an initial screen of their leaching potential. This screening procedure has been incorporated into the standard review procedure for all compounds proposed for registration. Table III is a list of the pesticides for which ground-water monitoring studies have been requested. This list includes pesticides that were flagged upon review of data submitted in response to the GWDCI, and those requested following review of data submitted to support the registration of a new chemical.

Monitoring may be required because of concerns about the leaching potential of the pesticide's active ingredient, it's degradates, metabolites, or contaminants. For example, monitoring for ethylene bisdithiocarbamate (EBDC) pesticides (including mancozeb, maneb, metiram, and zineb) was triggered largely because of concerns about a common contaminant, metabolite, and degradation product, ethylene thiourea (ETU).

All uses have been canceled for dalapon, and the cancellation of all uses of propazine and zineb have been proposed.

Table III Ground-water Monitoring Studies Required as of 12/20/90

COMMON NAME	STUDY TYPE	CHEMICAL NAME
dichloropropene	Small-scale retrospective	1,3 Dichloropropene
acifluorfen	Small-scale prospective, Small-scale retrospective	5-{2-chloro-4-(trifluoromethyl)phenoxy}-2-nitrobenzoic acid
alachlor	Large-scale retrospective	2-chloro-N-(2,6-diethylphenyl)-N-(methoxymethyl)acetamide
ametryn	Small-scale prospective	N-ethyl-N-(1-methylethyl)-6-(methylthio)-1,3,5-triazine-2,4-diamine
asulam	Small-scale prospective	methyl{(4-aminophenyl)sulfonyl} carbamate
chlorothalonil	Small-scale prospective	Tetrachloroisophthalonitrile
coumaphos	similar to Small-scale prospective	O,O-Diethyl o-(chloro-4-methyl-2-oxo-2h-1-benzopyran-7-yl) phosphorothioazte
cyanazine	similar to Large-scale retrospective	2-{[4-chloro-6-(ethylamino)-1,3,5-triazin-2-yl] amino-2-methylpropanenitrile
cyromazine	Small-scale prospective	N-cycloprpoyl-1,34,5-triazine-2,4,6-triamine
dalapon	Small-scale retrospective	2,2-dichloropropanoic acid
DCNA	Small-scale prospective	2,6-dichloro-4-nitroaniline
DCPA	Small-scale retrospective	dimethyl 2,3,5,6-tetrachloro-1,4-benzenedicarboxylate
DPX-M6316	Small-scale prospective	methyl 3-[[[[(4-methoxy-6-methyl-1,3,5-triazin-2-yl)-amino]carbonyl]amino]sulfonyl]-2-thiophencarboxylate
diclofop-methyl	Small-scale prospective	2-(4-(2',4'-dichlorophenoxy)-phenoxy)-methyl-propanoate
fluazifop-butyl	Small-scale prospective	butyl 2-(4-(5-trifluoromethyl-2-pyridyloxy)phenoxy) propionate

Table III (continued).

COMMON NAME	STUDY TYPE	CHEMICAL NAME
fluometuron	Small-scale retrospective	N,N-dimethyl-N'-[3-(trifloromethyl)phenyl] urea
fomesafen	Small-scale prospective	5-[2-chloro-4-(trifloromethyl)phenoxyl-N-(methylsulfonyl)-2-nitrobenzamide
Sodium tetrathiocarbonate	Small-scale prospective	carbono (dithioperoxo) dithioic acid, sodium salt
haloxyfop-methyl	Small-scale prospective	methyl 2-(4((3-chloro-5-(trifluoromethyl)-2-pyridinyl)oxy)phenoxy)propanoate
hexazinone	Small-scale retrospective	3-cyclohexyl-6-(dimethylamino)-1-methyl-1,3,5-triazine-2,4(1H,3H)-dione
lactofen	Small-scale prospective	(+/-)-2-ethoxy-1-methyl-2-oxoethyl 5-[2-chloro-4-(trifluoromethyl)phenoxy]-2-nitrobenzoate
mancozeb/ETU	Small-scale retrospective	coordination complex of zinc ion, manganese ethylene bisdithiocarbamate
maneb/ETU	Small-scale retrospective	manganese ethylenebisdithiocarbamate
metalaxyl	Small-scale retrospective	S-methyl-N-((methylcarbamoyl)oxy)-thioacetimidate
methomyl	Small-scale retrospective	S-methyl-N-((methylcarbamoyl)oxy)-thioacetimidate
metiram/ETU	Small-scale retrospective	tris{ammine{ethylenebis(dithiocarbamato)} zinc(2+1)}{tetrahydro-1,2,4,7-dithiadiazocine-3,8-dithione}, ploymer

Continued on next page

Table III (continued).

COMMON NAME	STUDY TYPE	CHEMICAL NAME
metolachlor	similar to Large-scale retrospective	2-chloro-N-(2-ethyl-6-methylphenyl)-N-(2-methoxy-1-methylethyl)acetamide
metribuzin	Small-scale prospective	4-amino-6-(1,1-dimethylethyl)-3-(methylthio)-1,2,4-triazin-5(4H)-one
oxamyl	Small-scale retrospective	methyl N',N'-dimethyl-N-{methylcarbamoyl)oxy}-1-thiooxamimidate
picloram	Small-scale prospective	4-amino-3,5,6-trichloro-2-pyridinecarboxylic acid
primisulfuron	Small-scale prospective	methyl 2-[4,6-bis(difluoromethoxy)-2-pyrimidinyl aminocarbonylaminosulfonyl]benzoate
prometon	Small-scale retrospective	6-methoxy-N,N'-bis(1-methylethyl)-1,3,5-triazine-2,4-diamine
prometryn	Small-scale retrospective	N,N'-bis(1-methylethyl)-6-(methylthio)-1,3,5-triazine-2,4-diamine
propazine	Small-scale retrospective	6-chloro-N,N'-bis(1-methylethyl)-1,3,5-triazine-2,4-diamine
tebuthiron	Small-scale retrospective	N-[5-(1,1-dimethylethyl)-1,3,4-thiadiazol-2-yl]-N,N'-dimethylurea
thiodicarb	similar to Small-scale prospective	dimethyl N,N' thiobis (methylimino) carbonyloxy bis ethanimidothioate
zineb/ETU	Small-scale retrospective	zinc ethylenebisdithiocarbamate

Evaluation of Leaching Potential. OPP utilizes both field and laboratory data to assess a pesticide's potential to contaminate ground water. Environmental fate studies required for registration are predominately laboratory studies, with the exception of the field dissipation study. In a field dissipation study, pesticides are applied to a field and soil samples are collected and analyzed to determine if pesticide residues persist in the soil at depth. The original protocol for this study did not standardize either the depth to which soil had to be sampled, or the frequency of sampling. This resulted in highly erratic detections, and produced data that was too inconclusive to make regulatory decisions. The field dissipation study protocol was revised to resolve these problems in 1989 (5). Data from the core studies (Table I) are evaluated using a weight-of-evidence approach to assess leaching potential.

Field Monitoring Data. In a laboratory, conditions are more easily controlled and data are more easily collected than in field study; however, all processes that control degradation and dissipation of the pesticide cannot be duplicated. Environmental fate parameters determined in a laboratory can be used in models to simulate the expected behavior of a compound in the field; but, these same parameters often differ significantly when measured in the field. These differences arise because of the scale at which processes are operating, and the influence of factors not considered by the model algorithms. Physical evidence that a pesticide leaches under a set of field conditions clearly establishes the leaching potential of the pesticide.

In 1987, when EPA attempted to alter pesticide labels to indicate concerns about possible ground-water contamination, the regulated industry argued that the presence of pesticide residues at a specified depth was not sufficient evidence upon which to base regulatory actions. The thought at that time was that pesticide residues in the vadose zone might degrade or adsorb to soils, and in that case would not pose a clear threat to ground-water quality. In line with this, pesticide residues detected in ground water were attributed to point sources of contamination rather than to normal agricultural use. Possible point sources resulting from mixer/loader activities, improper disposal practices, and cases of misuse must be differentiated from non-point sources resulting from field application. These activities technically do not constitute "normal agricultural use of pesticides," as required under FIFRA.

Criticisms that EPA's proposed label changes were based upon limited field data and monitoring data showing sporadic pesticide detections of questionable origin, led OPP to develop a new data requirement focusing specifically on pesticide residues in ground water. Thus, ground-water monitoring became an integral part of the pesticide registration process. A database of pesticide detections in ground water was developed in 1988 by OPP to provide additional monitoring information on pesticides that have leached to ground water as a result of normal agricultural use. This database identifies pesticides, the locations and frequency with which they are found, and classifies the origin of the pesticide contaminants as point source or normal field use. An initial report describing the Pesticides in Ground Water Data Base has been published (6), and an update is expected in 1991.

A small-scale ground-water monitoring study may be required for a pesticide if adequate field evidence exists to demonstrate that it can reach ground water from normal agricultural use. Information obtained from the field-scale study can be used to assure that the pesticide is not used in areas that are vulnerable to contamination.

In addition to providing information that is used to establish risk reduction measures, a second use of field data is to provide an estimate of the levels of residues to which people may be exposed, for example, in drinking water. This information is needed to perform risk assessments for chemicals undergoing Special Review. Studies included in the Pesticides in Ground Water Data Base do not provide this type of information because they are done according to different protocols, and are usually targeted to areas that are the most vulnerable to contamination. An exposure-based study is broader in scope than a field-based study.

Large-Scale Studies. In 1984, OPP required the registrant of alachlor to conduct the first large-scale retrospective ground-water monitoring study. Monsanto sampled drinking water wells in alachlor use-areas and analyzed for the parent compound. Also, in 1984–1985, EPA's Office of Drinking Water (ODW) and OPP began planning the National Pesticide Survey (NPS). The NPS is a nationwide survey designed to investigate the presence of pesticide residues in drinking water wells. Both the NPS and the alachlor studies were statistically designed and will estimate the number of people exposed to levels of pesticide(s) that exceed drinking water standards.

Approximately 1350 wells were sampled for the NPS over a two-year period and analyzed for residues of 126 pesticides and metabolites, and nitrate/nitrite. Certain categories of wells were over-sampled to account for differences in pesticide usage patterns and ground-water vulnerability in various parts of the country. Preliminary results indicate that about 10 per cent of community drinking water wells and about 4 percent of rural domestic wells in the United States contain detectable levels of at least one pesticide. The most commonly detected residues were of a degradate of DCPA, which has the trade name Dacthal. Atrazine was the second most commonly detected pesticide. Overall, less than one percent of wells contained pesticide concentrations exceeding Health Advisory levels (7).

Few large-scale studies have been required by OPP. Barrett et al. (8) discuss in more detail how large-scale studies are used in pesticide regulation and how early ground-water monitoring studies (for metribuzin and cyanazine) led to the design of the large-scale retrospective study.

Original Monitoring Study Designs. A ground-water monitoring study requirement was developed to provide OPP with solid evidence that a specific pesticide (or degradate) has the potential to move to the saturated zone as a result of normal agricultural use.   Originally, separate types of small-scale monitoring studies were developed for pesticides (1) proposed for registration or without a long history of use, or (2) for registered pesticides with an established use history.   The studies were conducted by pesticide registrants as a part of the reregistration process, with oversight by the Ground Water Section in the Environmental Fate and Effects Division of OPP.

Small-scale prospective studies were designed to evaluate recently registered pesticides.   The study tracks the fate of a pesticide from the time of application until it completely degrades.   It is conducted in a field where the pesticide has never been used, in a worst-case environment for leaching, by:

- applying the pesticide at highest label rates,

- where a shallow watertable aquifer exists,

- where soils in unsaturated zone are highly permeable, homogeneous and isotropic, and

- where climatic conditions are conducive to the persistence of the pesticide in question.

Soil, soil-water, and ground water are sampled for two years.   The original design was the result of a joint research project between the U.S. Geological Survey's Kansas District Office and OPP (9).   Results of theses studies link application during the study to detections of residues in ground water.   The prospective study design is the precursor of the small-scale field monitoring study.

Small-scale retrospective studies were designed to examine a pesticide's effect on local ground-water quality following years of use.   Several sites are monitored to evaluate the spectrum of pesticide uses.   The sites must be representative of a realistic worst case environment for leaching.   This means that they must be typical of sites on which the product is used, yet they must meet specified vulnerability criteria (e.g., a documented history of use of the pesticide, shallow watertable aquifer, highly permeable soils).   It was recommended that a tracer be applied with the pesticide under study, so that the impact of use during the study period could be evaluated.   Ground-water monitoring typically lasts for one year, and, importantly, only ground water is sampled.   A detailed description of retrospective study components and an example are given in (10).

The *large-scale retrospective study* was designed to examine the cumulative effect on ground-water quality from multiple pesticide applications. These studies are focused on areas where ground water is a major source of drinking water.   The study is statistically designed, and a large number of sites are sampled, so that results can be extrapolated to estimate the risk to the population exposed.   Monitoring typically lasts for several years, and only existing drinking water wells are sampled.

Recent Changes in Study Designs

Results from recently completed field studies have identified important considerations for the design of non-point source investigations. Some of the most important issues are discussed in the following section. A Guidance Document has been under development since 1987, and is expected to be released in 1991. Of the two original field-scale study types, prospective studies have provided OPP with the best information about the behavior of the chemical. These studies have resulted in geographic restrictions on usage for several chemicals in highly vulnerable environments such as Suffolk County, New York (an aquifer composed of sand and gravel in a terminal moraine), and in the Central Sands of Wisconsin (glacial outwash). Throughout the United States, intensive agricultural activity is taking place in areas highly vulnerable to ground-water contamination, as shown by Figure 1.

In contrast, results of retrospective studies have proven to be more difficult to interpret. Because only minimal site characterization is required, it is impossible to determine if residues of the pesticide dissipate in the vadose zone, or if the one-year sampling period is insufficient to observe movement. Also, if usage rates or agronomic practices change, it is difficult to separate the effects of years of usage on ground-water quality. This is complicated by the fact that tracers, although recommended, have been rarely used in retrospective studies.

OPP has determined that it is appropriate to first establish the behavior of a pesticide in a worst-case environment in an intensive field study, if laboratory data indicate a potential problem. The prospective field study gives the most useful information, for regulatory purposes, about both new (unregistered) chemicals, and old and new uses of a registered product.

A Tiered Approach. One change in OPP's approach has been to organize field studies into 2 tiers, as illustrated in Figure 2. Progressively higher tiers indicate EPA's increasing concern about the ground-water contamination potential of the compound. The approach is not highly structured to allow for flexibility in the type of study required, and in the study design. If sufficient data are available to indicate that detections of a pesticide are widespread, small-scale studies may not be required prior to conducting a large-scale study. For example, EPA was sufficiently concerned about the mobility and persistence of alachlor to require that a large-scale survey be undertaken to evaluate the extent of contamination without first requiring a field-scale study.

Building in this flexibility also enables study designs to be tailored to investigate compounds that have very different chemical properties and usage patterns. For example, sugarcane is typically grown in organic rich soils and has high water requirements. In part of Florida, unlined irrigation canals border the sugarcane fields. The sugarcane in these areas is irrigated by raising the water levels in the canals using a technique known

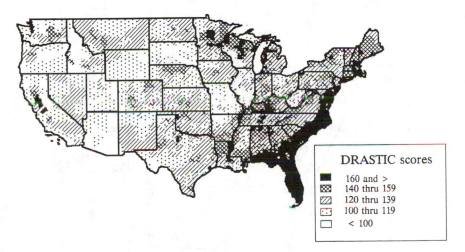

Figure 1. Susceptibility to ground-water contamination by pesticides.

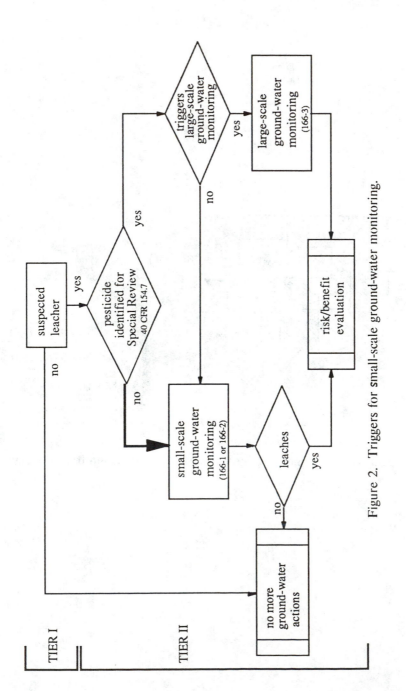

Figure 2. Triggers for small-scale ground-water monitoring.

as subsurface irrigation. Because of the hydraulic connection between ground and surface water in this area, both surface water and ground water must be monitored. At sites where the hydraulic connection is not as direct, surface-water sampling may not be as necessary.

The Tier I studies have been described previously, and are standard data requirements. Data from these studies are evaluated in a leaching assessment, and based on that assessment, further Tier II data may be required. Tier II is composed of several different types of monitoring studies that focus on ground-water sampling. Studies are designed to assess impacts of the pesticide and degradates at the field scale, or at a large (multi-state) scale, depending on the type of information required by EPA.

Study Types. Monitoring studies are conducted on different time-scales and focus on different issues relating to the leaching of pesticides. The *field dissipation study* is designed to evaluate physical and chemical processes that affect the loss of a pesticide from the field. Thus, monitoring focuses on the top meter of the root zone. The *small-scale field study* is designed to track a pesticide's movement in soils, soil-water, and ground water. The objective is to determine the leaching pattern of the pesticide and its degradation products in a specific usage environment. In most cases, a highly vulnerable environment would be selected as an initial study site. Results from this type of study help to establish a boundary condition for the leachability of the pesticide. The assumption is, if the chemical does not leach under extreme conditions it is not likely to leach under less extreme conditions. If a pesticide is shown to reach ground water in a highly vulnerable environment, but many use-areas are not as vulnerable, additional field studies may be required to gain information on how to best regulate the pesticide under typical conditions. The *large-scale retrospective study* requires extensive ground-water sampling, primarily from existing drinking water wells. Results are used to estimate exposure to different levels of pesticides resulting from a pesticide's past use. This type of study is required for registered chemicals in Special Review on a case-by-case basis.

Triggers for Studies. Triggers for Tier II studies are straight forward, and are based largely on results of the Tier I studies. The data requirements in Tier I consist of a battery of basic laboratory tests and one field test. The laboratory studies provide conservative estimates of how environmental processes influence the mobility and persistence of a pesticide. The field study is designed to assess the overall effect of these competing processes on the movement of a pesticide or its degradates moves through the upper portion of the soil profile.

EPA reviews the results of the Tier I tests to determine if the pesticide has a high probability of ground-water contamination. Further Tier II data is required if:

1. The weight of the evidence of all Tier I studies, taken as a whole, indicates that the pesticide has properties and characteristics similar to pesticides that have been detected in ground water (see Table II), and

2. *Either* a field dissipation study demonstrates movement of the parent or degradates 75 to 90 centimeters through the soil profile, *or* other monitoring studies report that the pesticide has been detected in ground water.

These criteria will lead to requirement of a Tier II study, except in cases where the substance is of such low toxicity that no food residue data or tolerance would be required for a major food use.

Small-Scale Field Study Requirements

This section describes requirements for field studies, and focuses on studies conducted in a highly vulnerable environment. Other types of small-scale field studies may be needed as a follow-up to studies in highly vulnerable environments, to establish the leaching potential under more realistic conditions. This may involve conducting a similar study in a different type of environment, or other types of studies, for example, a regional survey of existing wells in a high-use area.

Aquifer Vulnerability Assessments. Several methods for addressing aquifer vulnerability are available. Some vulnerability criteria are specifically for pesticides, others are for ground water in general. The most widely used screening tool is DRASTIC (*11*), developed by the National Water Well Association in conjunction with the EPA. This methodology was used, along with other criteria, to select sampling sites for the NPS. Many other methods are available to assess aquifer vulnerability. These techniques are described in a document soon to be published by EPA's Office of Ground Water Protection.

In the past, the selection of monitoring study sites has been determined by the usage history of the pesticide of concern at the site, hydrogeologic vulnerability, and locating a "cooperator," or land owner willing to participate in the study. A minimum amount of site characterization was required to verify soil and hydrologic attributes at the site prior to initiation of the study. The latest guidance document for the small-scale studies will require that more intensive site information be collected before and during a small-scale field study. Site maps, and the variability of parameters at the site will be defined more fully.

OPP's new guidance document will require that the following be identified or defined at each candidate site prior to final site selection:

- depth to the water table <10 meters,

- percentage of sand, silt, clay, organic matter, and cation exchange capacity of soils,
- definition of any natural drainage features (e.g., karst) affecting the gradient of ground-water flow,
- availability of irrigation equipment,
- surface slope less than or equal to 2 percent,
- definition of any man-made activities affecting the gradient of ground-water flow (pumping stations, well fields, industries nearby),
- definition of any potential restrictive layers in the soil profile (clay lenses),
- site owner cooperation and availability of site for 2–3 years, and
- reconnaissance for potential point sources.

This necessitates a thorough research through local sources of information on the geology and hydrology of the area and collection of soil cores at potential sites. From a set of candidate sites, appropriate sites are selected and approved by EPA prior to initiation of the study.

"Highly vulnerable" conditions for leaching must be independently determined for each pesticide based on the environmental chemistry of the compound. For example, pesticide degradation rates can be affected by climate and soil pH. The study site does not have to be one in which the compound is typically used.

History of Use Requirements for Small-Scale Studies. The study is performed in a field where the pesticide has never been used. Thus, if the pesticide is subsequently detected in soil, soil-water, or ground water it is clear that the source of the residues is from normal field use. The pesticide usage history of the proposed site must be carefully documented.

For some types of representative-use studies pesticide usage history is extremely important. This is especially the case when the study is designed to evaluate the cumulative effect of years of normal use of a pesticide on ground-water quality. These studies must have a documented pesticide use history. Typically, several sites are selected to represent the spectrum of uses, based on sales information provided by the pesticide registrant and based on the geological environments in which the product is used. Like the initial study, areas that are hydrogeologically vulnerable must be selected. While not the dominant concern, if large uses are associated with environments that are less conducive to leaching, for example heavier soils, one of the sites may be selected to characterize the behavior of the pesticide under these conditions. In this circumstance, the study requirements may be altered to include components of surface water monitoring, if pesticides are more easily transported off-site in runoff.

Detailed Site Characterization. Once appropriate sites are selected, detailed site characterization can begin.   This will include the development of detailed site maps identifying topographic and hydrologic features, contour intervals, latitude/longitude coordinates, and off-site features that may affect the local ground-water gradient.

Vadose Zone. The preferred site should have a single soil type (soil series) to insure that vadose zone media are relatively homogeneous.   Where this is not possible, the extent and location of different soil types (series) should be identified and mapped.   This will influence the placement of wells, soil–pore water samplers (suction lysimeters), and the location of soil cores.   Soils are then characterized as to:

- color and physical description,

- texture,

- organic carbon or matter percent,

- bulk density,

- permeability,

- cation exchange capacity,

- clay mineralogy, and

- background pesticide residues.

Much of this information may be available from local sources, e.g., state agricultural extension agents, USDA Soil Conservation Service Soil Surveys, and university research stations.   Available information should be compared to actual field values obtained through sampling and analysis. Hydraulic characteristics of unsaturated media should be assessed by performing infiltrometer tests at several locations at the proposed sites, and developed in sufficient detail that the data can be used in numerical models.

Saturated Zone. Piezometers will be used to establish the configuration of the water-table surface at the site.   Characterization of the local saturated flow conditions must include establishing hydraulic characteristics of the aquifer in sufficient detail that numerical models may be used to simulate pesticide movement in ground water.   Also, the direction of shallow ground-water flow and the hydraulic gradient must be established.   Any man-made activities that might influence the direction of the ground-water flow should be considered and documented.   Seasonal fluctuations in the depth to the water table are important when considering at which depth to place well screens.   By knowing how much the water table is expected to drop, because of seasonal fluctuations, and placing well screens accordingly, one can try to ensure that wells will not go dry during a study.

Water Chemistry. Basic chemical characteristics (major ions, pH, redox state, temperature, and conductivity) should be determined for ground-water samples. These parameters can affect the mobility and persistence of some pesticides. Also, irrigation water should be characterized to insure that ground-water chemistry is not significantly altered in the course of the study.

Data Reporting. Data presentation should include summaries of data collected, rather than just daily information. For longer time periods, daily information is unwieldy and trends can be seen better graphically (in plots of water levels versus time, for example). The daily data are important and should be included on magnetic media as well as data summaries.

Site maps should not differentiate between locations of cluster wells, suction lysimeters, and soils sampled for residue analysis. This information should be clearly shown on one map. Additional maps (using the same base map) should indicate wells that were installed in separate phases of sampling, locations of infiltration tests, locations of soils sampled for textural analysis, and locations of all other spatially distributed data.

Final maps showing the location of all soil series on the site, water table contours, stratigraphic cross-sections, and fence diagrams will provide information necessary to characterize the site hydrogeologically.

Instrumentation. The small-scale ground-water monitoring studies conducted in highly vulnerable or typical environments have more intensive field instrumentation requirements than do field dissipation studies regularly required as a part of registration.

Field dissipation studies, require only collection of soil cores for residue analysis. Small-scale field studies require collection of soil cores, soil-pore water, and ground water. In both cases the field is instrumented with soil-water samplers, and monitoring wells.

Soil Cores. The timing of sampling and the number of sampling dates should be adequate to describe the movement of the pesticide residues downward. Pesticides which are persistent and mobile (usually the case for compounds for which ground-water monitoring is required) should be sampled at weekly intervals for a month or more following application, then biweekly sampling for several months, then monthly sampling until the pesticide has dissipated. The sampling schedule is tailored to suit the conditions of the site, rate of water transmission, and the pesticide. The number of samples should be adequate to indicate the variability of residues in the field.

Soil Water. Soil-water sampling devices are installed on a site at varying depths depending on the depth to the water table. The suction lysimeters continue to qualitatively track the movement of pesticide residues in the soil-pore water at depths beyond the point at which soil sampling by hand becomes difficult, that is beyond 1 meter. It is emphasized here that data from suction lysimeters is only qualitative evidence of chemical movement through the soil profile; it is not quantitative. Soil-water samplers (suction lysimeters) can be placed at 1, 2, and 3 meters in the soil for a water table at the 4.5–9 meter depth. Tensiometers are used in conjunction with suction lysimeters to determine soil moisture. This information can be used to determine when to sample suction lysimeters.

Ground Water. Piezometers are used to establish the depth to the water table, direction of shallow ground-water flow, and the hydraulic gradient. The exact number of ground-monitoring wells installed on the site may vary, but should include a minimum number of "well clusters." A "well cluster" is defined as a group of 2 to 3 wells, located very near to one another, which penetrates an aquifer at 2 to 3 depths. The uppermost well-screen skims the water table, the second well-screen intercepts the aquifer 1.5 to 3 meters below the skimming well. Multiple wells, screened at different depths, allow wells to be sampled despite fluctuations in the elevation of the water table. Thus, if one well goes dry during sampling, a second (deeper) well can be sampled.

Climate. Climatic conditions during the study can affect the depth to which residues move into the subsurface. The study design requires that an irrigation source be available to make up any rainfall deficit, and that weather data be measured on site. The 30-year average rainfall is compared with the rainfall during the study period. Similarly, if surface water monitoring is also required, the characteristics of a large rainfall event should also be determined, for example, the 10-year, 24-hour storm.

Study Duration. The small-scale ground-water monitoring studies are expected to be 2–3 year studies, including site selection, site instrumentation (wells, lysimeters), sampling and analysis, and preparation of the final report.

The study duration varies depending on the climate during the study period, or results reported in a progress report. For example, if a distinct band of pesticide residue is identified in the initial soil sampling phase, and the year of sampling is unusually dry (and irrigation did not make up for the rainfall deficit), residues would be likely to move slowly. In this case, an extended period of sampling may be necessary. No study should be terminated before the questions which the study was designed to answer are adequately addressed.

Sampling Frequency. The effects of temporal variability of pesticide residues in ground water can be taken into account through sampling frequency. Ground water is sampled once a month for a minimum of one year in the small-scale studies, up to two years. Soil-pore water samples are collected when feasible; however, there can be no single schedule as the timing of sampling will vary depending on the soil moisture conditions during the study. Soil cores are collected regularly, depending on the half-life or persistence of the pesticide, but the sampling schedule should take into account the effects of recharge on pesticide residue movement.

Significant time lags exist between application of a pesticide to soil and plant surfaces and detection of the pesticide in shallow ground water in worst case environments. Intervals of 1 to 1.5 years are characteristic before sub-ppm concentrations are observed in highly vulnerable environments.

Tracers. For small-scale prospective type studies, it may be necessary to apply a tracer, such as bromide or chloride, at the same time as the pesticide is applied. Typical rates of application are 30–45 kilograms per hectare. Tracers define the depth of the water front as it moves downward, and identify possible points of preferential flow (*12*).

Conclusion

One of the most important findings of EPA's monitoring study program is that adequate site characterization and instrumentation of non-point source study sites is critical. In general, it has been problematic to derive conclusions from small-scale retrospective studies, conducted in areas with a history of pesticide use. This arises primarily from deficiencies in site characterization, and also difficulties in separating the impact of previous agronomic practices from current practices on ground-water quality. In short, because results of retrospective monitoring studies have often generated more questions than answers, OPP has developed a field-study design that is modeled on the original small-scale prospective study. Results from these small-scale field studies, conducted in highly vulnerable environments, and with a higher degree of instrumentation and site characterization, have provided useful information about the compound's fate.

Reviews of several completed monitoring studies have led OPP to revise the requirements for monitoring. These changes will be fully described in the final Guidance for Small-Scale Ground-Water Monitoring Studies, expected to be completed by the end of 1991.

Acknowledgment

The assistance of Henry Jacoby, Chief of the Environmental Fate and Ground Water Branch of OPP at EPA in Washington, DC, is gratefully acknowledged.

Literature Cited

1. *Code of Federal Regulations, Protection of the Environment;* Title 40, parts 150 to 189; Office of the Federal Register, National Archives and Records Administration: Washington, DC; July 1, 1990.

2. U.S. Environmental Protection Agency. *Pesticide Assessment Guidelines. Subdivision N. Chemistry: Environmental Fate;* EPA-540/9-82-021; Washington, DC, 1982.

3. Holden, P. H. *Pesticides and Ground-Water Quality — Issues and Problems in Four States.* National Academy Press: Washington, DC, 1986; pp 1–124.

4. Cohen, S. Z.; Creeger, S. M.; Carsel, R. F.; and C. G. Enfield.
    In *Treatment and Disposal of Pesticide Wastes;* R. F. Kruegar and J.
    N. Seiber, Eds.; ACS Symposium Series No. 259; American Chemi-
    cal Society: Washington, DC, 1984.

5. U.S. Environmental Protection Agency. *Environmental Fate and
    Effects Division Standard Evaluation Procedure: Terrestrial Field Dissi-
    pation;* EPA-540-0990-073; Washington, DC; December 1989.

6. Williams, W. M.; Holden, P. W.; Parsons, D. W.; and M. L.
    Lorber. *Pesticides in Ground Water Data Base: 1988 Interim Report;*
    U.S. Environmental Protection Agency, Office of Pesticides Pro-
    grams: Washington, DC; 1988.

7. U.S. Environmental Protection Agency. *National Pesticide Survey,
    Project Summary;* Office of Water; Fall 1990, pp 1–10.

8. Barrett, M. R.; Williams, W. M.; Lorber, M. N.; and E. Behl. In
    *Proceedings of the 1990 meeting of the Weed Science Society of Amer-
    ica;* 1990.

9. Perry, C. A.; Eiden, C.; and J. Tessari. *Designing and Conducting
    Investigations of Agricultural Chemical Leaching in the Unsaturated and
    Saturated Zones;* U. S. Geological Survey, Lawrence Kansas District
    Office, Water Resources Investigations Report, 1991.

10. DeMartinis, J. M. *Ground Water Monitoring Review* **1989**, *9*(No. 4),
    pp 167–176.

11. U.S. Environmental Protection Agency. *DRASTIC: A Standardized
    System for Evaluating Ground Water Pollution Potential Using Hydrolo-
    gic Settings;* EPA/600/2- 85/018; Washington, DC; 1985.

12. Davis, S. N.; Campbell, D. J.; Bently, H. W.; and T. J. Flynn.
    Ground Water Tracers, National Water Well Association: Dublin,
    OH; 1985; pp 1–200.

RECEIVED March 19, 1991

STATISTICAL DESIGNS

Chapter 4

Geostatistics for Sampling Designs and Analysis

Allan Gutjahr

New Mexico Institute of Mining and Technology, Socorro, NM 87801

Spatial variability and its affect on groundwater flow and transport is an active research field. The characterization of that spatial (and possible temporal) variability can often be done effectively by using geostatistical techniques. The methods used and the implications for designs and analysis of groundwater transport and pollution problems will be discussed and illustrated. Discussion will include the incorporation of soft-data and their utility.

Classical statistical procedures are mainly concerned with estimating mean values: variation is viewed as a nuisance that needs to be controlled. By way of contrast geostatistics deals with data and problems that include uncontrollable variation that also has structure. The data most often is taken in space (either two or three-dimensions) and is presumed to have some embedded connectedness for continuity.

The objectives of any analysis can vary and include explanation of the variability, construction of predictive models, interpolation and extrapolation of values, design of sampling plans and interrelationships between different properties like conductivity and concentration. The geostatistical approach views variation as part of an overall problem that can convey information about the phenomena being studied.

Another aspect to note about geostatistics is that often only one realization is available and any inference requires additional assumptions. For example, in making predictions about contamination only data from a single region may be available. In addition, predictions for that region, taking into account observations, are desired. While there may be multiple observations, they are generally not independent and consequently many of the classical statistical procedures are not applicable.

The statistical inter-relationship is summarized by a covariance function that is a measure of how observations at different locations are

0097–6156/91/0465–0048$11.75/0

© 1991 American Chemical Society

statistically connected. This covariance behavior needs to be inferred from the data. The result can then be used to make predictions for values at unsampled locations along with measures of uncertainty for those predictions.

In this paper the sections start with the fundamental terms and definitions and then move on to some ramifications and current developments. The key geostatistical definitions are summarized in a glossary at the end of the paper. The approaches will be compared to and contrasted with classical statistical ideas and important issues that arise in the application of geostatistics will be illustrated. In addition to the estimation problem concerns about prediction and the associated uncertainties will be addressed. The use of qualitative and "soft" data will be discussed and some simple examples will be used to show the utility of this approach. Several examples and a case study will be presented. The chapter closes with some open questions and a glossary.

An elementary and clear introduction for readers with a minimal background in statistics is given elsewhere (*1*). More advanced texts that require some greater knowledge of probability and stochastic processes are available in brief (*2*) and expanded (*3*) versions. For non-linear geostatistics and other extensions, the brief monograph (*2*) is highly recommended.

Geostatistical Concepts

Random Fields. The stochastic concepts underlying geostatistics come from the area of random fields, although Journal (*4,5*) also presents some of the ideas with a minimum of random field theory. Locations in space (1,2,3 or 4 dimensions where time may be included) are designated by x and a generic variable or function of x is designated by $V(x)$.

Definition: $V(x)$ is a random field if, for any fixed location x_0, $V(x_0)$ is a random variable.

$V(x)$ is also called a spatial stochastic process. The term random field is preferred here because it emphasizes the spatial nature of the region over which the process is defined. One point to note about this general definition is that a description of the statistical or probabilistic behavior of $V(x)$ involves not just the univariate distribution $P[V(x) \leq v_0]$, (the probability that $V(x)$ is less than or equal to v_0), but also the joint distributions $P(V(x_1) \leq v_1, V(x_2) \leq v_2,...,V(x_n) \leq v_n)$ for any set of locations $x_1...x_n$. An important aspect of the model is that $V(x_1)$ has some statistical relationship with $V(x_2)$. This contrasts, for example, with multiple regression trend surface models where $Y(x_j) = m(x_j) + e_j$; $m(x)$ is some non-random function and the e_j's are statistically independent.

The probability model postulates a structure where one could, in theory, have different outcomes "paths" or "realizations" at any location x. The operator E is used for expectation and $E[V(x)]$, is the expected value of V at location x, is taken over all possible "paths" or "realizations" at x. Namely $E[V(x)]$ is an ensemble average.

Stationarity. Because a complete joint probability distribution is virtually impossible to find, the first two moments are often the main focus:
The mean: $E[V(\underline{x})] = m(\underline{x})$
The covariance: $Cov[V(\underline{x}), V(\underline{y})] = E\{[V(\underline{x})-m(\underline{x})][V(\underline{y})-m(\underline{y})]\}$.

Here $\underline{x}$ and $\underline{y}$ denote two different locations in space. Even these, at this level of generality, may be hard to estimate and invariably further kinds of stationarity decisions or assumptions are made. The stationarity decision refers to the model (4-6) and is not one that can be tested in any statistical manner. Stationarity assumptions relate to inference questions (4,5). In standard statistical studies one has a population picked a-priori about which inferences are made: This population decision is analogous to the decision about stationarity. Stationarity may be reasonable to assume over one region studied (e.g. km²) but not over a larger region where subdivision or stratification may be required. Observations can throw doubt on the prior decision but can't be used in any direct statistical test.

In a conventional sampling study the variable of interest could be a variable like telephone usage. An initial study may not involve stratification but further examination of results could lead to a finer breakdown and use of special sampling techniques. Analogously, in geostatistics an initial study may involve a composite sample that after careful scrutiny could be divided into sub-divisions where the stationarity hypothesis is tenable. In another context stationarity can be taken as a working hypothesis.

Statistical Homogeneity. A common statistical assumption is that of statistical homogeneity or second-order stationarity.
Definition : $V(\underline{x})$ is statistically homogeneous if
 (i) $E(V(\underline{x})) = m$ is constant
 (ii) $Cov[V(\underline{x}),V(\underline{y})] = C_v(\underline{x}-\underline{y})$ only depends on the separation vector $\underline{x}-\underline{y}$.
Once again $\underline{x}$ and $\underline{y}$ denote two different locations in space.

The function $C_v(\underline{x}-\underline{y})$ is called the (auto-) covariance function for the field $V(\underline{x})$. If $C_v(\underline{x}-\underline{y})$ depends only on the distance $||\underline{x}-\underline{y}||$ and not the direction then $V(\underline{x})$ is called statistically isotropic. Figure 1(a) shows some typical forms of theoretical covariance functions and Table I summarizes associated formulas.

For example, the porosity in a region associated with a given layer of material can be taken as the random field $V(\underline{x})$. The covariance function at $\underline{0}$ is the variance of the field. The correlation function $\rho_v(x-\underline{y}) = C_v(\underline{x}-\underline{y})/C_v(\underline{0})$ is a measure of how well linear prediction can be made about $V(\underline{y})$ based on an observation $V(\underline{x})$. A correlation function that dies out slowly essentially corresponds to a "smoother" process that has a higher degree of predictability even if the variance is large.

Table I

Some Covariance Functions

$$\underline{s} = \underline{x} - \underline{y} = \text{separation vector}$$

Exponential: $C(\underline{s}) = \sigma^2 \exp\{-|\underline{s}|/\lambda\}$

Bell: $C(\underline{s}) = \sigma^2 \exp\{-|\underline{s}|^2/\lambda^2\}$

Triangular: $C(\underline{s}) = \sigma^2 [1 - |\underline{s}|/b]$, $|\underline{s}| \le b$
$C(\underline{s}) = 0$, $|\underline{s}| > b$

Exponential: $C(\underline{s}) = \sigma^2 \exp[-(\sum_{j=1}^{n} s_j^2/\lambda_j^2)^{1/2}]$

Spherical: $C(\underline{s}) = \sigma^2[1-1.5(|\underline{s}|/B) + 0.5(|\underline{s}|/B)^3]$, $|\underline{s}| \le B$
$C(\underline{s}) = 0$, $|\underline{s}| > B.$

Analogy to Nested Designs. The role of covariance or correlation is central in geostatistics and is one feature that sets it apart from classical statistics. Yet it is well to remember that in classical statistics correlation, too, plays an important role. For example in completely randomized (or nested) experiments (7), a model of the form

$$Y_{ijk} = \mu + A_i + B_{ij} + e_{ijk}; \quad i=1...I, \; j=1...J, \; k=1...K \tag{1}$$

is postulated. This model applies, for example, to measurements taken when I fields are selected at random, J plots within each field are selected and then K measurements made at random within each plot. Typically μ is assumed to be an unknown constant, the A, B and e random variables are independent and there is an intra-class correlation between Y_{ijk} and Y_{ijl}.

The nested model has several analogies with the stochastic models. The decision to write the data as in model (1) is similar to the stationarity decision - it is an underlying decision about the experiment. It can be modified based on the observations but it is not generally a testable question. An important aim in a nested experimental design is to characterize the variability associated with the various sources (e.g. fields, plots and measurements). In the random field model the estimation of the covariance function also has as its aim the characterization of variability.

Characterization of variability. What are some important features that characterize this variability? Two of these have been mentioned before: The variance of the field $V(\underline{x})$ and the covariance function. The covariance function, when rescaled by the variance, indicates predictability. The behavior of the covariance function is often summarized by a "scale" indicating a significant correlation distance. There is no rigorous definition of such a scale: here it is simply taken to be that value λ where $C_v(\lambda)/C_v(0) = e^{-1}$, assuming $V(\underline{x})$ is statistically isotropic.

Definition: The <u>scale</u>, λ, is that value where $C_v(\lambda)/C_v(0)=e^{-1}$.

Figures 1(b,c) illustrate some of these ideas. Figure 1(b) shows several paths from a process with small variance ($\sigma^2_v = \text{var}[V(x)] = 1$) and a short scale ($\lambda = 1.0$). Figure 1(c) shows paths with larger variance ($\sigma^2_v = 2$) but a long scale ($\lambda = 10$). In Figure 1(c) one can do a fairly good job of predicting $V(x + 0.5)$ from $V(x)$ - essentially the paths are smoother than in Figure 1(b) but there is a larger variability <u>between</u> paths. This again is analogous to the nested case. The variability within shorter path segments is small while the variability between paths is large.

In general, the functional form of the covariance function is not as important as the value of the variance and the scale.

Intrinsic Random Fields and Variograms. The concept of statistical homogeneity or second-order stationarity may be too restrictive and this has led to related stationarity constructs, most notably that of an intrinsic random field of order 0 (8,9).

Definition: $V(\underline{x})$ is an <u>intrinsic random field of order 0</u> (IRF-0) if

    (i) $E[V(\underline{x})] = m$

    (ii) $\gamma_v(\underline{x}\text{-}\underline{y}) = 1/2 \; E[V(\underline{x}) - V(\underline{y})]^2$ only depends on $\underline{x}\text{-}\underline{y}$

$$(iii) \quad var \; [\sum_{i=1}^{n} \alpha_i V(x_i)] \; \text{is finite if} \; \sum_{i=1}^{n} \alpha_i = 0$$

where var [ ] denotes the variance of the expression within [ ]. $\gamma_v(\underline{s})$ is called the (semi-) variogram, and $\underline{s} = \underline{x}\text{-}\underline{y}$ is the vector between the two locations $\underline{x}$ and $\underline{y}$. The modifier semi- is dropped and $\gamma_v(\underline{s})$ is simply referred to as the variogram.

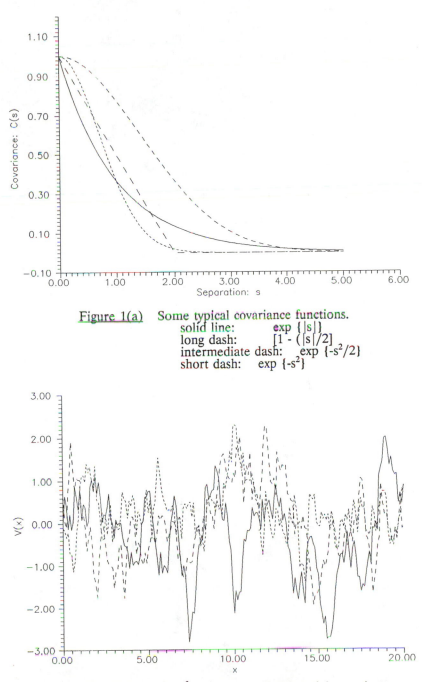

Figure 1(a)   Some typical covariance functions.
solid line: $\exp\{|s|\}$
long dash: $[1 - (|s|/2]$
intermediate dash: $\exp\{-s^2/2\}$
short dash: $\exp\{-s^2\}$

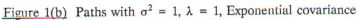

Figure 1(b) Paths with $\sigma^2 = 1$, $\lambda = 1$, Exponential covariance

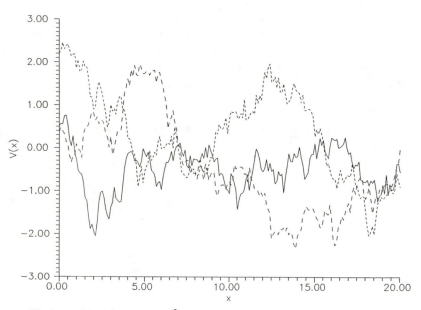

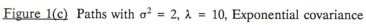

Figure 1(c) Paths with $\sigma^2 = 2$, $\lambda = 10$, Exponential covariance

A simple example of a random field in 1-dimension that is IRF - 0 is Brownian motion. The discrete version is of the form

$$V(m) = \sum_{k=1}^{m} W_k$$

where $E[W_k]=0$, var $(W_k) = \sigma^2 \Delta$ and the W_k's are independent. The covariance function is cov $(V(k) V(l)) = \sigma^2 \Delta$ min (k,l) and so $V(k)$ is not statistically homogeneous. On the other hand $\gamma_V(j) = \sigma^2 \Delta j/2$ and hence $V(k)$ is IRF = 0.

When $V(\underline{x})$ is statistically homogeneous then

$$\gamma_V(\underline{s}) = C_V(\underline{0}) - C_V(\underline{s}). \qquad (2)$$

The variogram is a measure of dissimilarity (2). For the statistically homogeneous case, from equation (2) it is seen that $\gamma_V(\underline{\xi})$ is the inverted covariance function. The limiting value of $C_V(\underline{0})$ is often referred to as the sill.

Data analysis and Geostatistics

Ergodicity. In most geostatistical studies the covariance function or variogram needs to be estimated along with other features of the observed data. This estimation is almost always based on observations from a single realization and invariably an assumption of ergodicity is made.

The assumption, loosely stated, says that if data from a single realization is taken over a large enough field then the sample mean and covariance/variogram will be close to the ensemble mean and covariance/variogram. In other words, space averages will converge to ensemble averages. This is a strong assumption and again is not amenable to statistical testing. For the field that has realization shown in Figure 1(c) one would need a sample over a large range to achieve the desired convergence. The ergodicity decision may be de-emphasized if the inference is for spatial averages on specific areas (5).

Data Analysis. Assume n observations, $V(\underline{x}_1)...V(\underline{x}_n)$ are available and the objective is to estimate quantities like the mean, covariance, and variogram. Prior to undertaking such estimation an exploratory data analysis should be carried out for the field. A good elementary reference for exploratory data analysis is the paper by Tukey (10).

Note that the data here is not like an independent random sample. Concepts like the empirical distribution function still can be used but distributional tests (e.g. chi-square or Kolmogorov-Smirnov goodness of fit tests) are not directly applicable. In effect the correlated nature of the data yields less information for certain purposes.

A declustering technique (*11*) can be used to account for the correlation where the estimator of the distribution function is similar to a probability weighted estimator of the type used in sampling theory. The idea is to overlay a regular grid (say of L squares) where the squares are large enough so all contain at least one point. The distribution function is then estimated by using averages of indicators within the squares, averaged over L.

In carrying out an exploratory data analysis, percentiles and quantiles (e.g. the median and deciles) can be especially useful. If there is sufficient data available the regions should be split into several subregions to see how these quantities change, whether transformations (eg. log, square root) are needed and whether other corrections would be useful.

Effective values. One concept that carries over simply from time series is that of an effective sample size. Let $V(\underline{x}_i)$, $i = 1, 2...n$ be a sample of size n from a statistically homogeneous random field with covariance function $C_v(\underline{x}-\underline{y})$. Consider estimating the expected value, $E[V(\underline{x})]$ by the sample

mean:
$$\bar{V} = \sum_{i=1}^{n} V(\underline{x}_i)/n.$$

Definition: Let

and
$$\sigma_1^2 = C_V(0)/n$$

$$\sigma_2^2 = var(\bar{V}) = \sum_{i=1}^{n} \sum_{j=1}^{n} C_v(\underline{x}_i - \underline{x}_j)/n^2.$$

The effective sample size, n_{eff}, is defined as

$$n_{eff} = n(\sigma_2^2/\sigma_1^2) \tag{3}$$

The ratio σ_2^2/σ_1^2 is a reduction factor due to the correlation structure of $V(\underline{x})$. If the $V(\underline{x}_i)$'s were independent the variance of $\bar{V}$ would be σ_1^2. However, the actual variance in the non-independent case is σ_2^2. The effective sample size gives the equivalent number of independent observations yielding the same variance for $\bar{V}$. Note that this reduction factor will depend on the data location and the covariance function. In Table II the reduction factor is shown for several designs in two dimensions, where the covariance function used was $\exp\{-|s|/\lambda\}$.

Table II
Reduction Factors for Sample Designs

Sample Network	Effective Sample Size	Reduction Factor σ_1^2/σ_2^2
10 x 10, Spacing = 0.5 (n = 100)	6.87	0.069
10 x 10, Spacing = 1 (n = 100)	19.62	0.196
10 x 10, Spacing = 2 (n = 100)	54.50	0.545
5 x 5, Spacing = 1 (n = 25)	6.38	0.255
5 x 5, Spacing = 2 (n = 25)	14.81	0.592

For example, if 100 samples are taken on a regular grid at spacing of .5 correlation scales, then the effective number of independent samples is only 6.87. Furthermore, for a spacing of 1 correlation length on a 5x5 grid, the effective sample size is 6.38 and for the same spacing on a 10x10 grid the effective sample size is 19.62.

The covariance structure consequently implies that, at least for averages, one can have considerably less information than in the independent case. The factor σ_2^2/σ_1^2 shows the reduction in the sample size needed to get n_{eff}.

Barnes (*12*) discusses a related concept and shows that the effective sample size has behavior of the type given in Table II by using a simulation approach. Other authors (*13*) have also examined estimation of means using geostatistical techniques and found similar results.

These considerations would imply that to get a greater effective sample size one would increase the intersampling distances. However we will see below that for accurate variogram estimation the reverse is true: In that case the intersample distances should be small.

Covariance and variogram estimation procedure. The data $V(\underline{x}_i)$ i = 1...n will ordinarily not be on equi-spaced or regular grids, but rather scattered across the field. For example, $V(\underline{x}_i)$ may represent concentration of a substance at location $\underline{x}_i$, where the $\underline{x}_i$'s are not on a grid.

The procedure for variogram/covariance estimation involves grouping the data into classes. Several standard programs (3,14) are available for the variogram calculations and for the non-ergodic covariance (*1,15*). The procedure described below allows for non-isotropic covariances: That is the covariance may depend on both distance and angle between points. In addition, it is restricted to a field in two dimensions.

The input to the estimation procedure involves a lag-increment parameter δ, a set of directions θ_i $i = 1...I$ and an angular tolerance parameter α,

Step 1 Form the set of data pairs

$A(r\delta,\theta_i)$ = the set of $\underline{x}_j$ and $\underline{x}_k$ with distance between $(r-1)\delta$ and $r\delta$, and angle between $\underline{x}_j$ and $\underline{x}_k$ in the interval $(\theta_i - \alpha, \theta_i + \alpha)$.

$$= \{(\underline{x}_j,\underline{x}_k):(r-1)\delta \leq ||\underline{x}_j-\underline{x}_k|| < r\delta; \theta_i-\alpha<angle(\underline{x}_j,\underline{x}_k)<\theta_i+\alpha\}$$

Let $N(r,i)$ = The number of pairs in $A(r\delta,\theta_i)$

Step 2 Form the estimates

$$\hat{C}[r\delta,\theta_i] = \Sigma[V(\underline{x}_j) - \hat{m}_r(\underline{x}_j)][V(\underline{x}_k) - \hat{m}_r(\underline{x}_k)]/N(r,i)$$

where the sum is over distinct points in $A(r\delta,\theta_i)$, for $r=1, ...R$. The estimator $\hat{C}[r\delta,\theta_i]$ is based on points that are separated by distances between $(r-1)\delta$ and $r\delta$ and by angles between $\theta_i-\alpha$ *and* $\theta_i+\alpha$. The quantity $\hat{C}[r\delta,\theta_i]$ estimates $C_V(\underline{s})$ where $\underline{s} = [(r+\frac{1}{2})\delta \cos\theta_i, (r+\frac{1}{2})\delta \sin\theta_i]$ in two dimensions.

In standard (ergodic) estimates, $\hat{m}_r(\underline{x}_j) = \hat{m}_r(\underline{x}_k) = \bar{V}$. A non-ergodic covariance where $\hat{m}_r(\underline{x}_j)$ is an average over a smaller set, has been proposed (14). Specifically if the isotropic case is considered, $\hat{m}_r(\underline{x})$ is the average over all points that are within distance $(r-1)\delta$ to $r\delta$ of some other sampling point, and estimates $E[V(\underline{x})]$. This estimator is especially useful if the observations seem to show a trend.

Generally the maximum separation used, $R\delta$, should not be too large. Although many authors calculate $\hat{C}$ out to very large separations, the author recommends being very wary of any estimates that are calculated when $R > 0.20$ max $||\underline{x}_j-\underline{x}_k||$.

Covariance and Variogram Estimation: Practice. Variogram or covariance estimation is one of the more important steps in geostatistical applications (*16*). It is complicated by the fact that assumptions like constancy of the mean may not be tenable in practice. In addition a phenomena can have variations on several scales necessitating the use of nested structures. For example, permeability will show one scale when a plot 100 m^2 is considered and display a second scale when it is examined on a larger domain. In the variogram there will then be several sills or limiting values as illustrated in Figure 2(a). One kind of multi-scale effect that shows up in many variogram estimates is the nugget effect; the variogram estimator at the origin is not zero. Two possible causes for this are:

(a) measurement error, uncorrelated from point to point

(b) extrapolation to the origin based on observations that are not spaced closely enough to detect shorter scales.

In the latter case there can be small scale variation that is not observed due to the sample spacing. The nugget effect can also be obtained by simple extrapolation to the origin and in that sense be artificially introduced.

Outliers can significantly influence variogram/covariance estimates (*17,18*) as can non-normality and trends (*19*). Simple trends will be exhibited by quadratic behavior in the variogram estimates (Figure 2(b)). Theoretical considerations show that true variograms can not behave quadratically at large separations.

The use of trend surfaces to first remove the effects of varying means leads to biased estimators. An iterative trend removal procedure that may seem logical has been proposed (20). However the statistical properties of the method are difficult to determine. Trends could also be accounted for by looking at directional variograms (*21*) where the variogram is estimated by examining the data in a direction perpendicular to the trend direction.

Alternative estimators like median polished estimators that are especially useful for partially gridded data have also been proposed (*22,23*). One of the more promising techniques is the non-ergodic covariance estimator (*1,15*) discussed above. It is easy to implement and seems to work well in a wide variety of situations.

Once an acceptable variogram estimator is selected, the general approach is to next fit a theoretical model. Except for the case when the data is presumed multivariate normal (*24*) the fit is generally "by-eye" and has no statistical basis. In practice one might be tempted to use least squares methods to fit the variograms with parametric functions. However the assumptions underlying standard least squares are severely violated by the non-linear variogram estimators. Specifically the variogram estimates are highly correlated while standard regression models assume independence. If weighted least squares are used the weighting factors should be related to the covariances for the estimators: these covariances are difficult to find. For these reasons the use of least squares estimators for variograms is

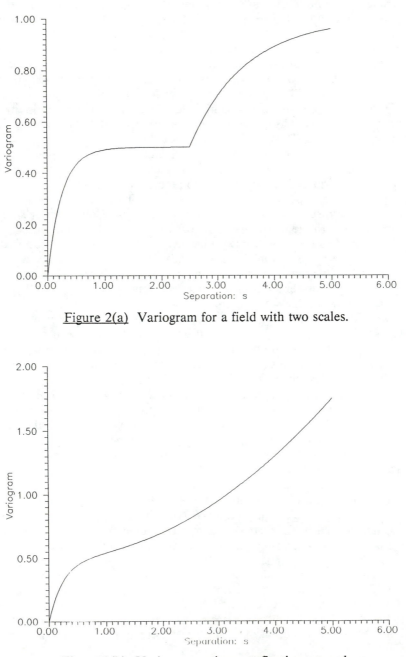

Figure 2(a) Variogram for a field with two scales.

Figure 2(b) Variogram estimate reflecting a trend.

discouraged. If it is used one can place no great credibility in confidence intervals and related quantities for one variogram parameters.

Some of the difficulties of variogram estimation are illustrated elsewhere (*25*) where data was generated using an exponential covariance and then variograms fit to the observations. The best fit variogram was linear rather than exponential in form with a nugget effect in one network. This poor fit was most likely due to the lack of data at small separations and further shows how spurious nugget effects can be introduced by extrapolation.

Estimation of variograms has been examined (*26,27*) using simulated fields. The estimation of scale and variances was difficult and to get reasonable estimates of the scale, λ, points are needed that are closer than $\lambda/2$. Furthermore, any trends in the realization will yield over-estimates of λ. It would be useful to re-run these experiments using the non-ergodic covariance (*15*).

Covariance and variogram estimation: Network Design. Most of the work on spatial sampling design has concentrated on estimation of means rather than covariances (*28*). A few studies, however, have examined estimation of structural properties like variograms and covariances.

A sample placement procedure that maintained the number of couples in a class at a fixed level (*29*) yielded an overall regular network with superimposed clusters. A systematic rather than random network has several advantages (*25*) though difficulties arise when multiple scales may be present. In that case, placement of regular samples (more or less) at large grid spacings with clusters around these locations may be preferred. The idea is to try to sample on several scales to capture any possible nesting.

Within geostatistics, sampling is often not point sampling (*30*): observations can be averages over zones of influence and that affects the variogram produced.

Covariance and variogram estimation: Summary. Geostatistical methods have had extensive applications in mining and other geological areas. In most of those cases the sample locations are pre-determined and that is perhaps one reason why there has been little work in sample design, especially in-so-far as variogram or covariance estimation is concerned. Essentially the samples used correspond to convenience samples, may occur preferentially in high pay-off zones and hence tend to be biased.

Some work (*31*) has been done on statistical modeling for correlated data using both parametric and non-parametric tests including extended t-tests. However the results assumed that the form of the covariance structure, with the exception of the variance, was known.

As indicated above, there is a conflict between sample size

placement for estimation of the mean, where larger sample spacing gives a higher effective sample size, and estimation of the covariance or variogram where smaller inter-sample distances are required to delineate the structure.

Other difficulties in the variogram/covariance estimation stage are
(1)   one needs considerable data (at least 50 points) to get reasonable estimates
(2)   It is difficult to separate trends from variability
(3)   The statistics of estimators and resulting reliability are hard to find.

The exact variogram form may not be important but even then variances and scales, which influence other properties of interest like transport, are hard to estimate.

Estimation and Kriging

As important as the characterization of variability is, in general it is not the final aim of a practical study. Invariably predictive statements about the process are desired. For example in studying a plume of material moving through the subsurface one wants to know
(1)   What is the location of greatest concentration?
(2)   How is the plume spread out?
(3)   What are the chances that at location x_0, level C_0 is exceeded?
(4)   Where will the plume be at a later time?

To answer these questions methods like kriging (8,9) can be used. It should be noted that relations between various different random quantities (e.g. between permeability and concentration contours) are often of interest and here either a model-based approach or a purely statistical approach could be applied (32).

Kriging: Theory. Kriging focuses on the following problem: Given values $V(x_i)$, i = 1...n, and an identified covariance function $C_v(x,y)$ = $Cov[V(x),V(y)]$ how can one best estimate $V(x_0)$ at a non-observed location x_0?

What is meant by best estimation? Generally this means that the mean squared error $E[V(x_0) - \hat{V}(x_0)]^2$ is minimized, though other loss-functions could also be considered (33,34). The general solution to this problem is $\hat{V}(x_0) = E[V(x_0)|V(x_1)...V(x_n)]$, the conditional expected value of $V(x_0)$ given the data $V(x_1)...V(x_n)$. This is usually hard to find and instead a <u>linear</u> estimator is used. This linear estimator can be viewed as the linear part of a McClaurin series approximation to the true conditional expected value:

$$E[V(\underline{x}_0)|V(\underline{x}_1)...V(\underline{x}_n)] \approx \sum_{j=1}^{n} \lambda_j V(\underline{x}_j).$$

The λ_j's are additionally chosen so that the expected value of the estimator equals the expected value of V at $\underline{x}_0$ (unbiasedness):

$$E[V(\underline{x}_0)] = \sum_{j=1}^{n} \lambda_j E[V(\underline{x}_j)]$$

If the $V(\underline{x})$ process is modelled as a statistically homogeneous or IRF - 0 field then

$$\sum_{j=1}^{n} \lambda_j = 1.$$

The mean squared error is quadratic in the λ_i's and the resulting problem can be solved by using a Lagrange multipliers and generating $n+1$ equations in $n+1$ unknowns.

This conventional approach leads to the $(n+1)$ equations

$$\sum_{j=1}^{n} \lambda_j C_v[\underline{x}_i\underline{x}_j] - \mu = C_v[\underline{x}_i\underline{x}_0]; \quad i = 1...n. \tag{4a}$$

$$\sum_{j=1}^{n} \lambda_j = 1 \tag{4b}$$

with minimum mean squared error (kriging variance)

$$\sigma_k^2 = var[V(\underline{x}_0) - \hat{V}(\underline{x}_0)] = var[V(\underline{x}_0)] - \sum_{i=1}^{n} \lambda_i C_v(\underline{x}_i\underline{x}_0) + \mu \tag{4c}$$

An alternate way to obtain this estimator is also instructive. The unknown mean can first be estimated using a minimum mean-squared error criterion and then the residuals (which have mean zero and lead to unconstrained optimization) can be estimated. The recombined estimators yield the same kriging estimator as those in equations (4). The mean estimation equations are (8,35)

$$\sum_{j=1}^{n} \lambda_{jm} C(\underline{x}_j,\underline{x}_i) = \mu_m: \quad i=1..n \tag{5a}$$

$$\sum_{j=1}^{n} \lambda_{jm} = 1 \tag{5b}$$

and the residual estimator equations are

$$\sum_{j=1}^{n} \lambda_{js} C_v (\underline{x}_j,\underline{x}_i) = C (\underline{x}_i,\underline{x}_0); \quad i=1..n \tag{6}$$

The original kriging weights, λ_j, are then

$$\lambda_j = \lambda_{js} + (1 - \sum_{i=1}^{n} \lambda_{is}) \lambda_{jm}. \tag{7}$$

The qualities μ_m are Lagrange multipliers determined from (5b). Specifically the mean is estimated using

$\sum_{j=1}^{n} \lambda_{jm} V(x_j)$. The residual at location x_i is $V(\underline{x}_i) - \sum_{j=1}^{n} \lambda_{jm} V(x_j)$ and it is these residuals which are used to estimate the residual at $\underline{x}_0$. To estimate the value at $\underline{x}_0$, the mean is added back onto the residual.

There is also a simple expression for the kriging variance using this approach. To see the advantages more clearly equations (5) and (6) are re-written in vector-matrix form where $\underline{e}^t = (1,...1)$, $\underline{C}$ is the n by n covariance matrix, $\underline{C}_0$ is the column vector $(C_v(\underline{x}_1,\underline{x}_0)...C_v(\underline{x}_n,\underline{x}_0))^t$, and $\underline{\lambda}_m = (\lambda_{1m}...\lambda_{nm})^t$. Then

$$\underline{C} \, \underline{\lambda}_m = \mu_m \, \underline{e} \tag{8a}$$

$$\underline{e}^t \, \underline{\lambda}_m = 1 \tag{8b}$$

$$\underline{C} \, \underline{\lambda}_s = \underline{C}_0 \tag{9a}$$

$$\underline{\lambda}_m = \mu_m \, \underline{C}^{-1} \underline{e} \tag{9b}$$

Hence only one matrix inversion is needed and furthermore this matrix, unlike the matrix of system 4(a,b), is positive semi-definite.

Another added advantage is that if moving neighborhoods are used, estimates of both the mean and actual value are obtained. This breakdown of the prediction into a mean and residual can be illuminating.

In the statistically homogeneous case $C_v(\underline{x},\underline{y})$ will only depend on $\underline{x}-\underline{y}$. In the IRF - 0 case the estimation equations are like those in equations (4), (5) and (6) except that $C_v(\underline{x},\underline{y})$ is replaced by $-\gamma_v(\underline{x}-\underline{y})$.

Note that the objective of kriging is to predict an actual value at $\underline{x}_0$ and not the mean. In that sense, as well as in the assumption of non-independence, it differs from trend surface or multiple regression analysis where the aim is to estimate a mean response function.

Kriging properties. Kriging estimators have several attractive properties *(9)*:

(1)   Kriging estimators agree with the data - i.e. they are exact interpolators

(2)   Kriging weights only depend on the locations and covariance functions and not on the actual $V(\underline{x}_i)$ values. A similar statement holds for the kriging variances.

(3)   Kriged values are correlated and are smoother than the actual field but more variable than the mean value.

(4)   The kriging estimators $\hat{V}(\underline{x}_0)$ can be viewed as an approximation to the conditional mean while the associated kriging variance $\sigma_k^2(\underline{x}_0)$ can be viewed as an average of the conditional variances.

(5) $\hat{V}(\underline{x}_0) \pm 2\sigma_k(\underline{x}_0)$ is an approximate 95% tolerance interval in the

case where $V(\underline{x})$ is a Gaussian (jointly normal) field. However this is not the same as a confidence interval.

Network design. Property (2) above can be used to carry out network design studies *(21,36)*. Because the weights and variance only depend on the covariance or variogram and location the effects of added measurements can be explored. This property can be further extended to the case of two or more fields where cross-covariances and co-kriging are used. For example, the inner data points can screen effects from farther points and show the effects of different types of data on the estimation procedure (37). Property (2) can be used to easily examine location of possible future samples and the effects of different designs. However it can be impractical to apply if many added points are desired.

Kriging does not require that the $V(\underline{x}_i)$'s be normally distributed if only the estimators are examined. However because normality leads to other useful properties the data is often transformed to get approximate n o r m a l i t y (e . g . v i a a n o r m a l s c o r e s transform: $Y_i = \phi^{-1}[\hat{P}_i] = F[V(\underline{x}_i)]$ where $\hat{P}_i$ = percentile of $V(\underline{x}_i)$ and ϕ is the standard normal).

Property (5) must be interpreted with care. It is not a confidence interval in the conventional sense; $\sigma_k(\underline{x}_0)$ in fact does not depend on the data $V(\underline{x}_i)$. Furthermore only in the multivariate normal case can probability content be ascribed to this interval. Otherwise the probability can be bounded by the Chebychev or Markov inequalities though it will still not be conditional on the data. For probability predictions as we discussed below, direct estimates via probability and indicator kriging may be preferable. The kriging variance is primarily a measure of data configuration and the role of different configurations on predictions (4).

Kriging can be validated by a jackknifing or "leave-out-one" procedure. This gives a global validation and here, as in regression analysis, graphs of the residuals are also useful for diagnostic purposes. Cross-validation is essentially an exploratory data analysis method and not a hypothesis-testing procedure (38).

Kriging variations. The stationarity assumptions require a constant mean but in practice this may often not be reasonable. As discussed in the variogram estimation section, an a-priori mean function could be fit to the data and the means removed (20) though that practice can introduce bias. Alternatively, several extensions - universal kriging and higher order intrinsic random functions (8,39,40) have been proposed. Journel and Rossi (41) argue that if moving neighborhoods and non-ergodic covariances are used these models are unnecessary in practice. The difficulty with universal kriging is that the problem becomes circular when it comes to covariance estimation.

The kriging procedure can be extended to estimate block averages or weighted averages. A variety of computer codes are available (3,14) (Geo-EAS available from USEPA, Evan Englund; Environmental Monitoring Systems Laboratory, P.O. Box 93478, Las Vegas, NV, 89193-3478) for carrying out kriging and geostatistical calculations.

Applications Review. Many non-model based and essentially purely statistical applications of kriging have appeared in the recent literature (42-48). It has been used to delineate high concentration areas in a study involving dioxin spills (42,43), to determine areas for additional conductivity and water surface levels (44) and to estimate concentrations and total contaminant in groundwater pollution problems (45-47) and chemical clean-up problems (48).

The results of kriging can be used for both risk evaluation and sampling design. Problems with the application of these geostatistical methods are; means may change throughout the domain thereby complicating variogram estimation (*43,45-47*); data may be censored (*42*) and taken preferentially in hot-spots (*48*); the fields may not be additive in nature (*45-47*); and, the studies ignored any physical interconnections between the fields. For concentration and probability estimates a useful alternative may be indicator kriging which is described below. Some kriging-based geostatistical design studies also have dealt with characterization and network design (*49*).

A comparison of kriging and regression analysis (*50*) led to the conclusion that with fewer than 50 points kriging has few advantages. Note, however, that the two techniques have different objectives as discussed above. Kriging in conjunction with conditional simulations can assist in determination of optimal sampling locations and in controlling the variability of the estimates (*51*).

Model Based Geostatistical Studies. The applications above along with many geostatistical applications in the literature, are non-model based in the sense that the field $V(\underline{x})$ is considered in isolation or when only statistical relationships between two fields are considered. However, many flow and dispersion phenomena are explained at least partially by physical and mathematical models. For example, transport is generally hypothesized to obey a convection-dispersion equation where $U(\underline{x})$, the velocity, may be random as well as the concentration $C(\underline{x})$. Similarly Darcy's law and continuity equations lead to partial differential equations involving both conductivity and head (*52*).

The added model information can be important for evaluating worth of data and for making predictions. Several model-based studies have been treated (*37,53-55*) and a more detailed discussion of these results is given elsewhere (*32*, Gutjahr, A.L., Math. Geol., In press). Other investigators (*56*) have partially validated theoretical stochastic models of transport (*52,57*).

Indicator Kriging and Soft Data

Indicator and probability kriging, along with indicator conditional simulation, (*2,9,33,58-60*) appear to be very promising for the kinds of problems that occur in groundwater transport.

Indicator Kriging. The objective in indicator kriging is to predict the conditional probability that $V(\underline{x}_0)$ is less than some value, v_0, given the data:

$$P[V(\underline{x}_0) \leq v_0 \mid V(\underline{x}_1)...V(\underline{x}_n)]. \tag{10}$$

This problem can be put into a kriging framework by encoding the data into indicator variables.

For a set of cut-off values, v_k the indicator functions

$$I[v_k:V(\underline{x}_j)] = 1, \text{ if } V(\underline{x}_j) \leq v_k$$
$$= 0, \text{ if } V(\underline{x}_j) > v_k \tag{11}$$

are introduced. Namely, $I[v_k:V(\underline{x}_j)]$ for fixed v_k simply indicates when the data value observed is less than v_k; hence a sequence of 0's and 1's is obtained in place of the previous data. As v_k changes different sequences occur though some order relations obviously will be satisfied: e.g.

$$I[v_1:V(\underline{x}_j)] \leq I\{v_2:V(\underline{x}_j)] \text{ if } v_1 \leq v_2$$

The expected value of $I[v_0:V(\underline{x})]$ is just $P[V(\underline{x}) \leq v_0]$

By treating the values of the indicators as the field of interest covariances and variograms can again be obtained. Note,

$$E\{I[v_1:V(\underline{x}_1)] \ I[v_2:V(\underline{x}_2)]\} = P[V(\underline{x}_1) \leq v_1, V(\underline{x}_2) \leq v_2] \tag{12}$$

Namely the covariance function for the indicators is related to the joint probability function for two values and two locations. While it is possible to examine cross-covariances based on these joint distributions in many cases it is sufficient to examine the covariances involving only $I[v_0:V(\underline{x}_j)]$ and $I[v_0:V(\underline{x}_i)]$.

After the covariances or variograms for the indicators are obtained the desired conditional probabilities can be predicted by kriging:

$$P(V(\underline{x})_0) \leq v_0 | I[v_0:V(\underline{x}_1)],...,I[v_0:V(\underline{x}_n)] \sim \sum_{j=1}^{n} \lambda_j \ I[v_0:V(\underline{x}_j)] \tag{13}$$

Again the kriging equations derived will be similar to those in equations (4). For K cut-off values $v_1...v_k$ there will be K values at each prediction location and these can be used to estimate the distribution of $V(\underline{x})$ at $\underline{x}_0$.

Transport Problems and Indicator Kriging. In the context of transport problems one can think about particles emplaced at an initial site and traveling through the domain. The number of particles at a particular location and time is proportional to the probability distribution discussed above and hence it appears that indicator kriging is well-suited to answer questions about concentrations at a specified location given data at other locations.

Once again the question of trends in the values has to be faced in both the covariance estimation stage as well as the prediction phase. In addition, inconsistencies may occur in the distribution functions and the estimated values may not be monotonic in the v-values. This generally is not serious and can be overcome by fitting monotonic functions to the estimated values.

"Soft" Data. The formalism of indicator kriging can be used to include qualitative or soft data of several types. At a location there may be no hard data but an assessment of the probability distribution given data in the surrounding area could be available. For example in examining permeability at a location, relatively simple information about geology and soil type could yield some rough information about the distribution of the non-measured value at the given location. Translated, this would give a prior distribution for $V(\underline{x}_i)$: $P[V(\underline{x}_i) \leq v] = G(v)$. Then at the indicated location $\underline{x}_l$ the data value can be encoded but this time using $I[v_0:V(\underline{x}_l)] = G(v_0)$, the probability value.

Other kinds of soft-data (e.g. interval constraints) can be included in this approach. Indicator kriging and use of soft data is very promising because it makes use of a wider range of data and directly predicts quantities of interest.

Conditional simulation. The previous geostatistical methods have estimation or prediction as their primary focus. The kriging procedure yields smoothed paths for the conditional means and conditional variance approximations. In many applications it is also important to know the variability that can occur in a particular region when the observations are given. One could argue that, in fact, this is the main aim - to predict what can happen in a specific instance rather than in general for related fields.

There are several procedures for doing this; in the standard case all of these involve some distributional assumptions. However, for indicator conditioning a method that is essentially non-parametric has been proposed (*2,61*).

Standard conditional simulations start by unconditionally generating a realization using the proposed mean and covariance estimates from the variability characterization phase of the study. One popular method for generating these unconditioned paths has been the Turning Bands Algorithm (*39,62*). However serious problems have been noted with this method (*63*) and instead a flexible alternative method called the Fast Fourier Transform Method is recommended.

The unconditioned path generated is next conditioned to agree with the data while still maintaining the specified variogram behavior. If $V_u(\underline{x})$ denotes the unconditioned generated realization, $\hat{V}_K(\underline{x})$ the kriged estimator based on the data and $\hat{V}_{uK}(\underline{x})$ a kriged estimator for $V_u(\underline{x})$ based on data taken at $\underline{x}_1...\underline{x}_n$ (the same corresponding location as in the actual field) then the conditioned path is

$$V_{CS}(\underline{x}) = \hat{V}_K(\underline{x}) + [V_u(\underline{x}) - \hat{V}_{uK}(\underline{x})]$$

This conditioned path has the following properties:

(i) $V_{CS}(\underline{x})$ agrees with the observations:
$$V_{CS}(\underline{x}_i) = V(\underline{x}_i) \quad i = 1...n$$

(ii) The mean of $V_{CS}(\underline{x})$ is $\hat{V}_K(\underline{x})$ the kriging estimator, and the variance is the   kriging variance

(iii) The variogram of $V_{CS}(\underline{x})$ is the initial variogram.

The conditioned path will be less variable than the true path but more variable than the kriged path. It also can be used for making other conditioned predictions.

Examples

In this section some examples are used to illustrate points discussed above. The first example deals with variogram estimation and sample spacing and the second two deal with indicator and soft kriging.

Example 1: Variogram Estimation. To illustrate the importance of closely spaced samples for variogram and covariance estimation several measurement networks are examined in a field of known covariance structure.

The realization is generated by using a Fast Fourier Transform generation algorithm (63). The covariance function is $C(\underline{s}) = \exp\{-|\underline{s}|\}$ so that the correlation scale is 1 as is the field variance. For the generated field, the actual variance was 0.869 and that is used as the target variance; Hence the true model variogram is $0.869 [1 - \exp\{-|\underline{\xi}|\}]$.

The networks used and the associated Figures are described in Table III. The networks had regular spacing and subgrids were located at the lower left corners of the larger grid. The original values were generated at a 0.2 spacing.

Table III. Networks and figures for a field of known covariance structure

Network 1:	10 x 10, spacing 1, 100 points	Figure 3(a)
Network 2:	5 x 5, 1.5 spacing	
	2 x 2 subgrids, 0.6 spacing,100 points	Figure 3(b)
Network 3:	4 x 4, 1.5 spacing	
	3 x 3 subgrids, 0.4 spacing, 144 points	Figure 3(b)
Network 4:	3 x 3, 3 spacing	
	4 x 4 subgrids, 0.4 spacing, 144 points	Figure 3(c)

For all the cases, variogram estimates were calculated as well as non-ergodic covariances. Generally the differences between the non-ergodic and ergodic covariances were small so only variogram results (corresponding to the ergodic covariance estimates) are given.

In Figure 3(a) the variogram is shown for the network 1 - a coarse network with measurements taken every correlation scale. Fitting a variogram model to the results in some ambiguity. In fact it would be tempting to fit a model with a nugget effect (eg. nugget variance = 0.3, correlation scale = 1.4) shown in the dashed curve in Figure 3(a) instead of the true curve (solid line) when the observations are more finely spaced like in network 2, the resolution near the origin is better and helps resolve that ambiguity. These results for networks 2 and 3 are shown in Figure 3(b) along with the true model. Figure 3(c) shows similar results for a coarser network with a fine superimposed network.

These examples show that to achieve scale identification, samples at a spacing of less than ½ the correlation scale are required and illustrate results given elsewhere (*26,27*).

Example 2: Soft Kriging; Theoretical Case. The same random field is used in this section to illustrate probability and soft kriging. For this purpose the data from network 2 is used.

Figure 4(a) shows the kriging estimates for predictions within a 5 x 5 region whose left corner is at (6,6): the sampling network is embedded within this region. The kriging map shows the smoothed map with a general low in the lower right hand corner, a high in the center left region and a valley trending from the lower left to the upper right corner. Away from the measured points the kriging standard deviations (map not shown) are around 0.5 or 0.6.

To illustrate indicator kriging a cut-off value of 0.5 was used. The encoded set of 0's and 1's yielded an isotropic spherical variogram estimate with variance 0.12 and maximum value B = 1.5: This was used in the kriging of the indicators and the contour map for the estimated probabilities is shown in Figure 4(b). As expected the contours have similarities to those of Figure 4(a), with values near 1 in the lower right corner (where the

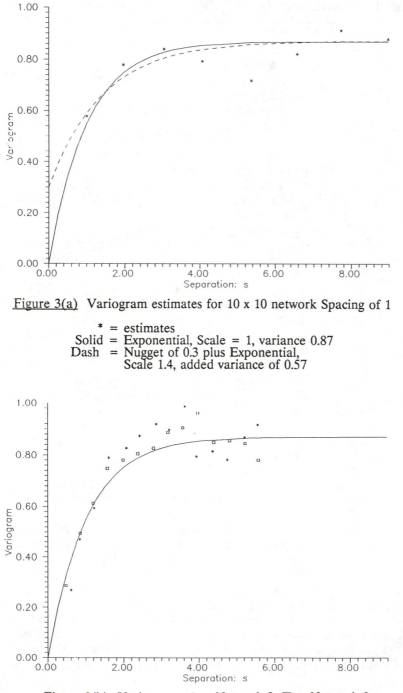

Figure 3(a) Variogram estimates for 10 x 10 network Spacing of 1

* = estimates
Solid = Exponential, Scale = 1, variance 0.87
Dash = Nugget of 0.3 plus Exponential,
Scale 1.4, added variance of 0.57

Figure 3(b) Variograms; * = Network 2, □ = Network 3

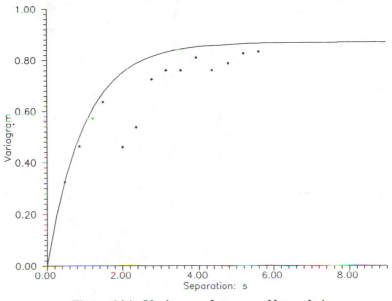

Figure 3(c) Variogram for coarse Network 4.

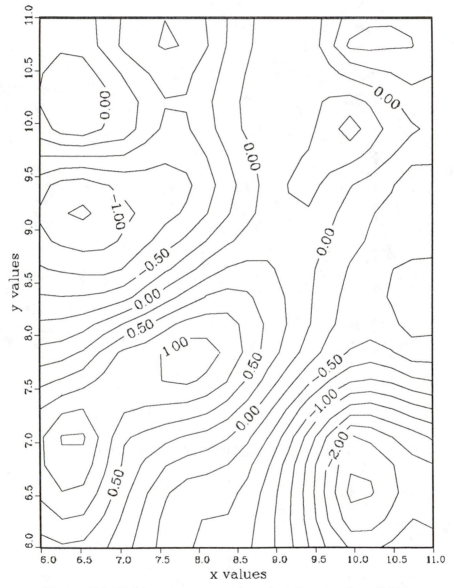

<u>Figure 4(a)</u> Kriging contour map for values from random field.

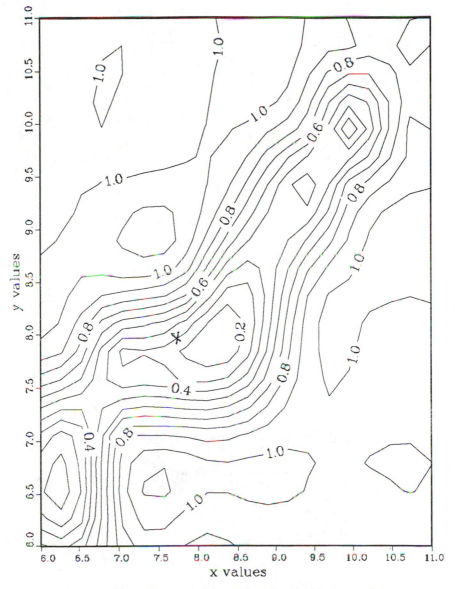

Figure 4(b) Indicator kriging V ≤ 0.5; All 100 data points

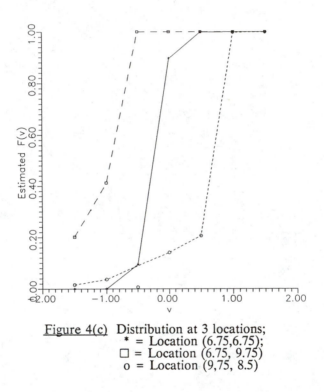

<u>Figure 4(c)</u> Distribution at 3 locations;
      * = Location (6.75,6.75);
      □ = Location (6.75, 9.75)
      o = Location (9,75, 8.5)

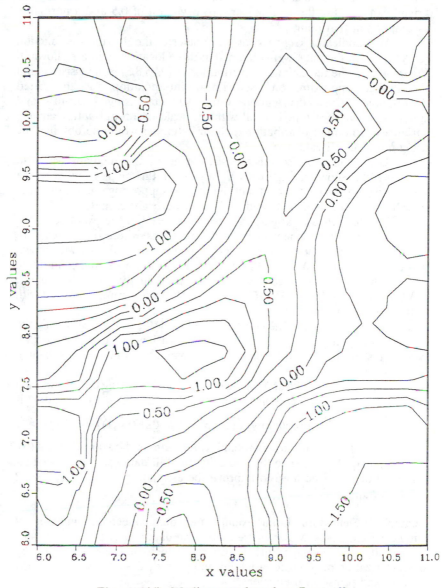

Figure 4(d) Median map based on 7 cut-offs

regular kriging estimator is low) and low values (on the order of 0.2) near the center left when the regular kriging estimator is higher. At the point indicated by an X in Figure 4(b), indicator kriging would estimate $P(V(\underline{x}) \leq 0.5)$ to be 0.30. This contrasts with the estimate of 0.16 using the regular kriged value of 1.0, the kriging standard deviation of 0.5 and a normality assumption at that point.

With indicator kriging, coarse estimates of the distribution function for $V(\underline{x})$, given the data, were obtained at three locations and are shown in Figure 4(c). Note the estimated distribution at (9.75,8.5) is inconsistent and the dashed approximation used linear interpolation for the fitted distribution. These distributions were estimated using 7 cut-offs and spherical covariance functions all with the scale of 1.5 but with changing variances. Smoother distribution function estimates would be obtained by using more cut-off values.

Using the same set of cut-offs, a map of the medians was also created and is shown in Figure 4(d). In general, this map is similar to the regular kriging map in the center of the region (supporting symmetry of the distributions) but slightly skewed near the lower right corner.

To illustrate soft kriging the 100 points used in the example above were split into 2 sets: one with 52 hard data points, where the values were known exactly, and the other with 48 data points where all that was known was that $V(\underline{x}) < 0$, $V(\underline{x}) > 1$ or $0 \leq V(\underline{x}) \leq 1$. In the latter set, the conditional probability that $V(\underline{x})$ is less than 0.5 was used in the indicator vector: $P[V(\underline{x}) \leq 0.5 \mid 0 \leq V(\underline{x}) \leq 1] = 0.515$, where $V(\underline{x})$ was assumed normal with mean 0 and variance 0.87.

Figure 5(a) shows the contour map for $P[V(\underline{x}) \leq 0.5]$ using only the 52 hard data points and the same spherical covariance structure as before. Notice it is substantially different from the map using all the data shown in Figure 4(a).

Continuing, the 48 "soft" data are added back in, using the information that $V(\underline{x}) < 0$ (in which case the indicator is 1), $0 \leq V(\underline{x}) \leq 1$ ("indicator" is

0.515 which is the probability $V(x) \leq .5$ given $0 \leq V(\underline{x}) \leq 1$)) or $V(\underline{x}) > 1$

(indicator is 0). The resulting probability estimate is shown in Figure 5(b). The map is quite similar to the one using the full data set and shows how non-hard data can be used to improve the estimate based on just the 52 hard data alone.

Example 3: Field Data. In this example field data collected at a geological site near Belen, New Mexico were used. The data is part of a larger study funded by the Department of Energy which has as its aim the characterization of geologic variability and to study the effects of variability on flow and transport predictions.

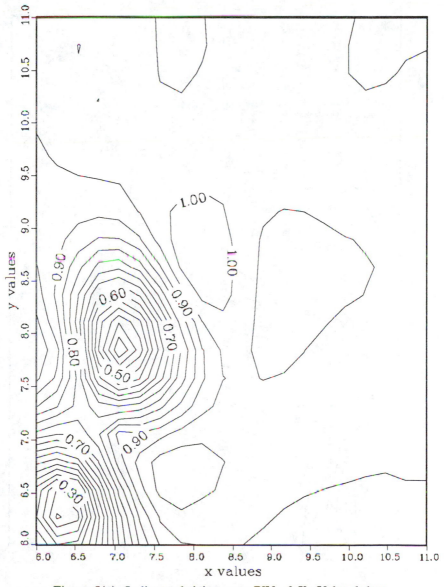

Figure 5(a) Indicator kriging map, P[V ≤ 0.5]; 52 hard data.

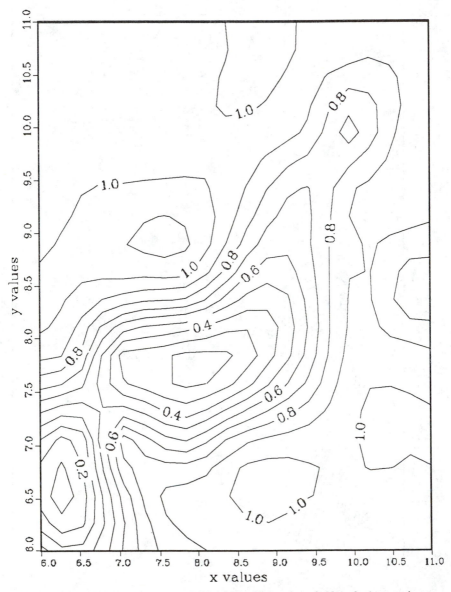

Figure 5(b)   Soft kriging map, P[V ≤ 0.5]; 52 hard and 48 soft data points.

Air flow permeameter instruments were taken on an outcrop of the Sierra Ladrones Formation in a region approximately 2.5 units wide and 6 units high (*64*). In addition other information like grain size distributions was collected. The objectives of this study were to see whether soft (non-quantitative) data could be used to predict permeability.

Three major grain size categories were used and the logarithm of the air flow measurements were related to the grain size types. In all cases the distribution appeared to be log-normal and the parameters are shown in Table IV.

Table IV
Log-airflow rate distribution

Grain size	mean	variance
fine	2.17	0.014
medium	3.35	0.157
coarse	3.96	0.131

Only the right half of the formation with 117 data points will be examined. The covariance-function estimates are anisotropic with a horizontal scale of about 17 cm and a vertical scale of 3 cm. An exponential covariance function with a variance of 0.25 was used for the raw data. The possibility does exist that there is a second larger scale but more data are needed to resolve that question. In addition to anisotropy there was a slight angular from the horizontal of about 10°.

The regular kriging estimates of the log-airflow data are shown in Figure 6(a). A layering is indicated and reflected in the geology.

To see whether the airflow data could be supplanted or supplemented by the grain size data (which is easier to obtain) indicator kriging was used with a cut-off value of 3.5. The contour map of the estimated probabilities is shown in Figure 6(b) and reflects the same trends as the original data.

The sample of 117 data points was next split into a set of 67 hard data points and 50 soft data points where only grain size data was used. For the soft kriging part of the example, the indicator used was the

probability that $V(\underline{x}) \le 3.5$ given the grain size at the location where again

normality was assumed for log-airflow rate. In all cases a covariance function with scales of 17 and 3 cm, and a variance of 0.22 was used. In addition a tilt of 10° was assumed.

Figure 6(c) shows the contour map where only hard data was used and Figure 6(d) where both hard and soft data were used. There seems to be little change between Figures 6(c) and 6(d) indicating the soft data in this case was not informative. This could be because most of the grain sizes

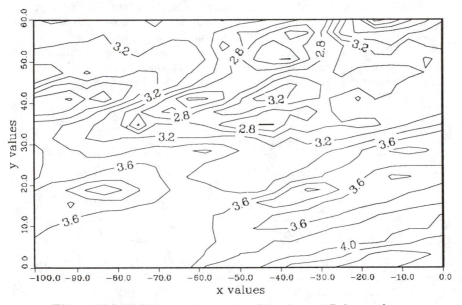

Figure 6(a) Kriging contours of Airflow data; 117 data points.

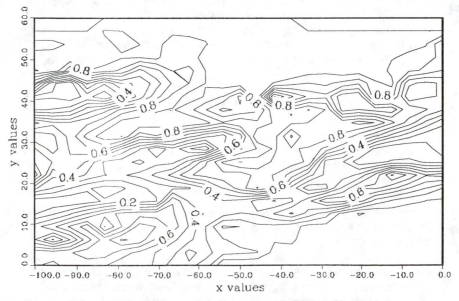

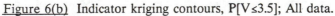

Figure 6(b) Indicator kriging contours, P[V ≤ 3.5]; All data.

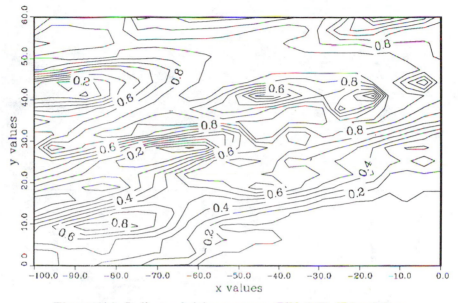

Figure 6(c) Indicator kriging contours, P[V ≤ 3.5]; 67 hard data.

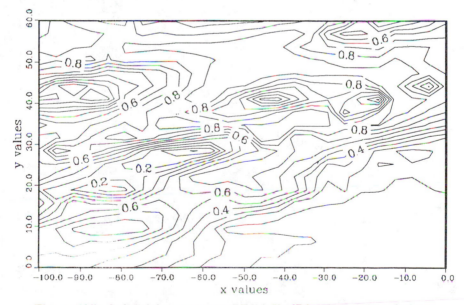

Figure 6(d) Soft kriging contours, P[V ≤ 3.5]; 67 hard, and 50 soft data.

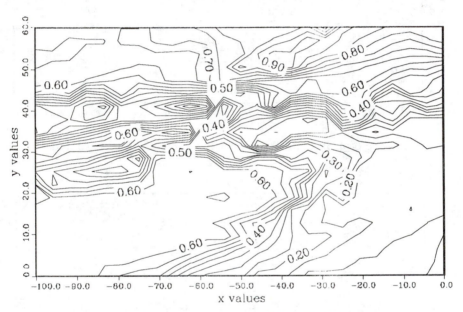

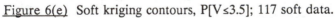

Figure 6(e) Soft kriging contours, P[V ≤3.5]; 117 soft data.

were coarse and medium, the cut-off of 3.5 was not informative and consequently a wider range of cut-offs could prove useful. In Figure 6(e) contours for __all__ soft data are presented and these have little resemblance to those of Figure 6(a). Clearly use of soft data must be supplemented by hard data to get realistic results.

Conclusions and Discussions

The exploratory data analysis phase of a geostatistical study is often of primary importance(*10*). Raw graphs of the data, graphs of distributions, checks for outliers with assignable causes, and examination for trends are examples of the kinds of calculations that are desirable.

Proceeding to variogram estimation one should try as much as possible to use robust techniques for variogram estimation. Cells with small numbers of data pairs need to be interpreted with care. New procedures like the non-ergodic covariance (*1,15*) may be especially useful.

Few guidelines exist for designing studies for variogram estimation. In general regular grids with random sub-networks superimposed could be useful. Spacing of observations should have a sufficient number of pairs close enough to get estimates of the variogram at one-half the scale. Use of a nugget effect in modeling a variogram can often be due to the fact that sample spacing is not fine enough in comparison to the scale. One difficulty is that initially quite often the scale is not well known. These guidelines also come in conflict with those for estimating mean values within regions where large sample spacings are desirable.

Kriging can be useful for prediction of the conditional mean value given the data. In many situations a variant of indicator kriging may, however, be a better procedure. This can be used, along with a moving neighborhood approach, to estimate probability values directly. It also allows for incorporation of "soft" data which can substantially improve the estimates.

Conditional simulations are useful for estimating the variability inherent in a specific field application. Such simulations should find increasing application within the domain of pollution and groundwater problems.

Model-based studies consonant with the data should be combined with the geostatistical approach and used whenever possible. The effect of spatial variability on predictions can be significant and should not be ignored.

There are several areas that need to be addressed from both a practical and theoretical standpoint. Optimal design of networks that can be used to characterize the heterogeneity within a system, as well as optimal designs for monitoring both need more research. The statistical uncertainties inherent in variogram estimation need further study. More studies tied to specific sites and including validation are needed for these procedures.

The geostatistical methodology is useful for studying many problems involving spatial variability which can influence properties like flow and transport. It is, however, not an automatic procedure and needs to be applied with caution, care and common sense.

Glossary

Random field or spatial stochastic process $V(\underline{x})$: For each fixed $\underline{x}$ (location in space) $V(\underline{x})$ is a random variable.

Probability density of $V(\underline{x})$: The function $g(v: \underline{x})$ such that

$$P(a < V(x) \leq b) = \int_a^b g(v: x)dv; \text{ where P denotes probability.}$$

Expected value or mean of $V(\underline{x})$: The expected value or mean (denoted by $E(V(\underline{x}))$ or $m(\underline{x})$ is the probability-weighted average,

$$E(V(\underline{x})) = \int_{-\infty}^{\infty} v\, g(v: \underline{x})dv.$$

Covariance function for a random field: Designated by $cov(V(\underline{x}), V(\underline{y})$ and defined as

$$cov(V(\underline{x}), V(\underline{y})) =$$

$$E[(V(\underline{x}) - m(\underline{x}))(V(y) - m(\underline{y}))]$$

It measures statistical relationship between field values at two different locations, $\underline{x}$ and $\underline{y}$.

Statistical homogeneity or second-order stationarity: The random field $V(\underline{x})$ is statistically homogeneous or second-order stationary if

(i) $E(V(\underline{x})) = m$, a constant; and

(ii) $cov(V(\underline{x}), V(\underline{y})) \equiv C(\underline{x} - \underline{y})$

only depends upon the separation vector $\underline{s} = \underline{x} - \underline{y}$ also called the lag vector.

Statistically isotropic process: A statistically homogenous process where $C(\underline{s}) = C(||\underline{s}||) \equiv C(s)$ depends on the separation distance, $s = ||\underline{s}||$.

Correlation function: $\rho(\underline{s}) = C(\underline{s})/C(\underline{0})$.

Scale: An average distance over which points are significantly correlated. For an isotropic covariance function this is sometimes taken as that value λ where $e^{-1} = \rho(\lambda)$, an e-fold drop.

Intrinsic random function of order 0: A random field $V(\underline{x})$ with constant mean where $E([V(\underline{x}+\underline{s})-V(\underline{s})]^2$ only depends on $\underline{s}$.

Variogram or semi-variogram: The function

$$\gamma(\underline{s}) = E\{[V(\underline{x}+\underline{s})-V(\underline{x})]^2\}/2$$

for an intrinsic random function of order 0.

Sill: if $\gamma(\underline{s})$ has a limiting value as y increases, the limit is called the sill and equals

$var(V(\underline{x}))$.

Effective Sample Size The sample size, based on independent observations, which would give the same variance for the sample mean as the correlated data:

$$\sigma_1^2 = C_V(\underline{0})/n, \quad \sigma_2^2 = \sum_{j=1}^{n} \sum_{i=1}^{n} C(\underline{x_i}-\underline{x_j})/n^2$$

$$n_{eff} = (\sigma_2^2/\sigma_1^2)n$$

Conditional Probability: $P[V(\underline{x}) \leq v|V(\underline{x_1})...V(\underline{x_n})]$ The probability that $V(\underline{x})$ is less

than or equal to v given the data $V(\underline{x_1})...V(\underline{x_n})$ at locations $\underline{x_1}...\underline{x_n}$.

Conditional Expected Value: The expected value of $V(\underline{x})$ given $V(\underline{x_1})...V(\underline{x_n})$ at locations $\underline{x_1}...\underline{x_n}$.

Linear Estimator: An estimator of the form

$$\hat{V} = \lambda_1 V(\underline{x_1}) + \lambda_2 V(\underline{x_2})+...\lambda_n V(\underline{x_n})$$

$$= \sum_{j=1}^{n} \lambda_j V(\underline{x_j})$$

where the $\lambda's$ are constants.

Kriging: The procedure that finds the best (minimum mean square error) linear unbiased estimator of $V(\underline{x})$ based upon observations $V(\underline{x_1})...V(\underline{x_n})$. For a statistically homogeneous process with covariance function n $C(\underline{s})$, this yields a set of linear equations for the "weights" λ_j;

$$\sum_{j=1}^{n} \lambda_j C(\underline{x_i}-\underline{x_j})- \mu = C(\underline{x_i} - \underline{x}), \quad i=1...n$$

$$\sum_{j=1}^{n} \lambda_j =1.$$

μ is a Lagrange multiplier. For an intrinsic random function of order zero the covariance function $C(\underline{s})$ can be replaced by $-\gamma(\underline{s})$, the negative of the variogram, to get the kriging equations.

<u>Kriging variance</u>: The variance associated with the kriging estimator, designated by σ_k^2. This is also the minimum mean squared error. For a statistically homogeneous random field it is

$$E([\hat{V}(\underline{x}) - V(\underline{x})]^2 = \sigma_k^2 = C(0) - \sum_{i=1}^{n} \lambda_i\, C(\underline{x} - \underline{x}_i) + \mu$$

<u>Co-kriging</u>: The extension of kriging to the case where $V(\underline{x})$ is estimated using observations from two random fields where now the cross-covariance, $\text{cov}(V(\underline{x}+\underline{s}),\, U(\underline{x}))$, also enter in.

<u>Indicator kriging</u>: Estimation of $P[V(\underline{x}) \leq v \,|\, V(\underline{x}_1)...V(\underline{x}_n)]$ using kriging applied to indicator data:

$$I(V(\underline{x}_i):\; v) = \begin{cases} 1 & if \quad V(\underline{x}_i) \leq v \\ 0 & if \quad V(\underline{x}_i) \leq v \end{cases}.$$

<u>Soft Data</u>: Data that is not a directly quantified variable which could just give a probability distribution or interval for the quantity of interest.

<u>Simulation</u>: Generation of paths or realizations with a prescribed mean and covariance structure.

<u>Conditional simulation</u>: Generation of paths or realizations with prescribed mean and variance structure that also agree with data observed at $\underline{x}_1...\underline{x}_n$.

Acknowledgments

The work in this paper was supported by Sandia National Labs Grant 4-R58-2690R-1 and Department of Energy Grant DE-FG04-89ER60843. The author also would like to thank Annette Aguilar and Monica Sivils for their skills in translating his scrawl, J. Matt Davis for making available the airflow data, and A.G. Journel for valuable discussions.

Literature Cited

1. Isaacs, E.H.; Srivastava, R.M., *Applied Geostatistics*, Oxford Univ. Press, New York, N.Y., 1989, 561 pp.
2. Journel, A.G., *Fundamentals of Geostatistics in Five Lessons*, American Geophysical Union, Short Course in Geology, 1989, Vol. 8, 40 pp.
3. Journel, A.G.,; Huijbregts, C., *Mining Geostatistics*, Academic Press, New York, 1978, 600 pp.
4. Journel, A.G., *Math. Geol.*, 1986, 18, 119-140.
5. Journel, A.G., *Math. Geol.*, 1985, 17, 1-15.
6. Myers, D.E.; *Math. Geol.*, 1989, 21, 347-362.
7. Hicks, C.R., *Fundamental Concepts in the Design of Experiments*, Holt Reinhard and Winston, New York, N.Y., 1964, 293 pp.
8. Matheron, G., *The Theory of Regionalized Variables and Its Applications*, Paris School of Mines, Cah. Cent. Morphologie Math, 5, Fontainebleau, France, 1971, 211 pp.

9. De Marsily, G., *Quantitative Hydrogeology*, Academic Press, Orlando, FL, 1986, 440 pp.
10. Tukey, P., Graphical Methods: In *Proceedings of a Symposium on Applied Mathematics*, Amer. Math. Soc., Providence, R.I., 1983, Vol. 28, Chapter 2.
11. Journel, A.G., *Math. Geol.*, 1983, 15, 445-468.
12. Barnes, R.J., *Geostatistics Newsletter*, 1989, 3(2), 10-13.
13. Olea, R.A., *Math. Geol.*, 1984, 16, 369-392.
14. Istok, J.D.; Cooper, R.M.; Flint, A.L., *Ground Water*, 1988, 26, 638-646.
15. Isaacs, E.H.; Srivastava, R.M., *Math. Geol.*, 1988, 20, 313-341.
16. Journel, A.G., *Geostatistics Newsletter*, 1989, 3(3), 5-6.
17. Armstrong, M., In *Geostatistics for Natural Resources Characterization*, D. Verly et. al., Eds.; Part 1, D. Reidel Pub., Hingham, MA, 1984, 1-19.
18. Armstrong, M., *Math. Geol.*, 1984, 16, 305-313.
19. Starks, J.H.; Fang, J.H., *Math. Geol.*, 1982, 14, 309-319.
20. Neuman, S.P.; Jacobson, E.A., *Math. Geol.*, 1984, 16, 499-501.
21. Delhomme, J.P., *Water Resour. Res.*, 1979, 15, 269-280.
22. Cressie, N.; Hawkins, D., *Math. Geol.*, 1980, 12, 115-125.
23. Cressie, N., *J. Amer. Stat. Assoc.*, 1986, 81, 625-634.
24. Kitanidis, P.K., *Water Resour. Bulletin*, 1987, 23, 557-567.
25. Russo, D., *Soil Sci. Am. J.*, 1984, 48, 708-716.
26. Russo, D.; Jury, W.A., *Water Resour. Res.*, 1987, 23, 1257-1268.
27. Russo, D.; Jury, W.A., *Water Resour. Res.*, 1987, 23, 1269-1279.
28. Barnes, R.J., *Math. Geol.*, 1988, 20, 477-490.
29. Warrick, A.; Meyers, D.E., *Water Resour. Res.*, 1987, 23, 496-500.
30. Zhang, R.; Warrick, A.W.; Meyers, D.E.; *Math. Geol.*, 1990, 22, 102-121.
31. Borgman, L.E., *Math. Geol.*, 1988, 20, 383-403.
32. Gutjahr, A.L., In *Studies in Ground Water Quality Studies*, Univ. of Nebraska Press, Lincoln, NE, 1990, 15-31.
33. Journel, A.G., In *Principles of Environmental Sampling*, Larry Keith, Ed., A.C.S., Washington, D.C., 1988, 45-72.
34. Journel, A.G., *Math. Geol.*, 1988, 20, 459-475.
35. Davis, M.W.; Grivet, C., *Math. Geol.*, 1984, 16, 249-265.
36. Delhomme, J.P., *Adv. Water Resour.*, 1978, 1, 251-266.
37. Gutjahr, A.L.; Wilson, J.L., *Transport in Porous Media*, 1989, 4, 585-598.
38. Davis, B., *Math. Geol.*, 1987, 19, 241-248.
39. Matheron, G., *Adv. App. Prob.*, 1973, 5, 439-468.
40. Delfiner, P., *The Intrinsic Model of Order k*, Centre de Geostatistique Report, Fontainbleau, France, 1978, 39 pp.
41. Journel, A.G.; Rossi, M.E., *Math. Geol.*, 1989, 21, 715-739.
42. Zirscky, J.H.; Keary, G.P.; Gilbert, P.O.; Middleborrks, E.J., *J. Environ. Eng.* 1985, 111, 777-789.
43. Zirscky, J.H.; Harris, D.J., *J. Environ. Eng.* 3, 1986, 112, 770-784.
44. Pucci, A.A.; Murashige, J.A.E., *Ground Water*, 1987, 25, 672-678.
45. Cooper, R.M.; Istok, J.D., *J. Envir. Eng. ASCE*, 1988, 114, 270-286.
46. Cooper, R.M.; Istok, J.D., *J. Envir. Eng. ASCE*, 1988, 114, 287-299.
47. Istok, T.D.; Cooper, R.M., *J. Envir. Eng. ASCE*, 1988, 114, 915-928.
48. Bras, R.L.; Vomvoris, E.G., *ASCE paper*, 1989, 6 pp.
49. Loaiciga, H.A., In *Computational Methods in Water Resources, Numerical Methods for Transport and Hydrology Processes*, M. Celia et. al. eds., Elsevier Pub., New York, N.Y., 1988, Vol. 2, 371-376.
50. Hughes, J.P.; Lattenmaier, D.P., *Water Resour. Res.*, 1981, 17, 1641-1650.
51. Virdee, T.S.; Kottegoda, N.T., *J. of Hydro. Sci.*, 1984, 29, 367-387.
52. Dagan, G., *Water Resour. Res.*, 1986, 22, 120-134.
53. Kitanidis, P.K.; Vomvoris, E.G., *Water Resour. Res.*, 1983, 19, 677-690.
54. Hoeksema, R.J.; Kitanidis, P.K., *Water Resour. Res.*, 1984, 20, 1003-1020.

55. Clifton, P.M.; Neuman, S.P., *Water Resour. Res.*, 1982, 18, 1215-1234.
56. Sudicky, E.A., *Water Resour. Res.*, 1986, 22, 2069-2082.
57. Gelhar, L.W.; Axness, C.L., *Water Resour. Res.*, 1983, 19, 161-180.
58. Journel, A.G., In *Geostatistics for Natural Resources Characterization*, Verly et. al., Eds., D. Reidel Pub., Hingham, MA, 1984, Vol. 1, 307-335.
59. Journel, A.G., *Math. Geol.*, 1986, 18, 269-286.
60. Journel, A.G., In *A Conference on Geostatistical Sensitivity and Uncertainty Methods for Groundwater Flow and Radionuclide Transport*, B. Buxton, ed., Battelle Press, Columbus, OH, 1988, 586-599.
61. Journel, A.G.; Alabert, F, *Terra Review*, 1989, 1, 123-134.
62. Mantoglou, A.; Wilson, J.L., *Water Resour. Res.*, 1982, 18, 1379-1394.
63. Gutjahr, A.L., N.M. Tech Report, *Fast Fourier Transform for Random Field Generation*, 1989, 106 pp.
64. Davis, J.M., *Variability of Permeability in the Sierra Ladrones Fm., Albuquerque Basin, MS Thesis in Hydrology*, N.M. Tech, Socorro, N.M., 1990, 110 pp.

RECEIVED January 25, 1991

Chapter 5

Minimum Cost Sample Allocation

Robert E. Mason[1] and James Boland[2]

[1]Research Triangle Institute, Research Triangle Park, NC 27709
[2]U.S. Environmental Protection Agency, 401 M Street SW,
Washington, DC 20460

A procedure for determining the minimum cost allocation of samples subject to multiple variance constraints is described. The procedure is illustrated using information developed for the National Pesticide Survey conducted by the United States Environmental Protection Agency.

Seldom are field studies conducted with but a single objective. More usually, the investigator is faced with the problem of designing a field study to satisfy multiple objectives, often with limited resources available. This paper addresses the problem of allocating field study resources to simultaneously satisfy an arbitrary number of objectives for the least cost.

Inferential Population

The first step in designing a field study is to develop a fully operational definition of the population (or universe) of inferential interest. Five points are addressed in the population definition.

- the spatial dimension of the population
- the temporal dimension of the population
- the units of observation that comprise the population
- eligibility criteria to differentiate between population units and otherwise similar units (of no interest to the study)
- the identification of domains (groups or subpopulations of units) that are of special interest to the investigation

The second step is to identify and define the population parameters that are to form the basis of the design, that is, the characteristics of the population that are the central subject of the investigation. These may be population totals, averages, proportions, regression relations, comparisons, and so on, and are defined as functions of observation or response variable values over the entire population.

The final design step is to specify the magnitudes of the variances that are to be associated with the identified parameter estimates. The specifications often take the form of quantities related to the variances rather than the variances themselves, such as relative standard errors, confidence intervals, or the power to be associated with a statistical test.

Population Concepts

The units comprising the population of inferential interest are denoted by u_g where,

$$g = 1, 2, ..., N .$$

Note the implications that,

- the population, although perhaps very large, is finite, there being N units in total, and,
- the population units are distinct, such that an individual is recognizable as the g-th unit.

Otherwise the units themselves may be anything, for example, rural domestic wells or ground water volumes defined within a three dimensional space. Arbitrary units such as the latter are constructed with the measurement technology in mind. That is, the units are constructed of such a size and shape that they can be accurately characterized by the measurement procedures planned for use. The objective is to construct units such that the measurement variability is small in relation to the variability among the population units.

The spatial dimension of the population definition defines the study site, for example, all rural domestic wells in the United States, or the total ground water volume to a specified depth underlying a specified field. Robust statistical inferences are, of course, limited to the selected study site. That is, statistical arguments supporting the validity of the conclusions reached are themselves valid only for the study site population.

If the population parameters of interest to the investigation are temporally varying quantities, then the population units, u_g, are defined in both time and space. The total data collection period defines the temporal reference for the study, and inferences are restricted to the corresponding time frame. The g-subscript in this case takes on the values,

$$g = 1, 2, ..., N_1, N_1+1, N_1+2, ..., N_t, N_t+1, N_t+2, ..., N ,$$

where the subscripted N-values denote the number of spatial units available for study at different times. The times are denoted by,

$$t = 1, 2, \ldots L \, ,$$

and the total population size is defined by,

$$N = \sum_{t=1}^{L} N_t \, .$$

The time intervals identified by the t-subscript are arbitrary. Like arbitrarily defined spatial units, temporal units are constructed such that measurement errors are kept small in relation to the variability that exists among the temporal units. That is, a temporal unit is of short enough duration that the variability of possible response variable values within a unit is small in relation to the variability that exists from one unit to another.

An observation or response variable value associated with the g-th unit in the population is denoted by y_g. Note the implication that every unit in the population is observable. The point has some importance in identifying the population parameters to form the basis of the design and in the subsequent data analysis.

A univariate population mean provides a familiar example of a population parameter. The quantity,

$$A_y = \tfrac{1}{N} \sum_{g=1}^{N} y_g \, ,$$

defines the mean. The population variance is defined by,

$$V_y = \tfrac{1}{N} \sum_{g=1}^{N} \left[y_g - A_y \right]^2 \, .$$

Two problems can arise. First, some information about the magnitudes of A_y and V_y is needed for design purposes. Sometimes the information is available from previous studies, but more commonly the information is not available, the purpose of the study being to provide it. Second, note that if y-values are not able to be obtained for some values of the g-subscript, then neither the parameter nor its variance is defined. If, for example, y_g is the observed concentration of a specified chemical in the g-th unit, then y-values are not observable for as many units as have concentrations below the method detection limit.

A convenient way around both problems is to design the study in the context of specifying the probabilities with which specified contamination frequencies will be detected. The exercise is equivalent to specifying the maximum values of the variances to be associated with sample estimates of specified domain sizes. The domain sizes to provide the basis for the design are determined based either on what is known about the actual state of nature, or on policy and program considerations. The sampling designs for both the EPA's National Pesticide Survey (Mason, R. E. and R. M. Lucas, Research Triangle Institute, report number RTI/7801/04-04F, 1988, unpublished) and Monsanto's agrichemical survey (Graham, J. A., presented at Groundwater Quality Methodology Workshop, Arlington, Virginia, November 1988) were developed along this line. Specifying the design problem this way has some generality and provides a useful surrogate for other parameters. Certainly parameters describing other domain characteristics are unlikely to be reliably estimated if the domain sizes themselves cannot be.

In this context, the observed chemical concentrations place the g-th unit in a specified concentration category or domain. Notationally, the indicator variable,

δ_{dg} = 1, if the g-th unit belongs to the d-th domain,

= 0, otherwise.

The indicator variable is observable for every unit in the population, assuming that 'below the detection limit' is one of the domains. The parameters of design interest become the relative domain sizes (population proportions) defined by,

$$P_d = \frac{1}{N} \sum_{g=1}^{N} \delta_{dg} \, ,$$

with the associated population variances,

$$V_d = \frac{1}{N} \sum_{g=1}^{N} \left[\delta_{dg} - P_d \right]^2 ,$$

$$= P_d \left[1 - P_d \right] .$$

Sampling Concepts

In designing a sample, the investigator assigns (relative) selection frequencies to each of the population units such that,

- linear statistics provide design unbiased estimates of corresponding parameters, and,
- the sampling variances of the parameter estimates do not exceed prespecified values.

Selection frequencies are denoted by,

$$\pi_g = n \frac{S_g}{S_+},$$

where,

$n =$ the sample size,

$S_g =$ the size measure associated with the g-th unit, and,

$$S_+ = \sum_{g=1}^{N} S_g.$$

In multi-stage sampling, the g-subscript is replaced by subscripts that identify the sampling units at each stage. The ranges of summation of these subscripts extend over the set of sampling units contained in each of the sampling units selected at the previous stage. That is, selection frequencies are assigned and samples are selected at each stage of sampling independently within the previous stage. If stratification has been imposed on the sampling frame, the ranges of summation extend over the set of sampling units contained in a stratum. That is, the selection frequencies are independently assigned and samples are independently selected within each stratum.

The size measure is (ideally) proportional to the value of the response variable associated with the unit, if information for the purpose is available, or can be set equal to one for all values of the relevant subscripts (equal probability sampling). Size measures can also be computed to simultaneously achieve specified sampling frequencies for multiple domains (*1*), if information for the purpose is available.

Similarly, in multi-stage, stratified designs the sample sizes (n-values) are determined for each stage of sampling within each of the design strata. The following section describes a procedure for determining sample sizes to satisfy arbitrary variance constraints for the least cost.

The Kuhn-Tucker Conditions

The sampling variances can be expressed as a function, $Var(\underline{n})_d$, of a vector of sample sizes, $\underline{n}$, selected from within each the design strata at each stage of sampling. The variable cost of the field study can be expressed as a function,

$C(\underline{n})$, of the same sample sizes. The sample allocation problem can then be stated in terms of minimizing the cost function, $C(\underline{n})$, subject to the inequality variance constraints given by,

$$Var(\underline{n})_d \leq K_d .$$

The values K_d are chosen by the investigator. The solutions sought, denoted by $^*\underline{n}$, are the sample sizes that minimize the objective function,

$$O(\underline{n},\lambda) = C(\underline{n}) + \sum_d \lambda_d \left[K_d - Var(\underline{n})_d \right] , \tag{1}$$

where λ_d is the Lagrange multiplier associated with the variance constraint imposed on the estimated size of the d-th domain.

Taking derivatives of the objective function with respect to the vector of sample sizes and equating to zero yields (gradient) equations of the form,

$$\frac{\partial(C(\underline{n}))}{\partial(\underline{n})} = \sum_d \lambda_d \frac{\partial(Var(\underline{n})_d)}{\partial(\underline{n})} . \tag{2}$$

If the variance constraints hold, then at $^*\underline{n}$ there must exist values of the Lagrange multipliers, $^*\lambda_d$, such that equation 2 evaluated at $^*\underline{n}$ is true and, additionally,

$$Var(^*\underline{n})_d \leq K_d , \tag{3}$$

$$^*\lambda_d \geq 0 , \tag{4}$$

$$^*\lambda_d \left[Var(^*\underline{n})_d - K_d \right] = 0 . \tag{5}$$

Equations 2 through 5 are the Kuhn-Tucker necessary conditions (see, for example, (2), pages 186 and 192). A general exposition of the application of Kuhn-Tucker theory to the problem of determining the minimum cost allocation of samples subject to multiple variance constraints is presented in (3).

For all but the simplest of sampling designs, the allocation solutions are found using iterative numerical procedures. If, in the iterative procedure, the initial values of the Lagrange multipliers, denoted by $^0\lambda_d$, are computed to equal the values that individually satisfy the variance constraints, then a comparison of the initial and final values will identify the relative importance of each constraint in determining the allocation solutions. Superfluous constraints, that is, those coincidentally satisfied with the imposition of other constraints, will have final Lagrange multiplier values

$$^{\bullet}\lambda_d = 0 \ .$$

The most important constraints will have final values

$$^{\bullet}\lambda_d \rightarrow {}^0\lambda_d.$$

The final values that most closely approach the initial values identify the variance constraints that are driving the field study costs. A small relaxation in the identified constraints can produce sizeable cost reductions.

An Example

The rural well component of the EPA's National Pesticide Survey (NPS) provides an example. The NPS design, data collection procedures and pilot implementation is described in Mason R. E., et al., Research Triangle Institute report number RTI/7801/06-02F, 1988, unpublished. A summary of the relevant sampling design information for present purposes is as follows.

Sampling Design. The sample was selected in three stages. A sample of counties was selected at the first stage. The county frame was stratified in two dimensions. The first dimension identified counties with quantifiably high, moderate, low and uncommon agricultural use of pesticides based on the use in 1982 of 63 targeted chemicals on 29 targeted crops. The second dimension identified those counties within use strata having the highest, intermediate and lowest potential for ground water contamination based on the distribution of county level DRASTIC scores (Alexander, W. J., et al., Research Triangle Institute unnumbered report, 1985, unpublished) over those counties in the same use stratum. First-stage strata are denoted by the subscript,

$$a = 1, 2, ..., 12 \ .$$

Second-stage sampling units were non-overlapping land area segments that, in the aggregate, accounted for the total rural land area in each sample county. The segments were constructed of a size convenient for counting and listing all domestic wells contained in a segment. The second-stage frame was stratified to identify those sub-county areas most vulnerable to ground water contamination and having the highest agricultural crop production. Second-stage strata are denoted by the subscript,

$$b = 1, 2 \ .$$

Third-stage sampling units were operable domestic wells. The number of wells in the b-th second-stage stratum and a-th first-stage stratum is denoted by N_{ab}, and the number of wells in the a-th first-stage stratum by,

$$N_{a+} = \sum_{b=1}^{2} N_{ab} .$$

The values shown in Table I, the numbers of households with wells, were used as surrogates for the values N_{a+} and N_{ab}.

Table II identifies the domains, the domain sizes, and associated precision requirements that form the basis of the design. The precision requirements were stated in terms of the relative standard errors to be associated with sample estimates of the specified domain sizes. The detection probabilities and approximate confidence intervals shown in the table were computed from the standard errors.

The first specification in Table II, for example, says that the relative standard error to be associated with a sample estimate of any domain of wells that comprises one percent or more of all wells nationally is not to be greater than 100 percent of the domain size. Equivalently, the survey is required to have at least a 63 percent chance of detecting any domain of wells that comprises one percent or more of the total, or, that the confidence interval about the sample estimate of a domain of this size have the limits indicated in the table. The specifications for the remaining domains have a similar interpretation, except that one percent of the wells in stratum 1, 2, and 3 (domain 2 in Table II), translates into 0.14 percent of wells nationally (and so on for domains 3, 4 and 5).

Other interpretations of the precision requirements shown in the table and, indeed, other equivalent specifications can be developed. The essential point of the exercise in developing the table is to provide,

- with pre-specified reliability,
- estimates of parameter values that have policy and program importance,
- within the resources available for the study.

Variance Model. If P_{dab} denotes the relative size of the d-th domain in the b-th second-stage and the a-th first-stage stratum, then the parameters of interest are given by,

$$\mathbf{P}_d = \sum_{a=1}^{12} \frac{N_{a+}}{N_{++}} \sum_{b=1}^{2} \frac{N_{ab}}{N_{a+}} P_{dab}$$

$$= \sum_{a=1}^{12} \frac{N_{a+}}{N_{++}} \mathbf{P}_{da} ,$$

Table I. Stratum Sizes

| First Stage Strata | Households With Wells (thousands) | |
| Second Stage Strata | | |
a b	N_{a+}	N_{ab}
1. High average use, high average vulnerability	455	
1. Most heavily cropped and vulnerable 25 percent		114
2. Remaining areas		341
2. High average use, moderate average vulnerability	916	
1. Most heavily cropped and vulnerable 25 percent		229
2. Remaining areas		687
3. High average use, low average vulnerability	440	
1. Most heavily croppped and vulnerable 25 percent		110
2. Remaining areas		330
4. Moderate average use, high average vulnerability	684	
1. Most heavily cropped and vulnerable 25 percent		171
2. Remaining areas		513
5. Moderate average use, moderate average vulnerability	1,417	
1. Most heavily cropped and vulnerable 25 percent		354
2. Remaining areas		1,063
6. Moderate average use, low average vulnerability	671	
1. Most heavily cropped and vulnerable 25 percent		168
2. Remaining areas		503
7. Low average use, high average vulnerability	1,154	
1. Most heavily cropped and vulnerable 25 percent		289
2. Remaining areas		866
8. Low average use, moderate average vulnerability	2,270	
1. Most heavily cropped and vulnerable 25 percent		568
2. Remaining areas		1,702
9. Low average use, low average vulnerability	1,170	
1. Most heavily cropped and vulnerable 25 percent		293
2. Remaining areas		878
10. Uncommon average use, high average vulnerability	1,043	
1. Most heavily cropped and vulnerable 25 percent		261
2. Remaining areas		782
11. Uncommon average use, moderate average vulnerability	1,894	
1. Most heavily cropped and vulnerable 25 percent		474
2. Remaining areas		1,421
12. Uncommon average use, low average vulnerability	997	
1. Most heavily cropped and vulnerable 25 percent		249
2. Remaining areas		748

Table II. Precision Requirements

Domain Description d	Item	Value
1. All wells nationally	Relative domain size	0.01
	Relative standard error	1.0
	Detection probability	0.63
	Confidence interval	0.0 - 0.30
2. Wells in counties with highest average use (a=1, 2, 3)	Relative domain size	0.0014
	Relative standard error	0.85
	Detection probability	0.75
	Confidence interval	0.0 - 0.004
3. Wells in counties with highest average vulnerability (a=1, 4, 7, 10)	Relative domain size	0.0025
	Relative standard error	0.85
	Detection probability	0.75
	Confidence interval	0.0 - 0.007
4. Wells in the cropped and vulnerable parts of counties (b=1)	Relative domain size	0.0025
	Relative standard error	0.525
	Detection probability	0.97
	Confidence interal	0.0 - 0.005
5. Wells in counties with highest average use and vulnerability (a=1)	Relative domain size	0.0003
	Relative standard error	1.25
	Detection probability	0.47
	Confidence interval	0.0 - 0.011

where,

$$N_{++} = \sum_{a=1}^{12} \sum_{b=1}^{2} N_{ab} \, .$$

The sampling variance, $Var(\underline{n})_d$, is made up of three components, one for each stage of sampling, divided by the (to be determined) sample sizes selected at each stage. Notationally,

$$Var(\underline{n})_d = \sum_{a=1}^{12} \left[\frac{Vcp_{1da}}{n_{1a}} + \sum_{b=1}^{2} \left[\frac{Vcp_{2dab}}{n_{1a} \, n_{2ab}} + \frac{Vcp_{3dab}}{n_{1a} \, n_{2ab} \, n_{3ab}} \right] \right], \qquad (6)$$

where,

n_{1a} = the number of sample counties (to be) selected from the a-th first-stage stratum,

n_{2ab} = the number of sample sub-county segments (to be) selected from within the b-th second-stage stratum constructed within each of the sample counties,

n_{3ab} = the number of sample wells (to be) selected from within each sub-county segment classified into the b-th second-stage stratum and a-th first-stage stratum.

The variance components themselves are functions of population variances and (intracluster) correlations. The correlations, denoted by R_{1da} and R_{2da}, arise respectively because of,

- selecting segments within the same county, and,
- selecting wells in the same segment.

The population variances are the binomial quantities,

$$V_{da} = P_{da} \left[1 - P_{da} \right],$$

$$V_{dab} = P_{dab} \left[1 - P_{dab} \right],$$

and are computed using the stratum sizes in Table I and the domain sizes in Table II. Quantitating the intracluster correlations is more problematical. For the NPS, relevant literature sources, largely well water surveys conducted by various States and by the private sector, were consulted. However, no quantitative information concerning geographic correlations among pesticide residues in wells was found.

Considering the correlations, an argument can be made that an area as large as a county contains, on average, wide ranges of variability both with respect to patterns of agricultural pesticide use and hydrogeologic features affecting and effecting ground water vulnerability to pesticide contamination. Segments within a given county might be expected to exhibit a range of variability nearly as wide as that exhibited by segments in different counties.

Wells within a segment, on the other hand, might be expected to be quite strongly correlated. The segments themselves, although variable in size, encompass a small geographic area (constructed to average about 25 housing units). Houses within a segment might tend to have wells of similar depths, with similar construction characteristics, that tend to draw from the same aquifer. Wells within a segment would be located at about the same proximity to the same type of agricultural activity and to ground water recharge areas.

Following an ad hoc sensitivity analysis, the values,

$$R_{1da} = 0.01 \, ,$$

$$R_{2da} = 0.10 \, ,$$

for all values of the d- and a-subscripts, were chosen for design purposes.
The variance components are given by,

$$Vcp_{1da} = \left[\frac{N_{a+}}{N_{++}} \right]^2 V_{da} \, R_{1da} \, ,$$

$$Vcp_{2dab} = \left[\frac{N_{ab}}{N_{a+}} \right]^2 V_{dab} \, R_{2da} \left[1 - R_{1da} \right] \, ,$$

$$Vcp_{3dab} = \left[\frac{N_{ab}}{N_{a+}} \right]^2 V_{dab} \left[1 - R_{2da} \right] \left[1 - R_{1da} \right] \, .$$

Cost Model. A cost function that is compatible with equation 6 and that facilitates taking the derivatives in equation 2 is,

$$C(\underline{n}) = \sum_{a=1}^{12} \left[n_{1a}C_{1a} + \sum_{b=1}^{2} \left[n_{1a}n_{2ab}C_{2ab} + n_{1a}n_{2ab}n_{3ab}C_{3ab} \right] \right] . \quad (7)$$

Equation 7 describes that part of the total field study cost that depends on the sample size and allocation. Fixed costs, those that do not depend on the sample size, are arbitrarily excluded, although a fixed cost coefficient, C_0, could be added to the equation to compute the overall field study cost. With respect to the sample allocation problem, the fixed cost coefficient, if included in equation 7, disappears upon taking the derivatives in equation 2.

The cost coefficients in equation 7 express the average per sampling unit costs for each stage of the design. That is,

$C_{1a} =$ the per county cost of sampling frame construction and stratification, sample selection, data collection and data processing, averaged over all counties in the a-th first-stage stratum,

$C_{2ab} =$ the average per sub-county segment cost, as above, for segments contained in the b-th second-stage stratum within the a-th first-stage stratum, and,

C_{3ab} = the average per well cost, as above.

For the NPS, the cost coefficients were quantitated by listing all of the planned sampling, data collection and analysis activities and estimating the cost of each activity, assuming a likely allocation of the sample. The activity level costs were then partitioned into components associated with the relevant stages of sampling. For example, costs associated with construction and stratification of the first-stage frame are fixed costs because they remain the same regardless of the sample size selected. Costs associated with sampling frame construction and stratification at the second-stage, on the other hand, are largely determined by the size of the first-stage sample and therefore contribute to the coefficient C_{1a}. Once the component costs are determined, they are summed over all activities and divided by the number of sampling units assumed for the costing exercise. The cost coefficients used in the NPS design are,

C_{1a} = \$3,947 per sample county,

C_{2ab} = \$678 per sub-county segment,

C_{3ab} = \$2,832 per sample well,

for all values of the a- and b-subscripts. The dollar values themselves are largely uninformative when presented, as above, without reference to the data collection and other planned activities. They have no general applicability, although the procedure used to determine them is quite routine.

Allocation Solutions

The design specific form of the objective function, equation 1, is provided by equations 6 and 7. Substituting these equations into equation 1 and taking the derivatives with respect to the sample sizes, n_{1a}, n_{2ab} and n_{3ab} (equation 2), equating to zero and solving, yields allocation solutions of the form,

$$n_{1a} = \left[\frac{\sum_d \lambda_d \, Vcp_{1da}}{C_{1a}} \right]^{\frac{1}{2}},$$

$$n_{2ab} = \left[\frac{C_{1a} \sum_d \lambda_d \, Vcp_{2dab}}{C_{2ab} \sum_d \lambda_d \, Vcp_{1da}} \right]^{\frac{1}{2}},$$

$$
n_{3ab} = \left[\frac{C_{2ab} \sum\limits_d \lambda_d \, Vc_{p_{3dab}}}{C_{3ab} \sum\limits_d \lambda_d \, Vc_{p_{2dab}}} \right]^{\frac{1}{2}} .
$$

The solutions are obtained using an iterative numerical procedure. One way to proceed is to simply multiply values of the Lagrange multipliers at successive iterations by the ratios of the corresponding variances, given the sample sizes at that iteration, to the corresponding variance constraints. The ratio increases a Lagrange multiplier value when the variance exceeds the constraint (i.e., the sample sizes are too small) and decreases it when the variance is smaller than the constraint. Sufficient accuracy is ensured by continuing the process until equation 5, squared and summed over the d-subscript, is less than some arbitrarily small amount such as 10^{-6}.

Informative initial values for the Lagrange multipliers can be obtained by first computing starting values for the third stage sample sizes,

$$
{}^0 n_{3ab} = \left[\frac{C_{2ab} \left[1 - R_{2da} \right]}{C_{3ab} \, R_{2da}} \right]^{\frac{1}{2}} ,
$$

and the second stage sample sizes,

$$
{}^0 n_{2ab} = \left[\frac{C_{1a} \left[1 - R_{1da} \right] R_{2da}}{C_{2ab} \, R_{1da}} \right]^{\frac{1}{2}} ,
$$

and then the initial Lagrange multiplier values,

$$
\left[{}^0\lambda_d \right]^{\frac{1}{2}} = \frac{1}{K_d} \sum_{a=1}^{12} \left[\left[{}^0 Var({}^0\underline{n})_{da} \right] \left[{}^0 C({}^0\underline{n})_a \right] \right]^{\frac{1}{2}} ,
$$

where,

$$
{}^0 Var({}^0\underline{n})_{da} = Vc_{p_{1da}} + \sum_{b=1}^{2} \left[\frac{Vc_{p_{2dab}}}{{}^0 n_{2ab}} + \frac{Vc_{p_{3dab}}}{{}^0 n_{2ab} \, {}^0 n_{3ab}} \right] ,
$$

$$
{}^0 C({}^0\underline{n})_a = C_{1a} + \sum_{b=1}^{2} \left[{}^0 n_{2ab} C_{2ab} + {}^0 n_{2ab} \, {}^0 n_{3ab} C_{3ab} \right] .
$$

Although other $^0\lambda_d$ values could be chosen to start the iterative procedure, the above calculations provide the values of the Lagrange multipliers that individually satisfy the variance constraints, considering them one at a time. Note that the initial ^{0}n-values are equal to the corresponding allocation solutions if there were but one value of the d-subscript. Comparison of the final λ-values with these particular initial values identifies those constraints that essentially determine the field study costs.

The allocation solutions for the NPS are shown in Table III. The fractional sample sizes at the second and third stage of sampling are obtained in expectation. For example, a sample of 1.468 wells is achieved in expectation by selecting one well with probability 0.532 and two wells with probability 0.468. The total sample sizes in Table III are,

$$n_{1+} = 90 \, ,$$

$$n_{2++} = 500 \, ,$$

$$n_{3++} = 734 \, .$$

Recall that the sample sizes in Table III are those that satisfy the stated objectives of the study (Table II) for the least cost. The most costly objective was that of requiring a 47 percent chance of detecting contamination in wells in counties classified as having above average use of pesticides and above average vulnerability, if, in fact, 1 percent of these wells were contaminated (a domain size of 0.03 percent of all wells in the nation, labelled domain 5 in Table II).

Higher probabilities of detection are, of course, coincidentally afforded higher contamination frequencies. The detection probabilities increase most rapidly in response to increases in the size of domain 4, wells in the most heavily cropped and most vulnerable parts of counties. The sample allocation in Table III will detect contamination frequencies of 1 percent of the wells in the domain with virtual certainty (probability of 0.98). Comparable probabilities are obtained for domain 1 (all wells nationally) for domain sizes as small as 2 percent; domains 2 and 3 (wells in counties with above average pesticide use and wells in counties with above average vulnerability), 2.5 percent each; and wells in domain 5 (wells in counties with both above average use and above average vulnerability), 5 percent. Although the choices of precision constraints might change with different investigators, virtually no chance exists for long term widespread contamination by any of the chemicals tested to escape detection.

Table III. Sample Allocation

| First Stage Strata | Sample Allocation | | |
| Second Stage Strata | Stage of Sampling | | |
a b	n_{1a}	n_{2ab}	n_{3ab}
1. High average use, high average vulnerability	7		
1. Most heavily cropped and vulnerable		2.322	1.468
2. Remaining areas		5.196	1.468
2. High average use, moderate average vulnerability	8		
1. Most heavily cropped and vulnerable		3.038	1.468
2. Remaining areas		3.919	1.468
3. High average use, low average vulnerability	4		
1. Most heavily cropped and vulnerable		3.038	1.468
2. Remaining areas		3.919	1.468
4. Moderate average use, high average vulnerability	5		
1. Most heavily cropped and vulnerable		3.495	1.468
2. Remaining areas		2.512	1.468
5. Moderate average use, and vulnerability	9		
1. Most heavily cropped and vulnerable		3.776	1.468
2. Remaining areas		1.000	1.468
6. Moderate average use, low average vulnerability	4		
1. Most heavily cropped and vulnerable		3.776	1.468
2. Remaining areas		1.000	1.468
7. Low average use, high average vulnerability	8		
1. Most heavily cropped and vulnerable		3.495	1.468
2. Remaining areas		2.512	1.468
8. Low average use, moderate average vulnerability	14		
1. Most heavily cropped and vulnerable		3,776	1.468
2. Remaining areas		1.000	1.468
9. Low average use, low average vulnerability	7		
1. Most heavily cropped and vulnerable		3.776	1.468
2. Remaining areas		1.000	1.468
10. Uncommon average use, high average vulnerability	7		
1. Most heavily cropped and vulnerable		3.495	1.468
2. Remaining areas		2.512	1.468
11. Uncommon avg. use, moderate avg. vulnerability	11		
1. Most heavily cropped and vulnerable		3.776	1.468
2. Remaining areas		1.000	1.468
12. Uncommon average use, low average vulnerability	6		
1. Most heavily cropped and vulnerable		3.776	1.468
2. Remaining areas		1.000	1.468

Summary

The steps in the allocation procedure are summarized in the following points.

- Identify the key parameters to provide the basis of the design.
- Quantitate the maximum values of the variances to be associated with the sample estimates of the key parameters.
- Develop equations to describe the variances of the parameters in terms of the sampling design (constants in the equations) and the sample sizes (unknowns in the equations).
- Develop equations to describe the per sampling unit costs of frame construction and stratification, sample selection, data collection and data processing.
- Simultaneously solve the equations subject to the imposed variance constraints.

Often the procedure tends to be iterative, in that budget realities act to modify the initially determined precision requirements.

Some emphasis is given the fact that questions of precision and cost cannot be sensibly addressed in the absence of a well specified field study design, including the sampling design and the design of the data collection procedures. In general, the information required to address the sample size and allocation question, with cost following as a consequence, is derived from the design specifications.

Literature Cited

1. Folsom, R. E., F. J. Potter and S. R. Williams, *Proc. Sect. Survey Res. Methods Am. Statistical Assoc.,* 1987, p 792.
2. Simmons. D. M., *Nonlinear Programming For Operations Research,* Prentiss-Hall, Inc., Englewood, N. J. 1975.

3. Chromy, J. R., *Proc. Sect. Survey Res. Methods, Am. Statistical Assoc.,* 1987, p 194.

RECEIVED October 18, 1990

CURRENT DESIGNS

Chapter 6

Regional and Targeted Groundwater Quality Networks in the Delmarva Peninsula

Michael T. Koterba, Robert J. Shedlock, L. Joseph Bachman, and Patrick J. Phillips

Water Resources Division, U.S. Geological Survey, 208 Carroll Building, 8600 La Salle Road, Towson, MD 21204

A multi-network monitoring and quality-assurance program was designed to assess the occurrence and distribution of selected pesticides and nutrients in groundwater in the Delmarva Peninsula in Delaware, Maryland, and Virginia.   As part of the U.S. Geological Survey's National Water-Quality Assessment (NAWQA) Program, four interrelated networks were established with wells distributed regionally across the peninsula and locally in small watersheds.   Data from these networks are being used to assess groundwater quality relative to differences in soil, land use, geomorphology, physiography, and hydrogeology at regional and local scales. An accompanying quality-assurance program was designed to help ensure accurate data and determine whether differences in water quality among network samples result from changes in hydrologic setting or are from sampling design.

In 1986, the U.S. Geological Survey (USGS) began a pilot National Water-Quality Assessment (NAWQA) Program. The long-term goals of this program are (1) to describe the quality of our nation's water resources and (2) to provide an understanding of how natural and human factors affect the quality of these resources (*1*).   Approximately 60 regional study units ranging from several thousand to tens of thousands of square kilometers in area will be investigated to achieve these goals.

Currently (1990), seven regional pilot projects are being used to test and refine concepts and approaches for the fully implemented NAWQA Program (*2*). Each project includes an analysis of existing information, hydrochemical measurements, and regional, targeted, and long-term sampling. The projects also follow similar sampling and quality-assurance guidelines to ensure consistency in data collection and storage (3). This paper describes the regional and targeted sampling designs and the quality-assurance program for the Delmarva Peninsula groundwater project. Regional sampling provides data for a broad range of chemical constituents in ground water across the study area (1). Targeted sampling provides data in relation to selected natural and human factors that can affect water quality in the study area. For example, in the Delmarva Peninsula, agriculture was targeted for investigation because it is the predominant land use. To evaluate the effects of agriculture on groundwater quality, samples are being collected from agricultural areas in various hydrological settings. These samples are analyzed to determine the concentrations of major ions, selected trace elements, nutrients, dissolved-organic carbon, and selected pesticides used in the Delmarva Peninsula.

A quality-assurance (QA) program is an integral part of the Delmarva regional and targeted sampling designs. The goals of this QA program are to provide accurate water-quality data, and estimate the variability in measuring selected water-quality constituents for use in data interpretation.

Description of the Study Area

The Delmarva Peninsula includes most of Delaware and the sections of Maryland and Virginia east of the Chesapeake Bay. The Peninsula is an oval-shaped land mass of about 15,700 km^2 in area that extends 240 km from north to south and about 115 km from west to east into the Chesapeake and Delaware Bays and Atlantic Ocean (Figure 1). The study area consists of that part of the Peninsula south of the Fall Line and lies in the Atlantic Coastal Plain physiographic province (4).

Physiography, Geomorphology, and Drainage Characteristics. The Peninsula is a coastal lowland formed by the deposition of fluvial, estuarine, and marine-marginal sediments. It is also an area of low relief; the maximum elevation is about 60 m above sea level and most of the area ranges from 15 to 25 m in elevation.

The geomorphology and drainage features of the Peninsula differ areally (4). In the northern part of the Peninsula, sandy, fluvial deposits form a broad, flat to gently rolling central upland area flanked by estuarine lowlands that gradually slope toward the Chesapeake and Delaware Bays (Figure 2). In the southern part of the Peninsula, the upland area is covered by a thin veneer of aeolian sand overlying fluvial and marine-marginal deposits. The upland area is flanked by lowlands containing broad tidal wetlands along the Chesapeake Bay and tidal wetlands and back-barrier lagoons and beaches along the Atlantic Ocean (Figure 2).

Upland and lowland areas can be divided into subregions on the basis of differences in soil, geomorphology, geology, drainage, and other hydrogeomorphic features (Figure 3). The hydrogeomorphic subregions can reflect differences in

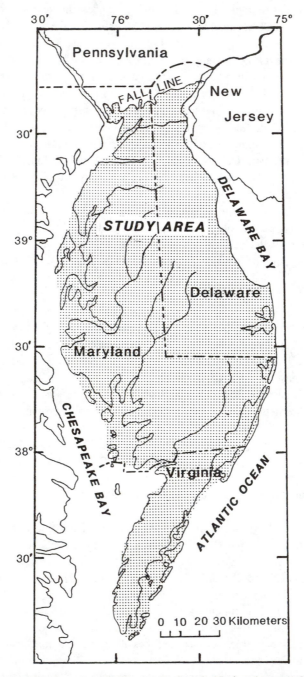

Figure 1. Location of the Delmarva Peninsula National Water-Quality
          Assessment study area

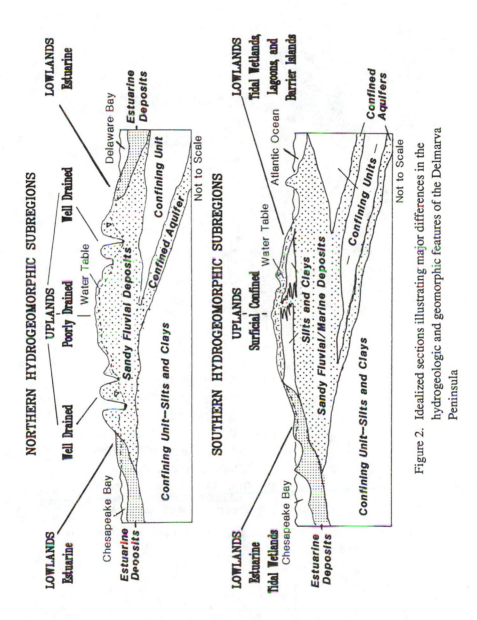

Figure 2. Idealized sections illustrating major differences in the hydrogeologic and geomorphic features of the Delmarva Peninsula

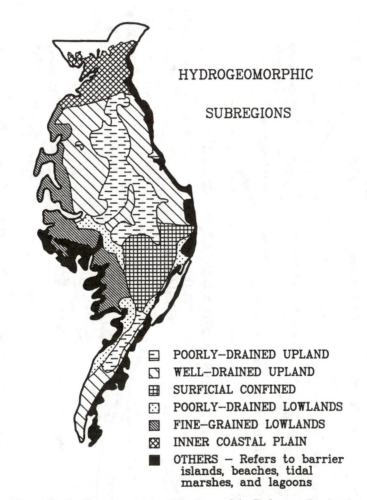

HYDROGEOMORPHIC

SUBREGIONS

⊟ POORLY–DRAINED UPLAND
◨ WELL–DRAINED UPLAND
⊞ SURFICIAL CONFINED
⸬ POORLY–DRAINED LOWLANDS
▨ FINE–GRAINED LOWLANDS
⊠ INNER COASTAL PLAIN
■ OTHERS – Refers to barrier
     islands, beaches, tidal
     marshes, and lagoons

Figure 3. Hydrogeomorphic subregions of the Delmarva Peninsula
            (Adapted from ref 4)

groundwater quality across the study area. These subregions are described by Hamilton and others (4) and have been used in evaluating existing groundwater-quality data in the study area.

Hydrogeology and Groundwater Flow. The study area is underlain by a wedge of unconsolidated sediments that range in thickness from 0 m at the Fall Line to greater than 2,000 m near the Atlantic Ocean (Figure 4). Sandy layers in these sediments form a series of confined aquifers that dip toward the Atlantic Ocean (5). The confined aquifers are overlain by a surficial aquifer that is under water-table conditions everywhere except in the "surficial confined" subregion (Figure 3).

The hydrogeologic characteristics of the surficial aquifer differ areally. Depths to the water table generally range from 0 to 6 m in most hydrogeomorphic subregions, but can exceed 10 m in the well-drained subregion. The saturated thickness of the surficial aquifer also ranges from about 6 to 15 m in the northern half of the study area to about 12 to 30 m in the central and southern parts of the Peninsula (Figure 2). This variability results from differences in the thickness of aquifer sediments and the depth to the water table.

The hydrogeology of the confined-aquifer system differs from the surficial-aquifer system. Although the surficial aquifer covers most of the study area, the confined aquifers are limited in areal extent. The confined aquifers also differ in thickness, age, and depositional environment (4), leading to regional changes in the depth to groundwater and areal extent of each confined aquifer.

All of the confined aquifers except one (Piney Point, Figure 4) make geo-logic contact with the surficial aquifer. This area of contact between the surficial aquifer and an underlying confined aquifer is called the subcrop zone. Subcrop zones are potential areas of transport for agricultural and other non-point-source contaminants to the deeper aquifers.

Recharge-discharge and groundwater-flow patterns differ in the surficial and confined aquifers. Determining these patterns is important because they can affect the quality of groundwater. In the surficial aquifer, recharge is pri-marily by infiltration of rainfall or snowmelt and is seasonal, occurring mainly from late fall to early spring when vegetation is dormant. Beginning with the growing season in the spring and continuing on until the fall, the depth to the water table usually increases as a result of evapotranspiration and groundwater discharge (6). Temporal variations in water-table depths and groundwater flow also can occur in an area as a result of groundwater pumping (7).

In the shallow surficial aquifer the length of groundwater flow-paths, or the distance water travels between its point of recharge and its point of discharge, ranges from a few meters to several kilometers. The variation in flowpath length is a result of differences in subregional and local drainage conditions (6,7, and J.M. Denver, in this volume).

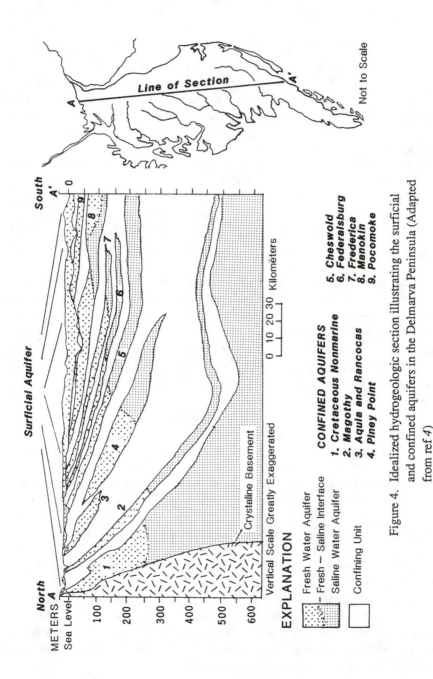

Figure 4. Idealized hydrogeologic section illustrating the surficial and confined aquifers in the Delmarva Peninsula (Adapted from ref 4)

Groundwater flow patterns also differ in the confined aquifers. The subcrop and adjacent zones of a confined aquifer can be part of a local, shallow-flow system in the surficial aquifer, where flowpath lengths are relatively short. In the intermediate and deep parts of the confined aquifers, groundwater often moves in regional flow systems and the length of flowpaths can range from a few to several hundred kilometers based on estimates from model simulations (*8-10*).

In a confined aquifer, differences in the length of flowpaths, and hence, groundwater flow patterns, arise because recharge to the aquifer can be through a subcrop zone, along an outcrop area, and by leakage through an adjacent confining layer. The corresponding point of discharge also can vary and be direct--to local streams, rivers, or the Chesapeake and Delaware Bays--or indirect--by upward leakage through an overlying confining bed and into another aquifer. Model simulations also show that groundwater flow in some areas of the Peninsula has been altered, or even reversed, near large pumping centers (*8-10*).

Population, Land Use, and Water Use. The population of the study area is about 600,000 (*11*). Most residents live in small, rural communities and are employed by agribusinesses and a few light industries.

The most significant agribusiness in the study area is the poultry industry (*4*). The importance of this industry is reflected in land-use and cropping patterns. About half the land use in the Peninsula can be classified as agricultural (Table I). Most agricultural land use is in corn rotated with soybeans, which provide feed to poultry growers.

The poultry industry produces about 5 tons of litter per 1,000 birds (*12*). This litter contains nitrogen, phosphorus, and soluble metals (*13*), and is recycled in the poultry areas, spread on cropland, or disposed of locally.

Table I.-- Land Uses on the Delmarva Peninsula
(Adapted from ref. *4*)

Land Use	Area (km^2)	Relative Proportion (%)
Agricultural	7,555	48
Woodland	4,861	31
Wetlands	2,038	13
Urban	1,044	7
Barren	192	1
Total	15,690	100

In addition to grain crops, a small number of farms produce fruits, vegetables, and nursery stock for local and regional markets. State and private surveys (*14,15*) indicate that fertilizer and pesticide use is widespread in the production of both vegetables and grains in the study area. These surveys also indicate the most commonly used pesticides are herbicides, such as the S-triazines, and insecticides, such as the carbamates.

Although agriculture is the dominant land use, cultivated fields differ in area and are interspersed with woodlands and wetlands in the study area. The differences in the area of agricultural fields, the relative proportion of different land uses (agriculture, forest, or wetland) found per unit area, and the location of agricultural lands relative to landscape features such as ridges, hilltops, bottomlands, ponds, and streams appear to reflect differences in drainage conditions among the hydrogeomorphic subregions (*4*).

The widely dispersed nature of both the population and agricultural activities have made the surficial- and confined-aquifer systems important sources of water supplies. Although the amount of groundwater pumped from each system can vary locally, total pumpage from the surficial aquifer is about equal to that from the confined aquifers (*4*).

In the Delaware and Maryland parts of the Peninsula, most groundwater for agricultural needs and rural domestic supplies is pumped from the surficial aquifer. Water needs of small towns and larger cities are mostly met by pumping from the confined-aquifer system; an exception is in southern Delaware and adjacent parts of Maryland where the surficial aquifer is the main source of water supply.

In Virginia, about two-thirds of the groundwater is pumped from the confined aquifer. This aquifer also supplies most domestic water needs. The surficial aquifer supplies most agricultural water needs.

Groundwater Quality. Available data on groundwater quality in the study area were compiled and analyzed by Hamilton and others (*4*). They found that natural factors seemingly accounted for only a small part of the variability in the concentration of each of a number of water-quality constituents.

Most of the variability in the existing data was attributed to inconsistencies and limitations among the existing data bases, which were compiled from several sources. Inconsistencies were found in relation to (1) well location, methods and materials used in well construction, and site-identification data; (2) quality control in sampling and analytical methods; and (3) sample-site selection (commonly biased to known or expected water-quality conditions). Analyses of trace elements and radiochemical concentrations, which are useful in assessing the effects of natural processes on water quality, were limited to a few samples. Data on pesticides and other potential, nonpoint-source contaminants also were sparse, a critical limitation given water resources within and around the Peninsula have been classified as susceptible to this type of contamination (*16,17*).

Despite the above inconsistencies and limitations, data for some constituents indicate there are differences in the quality of groundwater between the surficial and confined aquifers (*4*). Analyses of samples from wells completed in the upper part of the surficial aquifer showed that groundwater not affected by human activities has low pH and small concentrations of alkalinity, total dissolved solids, sodium, and nutrients such as nitrogen and phosphorus. The low values were attributed to rainfall chemistry and the weak solubility of the minerals, such as quartz and feldspar, in the upper sandy sediments of the surficial aquifer.

Waters in the deep surficial and confined aquifers generally vary in chemical composition, have large concentrations of dissolved solids, and a high pH. The large dissolved solids, high pH, and differences in the relative proportions of the major cations and anions in groundwater within and among these aquifers have been attributed to long groundwater residence times and different suites of minerals found in these aquifers (*4*).

Hamilton and others (*4*) also found that water quality, particularly in the surficial aquifer, is affected by human activities. For example, although nitrate-N concentrations seldom exceeded a few tenths to several milligrams per liter in forested areas, elevated nitrate-N concentrations were found in agricultural and urban-residential lands. These results are consistent with previous studies, in which elevated nitrate-N concentrations in groundwater were related to applications of fertilizers, sewage effluents, and animal wastes, or to leaking septic systems (*17-19*).

Differences were also found in nitrate-N concentrations among hydrogeomorphic subregions for a specific land use (*4*). In hydrogeomorphic subregions (such as the well-drained uplands) where nitrogen is readily oxidized, large concentrations of nitrate-N are found in agricultural and residential areas. These nitrate-N concentrations are significantly larger than those found in the same land use in hydrogeomorphic subregions (such as the poorly drained lowlands and surficial confined uplands) where nitrogen is not readily oxidized.

Sampling Design

The design for the Delmarva NAWQA sampling program is intended to reflect features and water-quality conditions in the study area that were ubiquitous and persistent over time. This meant the design had to: (1) include both the surficial and confined aquifers, (2) reflect current and potential water use, and (3) incorporate current water-quality concerns in addition to examining general water-quality conditions. The design also had to ensure that water quality could be assessed relative to differences in (1) study-area features (for example, soils, land use, hydrogeomorphic conditions, and geology), (2) groundwater conditions (such as positions of and depths along flowpaths in the surficial and confined aquifers) and (3) human activities (such as agriculture).

The sampling design also had to consider that relations between water quality and the above features, conditions, and activities can depend on the spatial scale over which data are collected and examined. A multi-network sampling design was conceived and developed to describe and assess water quality at several spatial scales. In addition, a quality-assurance (QA) program was developed to ensure that (1) unbiased procedures would be used in selecting sites, constructing wells, and collecting this data and (2) that QA data would be used to eliminate erroneous water-quality data and determine the affect of the sampling design on the measurement of selected water-quality constituents.

Sampling Networks. Four networks were developed as part of a multi-network sampling design (Figure 5). Two networks provide data for regional descriptions and assessments of groundwater quality. Two additional networks target hydro-geomorphic subregions and local study-area features that could affect water quality.

Regional Sampling. The areal and confined-aquifer networks were designed to be regional in scope. They provide the data for a broad, spatial assessment of groundwater quality in the surficial and confined aquifers.

The areal network was designed for the surficial aquifer. Consisting of two wells at different depths near each of 35 sites, it is the most widely distributed network and provides a relatively uniform coverage of the study area (Figure 5). The broad distribution of areal sites was achieved with minimal selection bias by first subdividing the study area into 12.5 minute by 12.5 minute grid cells. The boundaries of these cells were adjusted to divide the study area into 35 polygons of approximately equal area. A random site was chosen in each polygon, and state and county records were searched to find wells near each site that were suitable for sampling (3). If a suitable well could not be found, one was drilled according to NAWQA protocols. Existing wells , which reflect current water use, were chosen whenever possible.

In most cases, areal-network sites consist of a shallow, newly drilled, well and a deeper, existing well completed in an unconfined part of the surficial aquifer (Figure 6). At a few sites, such as some of those in the surficial-confined subregion (Figure 3), wells were located at different depths in the confined aquifer closest to the land surface when no significant thickness of aquifer material could be found at the water table.

The shallow wells in the areal network are screened within a few meters below the water table in that part of the surficial aquifer considered most vulnerable to non-point-source contaminants. The deeper wells are screened at least 5 to 10 m below the water table in that part of the surficial aquifer that is used for many rural and some urban water supplies.

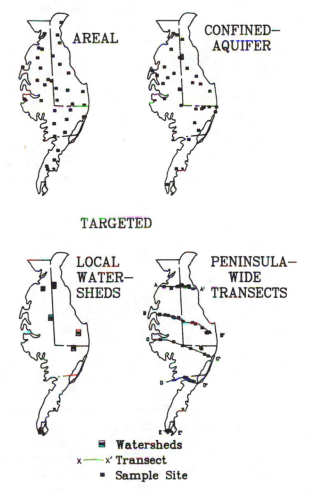

REGIONAL

AREAL

CONFINED–AQUIFER

TARGETED

LOCAL WATER–SHEDS

PENINSULA–WIDE TRANSECTS

Watersheds
x——x′ Transect
Sample Site

Figure 5. Sampling-site locations for each of the four water-quality networks of the Delmarva Peninsula National Water-Quality Assessment pilot-project

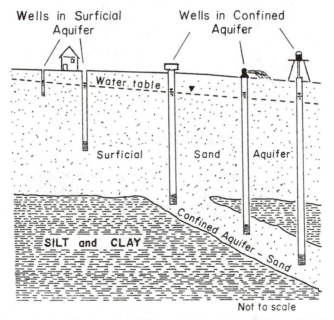

Figure 6. Idealized section illustrating depth distribution of wells
in the surficial and confined aquifers

The confined-aquifer network was designed to provide a broad spatial assessment of water quality in the confined-aquifer system. It also was designed to identify major changes in freshwater quality from the shallow to deep parts of each confined aquifer.

At least two sampling sites were chosen in each confined aquifer along flowpaths inferred from simulations and measured potentiometric surfaces (5,10). The first site was located near the base of the surficial aquifer and the upper part of the subcrop zone in each confined aquifer (Figure 6). A second site was chosen downgradient in the confined part of each aquifer. A third site was sometimes chosen in the deepest freshwater part of the aquifer tapped by an existing well. Two different sets of flowpaths and up to six sites were used to improve coverage in the more extensive confined aquifers or in aquifers undergoing more rapid development (such as the Pocomoke and Manokin--Figure 4).

A total of 35 sites were selected for the confined-aquifer network. Existing wells suitable for sampling were found near each site. These wells are distributed throughout the study area (Figure 5) and are screened at depths from about 5 m to greater than 350 m below land surface. About half of the wells are water-supply wells for major urban areas. The remaining wells are existing observation wells or domestic supply wells.

Targeted Sampling. Two networks were designed to target specific study-area features that could affect groundwater quality, particularly in the surficial aquifer (Table II). The networks were also designed to help evaluate regional water-quality.

The Peninsulawide transect network was designed to assess differences in water quality patterns among the major hydrogeomorphic subregions. These subregions were targeted for study because they differ in landforms, land uses, drainage characteristics, and shallow groundwater-flow patterns, all of which could affect water quality. This network consists of sites located along five transect lines that cross the study area from east to west (Figure 5). The transects were spaced at intervals that provide a broad north-south coverage of the study area.

Eight to twenty sites are located along each transect, depending on the number of subregions and transition zones the transect crosses. Two to five sites were chosen in each major hydrogeomorphic subregion along a transect (Figure 3). Several additional sites also were selected in transition zones between subregions. The five transects in this network contain a total of 80 sites.

Most of these sites have one newly drilled well, screened within a few meters below the water table. Some sites, where the surficial aquifer is thicker, have two wells screened at different depths. All sites initially were used to obtain geologic data and depths to the water table. Water levels are measured seasonally at some sites. About 40 wells in this network, or from 6 to 12 wells per transect, were used for water-quality sampling.

Table II.-- Water-Quality-Sampling Objectives, by Network Design, for the Delmarva National Water Quality Assessment Pilot Project

Water-quality-sampling objective	Network design
To describe and assess the quality of groundwater throughout the surficial aquifer.	Regional, Areal Network: Consisting of 35 sites, areally distributed over the study area, randomly selected, with a shallow and deep well being sampled at each site.
To describe and assess the quality of groundwater in the freshwater portion of the confined-aquifer system.	Regional, Confined-Aquifer Network: Consisting of 35 sites, distributed among 10 confined aquifers, along inferred flowpaths in each aquifer, with at least 2 to 3 sites along each flowpath. Sampled one well at each site.
To describe and assess the quality of groundwater in the surficial aquifer and in different hydrogeomorphic subregions in the study area.	Targeted, Peninsulawide Transect Network: Consisting of about 40 sites, distributed along 5 east-west transects across the Peninsula, with 2 to 5 sites in each hydrogeomorphic subregion crossed by a transect. Sampled 6 to 12 wells (at 5 to 10 sites) along each transect.
To describe and assess the quality of groundwater in local, shallow-flow systems in different hydrogeomorphic subregions, and to assess temporal variations in water quality and the effect of land use on water quality.	Targeted, Local-Watershed Network: Consisting of over 40 sites, distributed among 7 small watersheds in different hydrogeomorphic subregions, where wells are distributed among different agricultural and forested settings and along shallow groundwater flowpaths. Sampled 4 to 8 wells (at 4 to 5 sites) in each watershed seasonally.

The second targeted network consists of a series of local well networks in small watersheds, from about 5 to 10 km^2 in area, located in different hydrogeomorphic subregions along the Peninsulawide transects (Figure 5). Six of these watersheds are in predominantly agricultural settings. A seventh watershed was chosen in the poorly drained upland subregion to help evaluate changes in water-quality patterns in a forested wetland area (5).

Land use is being studied in each watershed in relation to the local groundwater flow system (4,6). Ten to twenty wells were placed at different locations and screened at various depths within the surficial aquifer to investigate local hydrochemical changes in water quality in each watershed. These wells are used to identify shallow groundwater flowpaths and to monitor seasonal changes in water levels and flowpaths. Four to eight wells are sampled seasonally to monitor and assess temporal variations in water quality.

The design and objectives of each networks are summarized in Table II. The data from the local-watershed, Peninsulawide transect, and areal networks provide a means of describing and assessing water-quality patterns in the surficial aquifer at local, subregional, and regional scales. The results obtained from these three networks, combined with those for deep aquifers using the confined network, provide a broad description and assessment of groundwater quality in the major aquifers of the study area.

Network Implementation and Sampling Strategies. The four sampling networks required the locating of sites and the identification or installation of about 230 wells. Because of the size of the study area and other developmental logistics, about 2 years were needed to install, develop, and sample the wells in all four networks.

Field activities within the study area were generally restricted by weather conditions to the period of April to December. Wells for each network generally were located or installed and developed in the spring. Sampling generally took place during the late spring to early winter.

Wells for the areal network and four of the local watersheds were in place by late spring of 1988. Wells in the other three local watershed and Peninsulawide transect networks were installed by the spring of 1989. The confined-aquifer network was established during the spring to fall of 1989.

Water samples from the areal network and the first seasonal samples from wells in four of the local watershed networks were collected in the summer of 1988. The Peninsulawide transect samples and the first seasonal samples from wells in the remaining watersheds were collected in the summer of 1989. Wells in the confined-aquifer network were sampled from the fall of 1989 through the early winter of 1990.

Wells were pumped to remove water that had been standing in the casing, and sampled for water quality according to NAWQA protocols (3). Analytical work not done in the field was performed or supervised by the USGS National Water Quality Laboratory (NWQL).

For the areal and Peninsulawide networks, the analyses included field measurements (Table III) and concentrations of inorganic constituents, nutrients, radiochemicals, stable isotopes, and organic compounds. Confined-aquifer water samples were analyzed for the same field measurements and concentrations of inorganic constituents and radiochemicals. Upgradient-well water samples also were analyzed for organic compounds. Analyses for local watershed samples included the same field measurements and concentrations of inorganic constituents. Organic analyses were customized for each local watershed based on known, or suspected, pesticide use.

Quality Assurance

The quality-assurance (QA) program for the Delmarva project complements existing QA programs at NWQL (20,21) and expands on the QA guidelines and protocols for the NAWQA program at the national level (3,21). Under these programs and guidelines, the extent to which QA sampling and data are employed by an individual NAWQA project are flexible because water-quality conditions and concerns differ among the projects. The objectives of the Delmarva QA program are to (1) provide accurate and representative water-quality data for each sampling network, and (2) estimate the variability in the measuring selected water-quality constituents and use these estimates as constraints in data interpretation.

Quality-assurance Measures and Criteria. Individual samples and all samples in each network are evaluated using a combination of quality-assurance measures and criteria (Tables IV and V).   The measures and criteria are used to: (1) monitor, detect, document, and when possible, eliminate erroneous data, (2) classify data for assessments, and (3) provide estimates of the variability in measuring the concentrations of selected constituents.

Detecting Erroneous Data. Potentially erroneous data are identified by (1) instabilities in field measurements while sampling, (2) bias or extreme errors in the electroneutrality balance among major dissolved cations and anions, (3) contamination of field blanks, (4) routine checks for transcription errors, and (5) trends or patterns in sample data that appear inconsistent with expected results (such as volatile organic compounds appearing in a samples collected from areas where there was no logical source for these compounds). Special efforts are made in the case of organic compounds to verify (disavow) low-level detections of pesticides and identify contaminants introduced during or after sample collection. This is done through the use of field blanks, modified ruggedness tests (23), and verification

Table III.-- Water-Quality Analyses for the Delmarva National Water Quality Assessment Pilot-Project (Adapted from ref 3)

Group	Class	Constituents and properties
Field measurements	Properties	pH, specific conductance, alkalinity, and bicarbonate.
Inorganic constituents	Major ions and metals and trace elements	Calcium, magnesium, potassium, sodium, chloride, fluoride, bromide, sulfate, silica, aluminum, antimony, arsenic, barium, beryllium, boron, cadmium, chromium, cobalt, copper, iron, lead, lithium, manganese, mercury, molybdenum, nickel, selenium, silver, strontium, vanadium, and zinc.
Nutrients	Nitrogen	Ammonium, nitrate-nitrite, nitrite, kjeldahl nitrogen (ammonium and organic).
	Phosphorus	Soluble-reactive phosphorus.
	Carbon	Dissolved-organic carbon; includes natural and synthetic.
Radiochemicals and isotopes	Radionuclides	Gross-alpha, gross-beta, radon-222, and tritium.
	Stable isotopes ratios	Deuterium/protium and oxygen-18/oxygen-16.
Organic compounds	Pesticides	Over 20 different herbicides including S-triazine- and chlorophenoxy-acid-based compounds. Over 15 different insecticides including carbamate-based compounds.
	Volatile organic compounds	Approximately 40 compounds, including simple, (multi-) halogenated alkanes, alkenes, and aromatics. Some of these may be used in agricultural areas, for example, 1,2-Dichloropropane and 1,3-Dichloropropene.

Table IV.-- Quality-Assurance Measures, Criteria, and Comments Related to Screening
Water-Quality Data for Individual Samples

Measures	Project criteria	Comments
Successive temperature, pH, and specific measurements are taken just prior to and during the sampling of each well.	Stable sampling conditions are achieved when the temperature, pH, and specific conductance differ by less than 0.2 degrees Celsius, 0.1 pH units, and 5%, respectively, over time.	At each well, water samples for different chemical analyses are collected sequentially. Instability in field measurements and indicate a change in the chemistry of water being sampled.
Successive field determinations of alkalinity as calcium carbonate, mg/L, and bicarbonate, mg/L, are made on at least two replicate samples from each well.	Resulting values for field alkalinity and bicarbonate are satisfactory if replicate values differ by less than 10%.	Bicarbonate can be a significant part of the total dissolved solids and errors in measurement can lead to large ionic balance errors.
The magnitude and sign of the error in the ionic balance between cations and anions (expressed as the ratio in percent of their difference divided by their sum) is estimated for each well.	The error in ionic balance is acceptable if it falls within a range determined by all the samples within a given network (see Table V).	A sample should theoretically have an ionic balance of zero. In practice, errors should be near-zero and a random function of sampling design.

tests (Table V). Potentially erroneous data are corrected or eliminated when it is possible to identify the source of the error.

Data falling under scrutiny and any action taken regarding that data are documented in the project's computerized data-management system. The documentation for each network also summarizes QA results for individual samples and informs users as to the general quality and limitations in the sample data. An example of the summary for inorganic constituents in areal-network samples is given in Table VI.

Sources of Variability in Analytical Measurements. Most of the variability in the concentration of a water-quality constituent is generally assumed to be from natural processes or human activities in the study area. However, some variability in measurements can be attributed to differences in the following: (1) ambient field, shipping, handling, and laboratory conditions, (2) sampling and laboratory methods and procedures and (3) possible analytical interferences caused by differences in the background hydrochemistry of water samples. To estimate this variability, which hereafter is referred to as the variability in measuring a constituent due to sample design, QA sampling was integrated into each network sampling schedule. This was done by distributing similar QA samples as follows: (1) according to network sampling objectives, (2) among the major hydrochemical types of groundwater sampled, (3) over the time it took to sample all network wells, and (4) over the range in field conditions encountered while sampling network wells.

Cost considerations precluded collecting a QA water sample at every well in a network. Thus, the chief problem in integrating QA sampling was ensuring that a small number of QA samples (about 10 to 15 percent of the total) would represent the range in hydrochemical variability found in a larger number of network water-quality samples. The variability in the hydrochemistry of groundwater in the study area is large (*4*) and the chemical characteristics of groundwater at many of the newly installed wells was completely unknown.

To overcome this uncertainty, QA sites were selected on the basis of the network design and the inferred relations between water-quality, study-area features, and human activities reflected in each sampling network (Table VII). For example, under these guidelines at least one QA site was selected along each transect and in each major hydrogeomorphic subregion in the Peninsulawide transect network. This network-based approach to QA-site selection reduced the problem of trying to select 7 QA sites from the over 40 wells in this network to one of choosing 1 or 2 QA sites along each transect from a smaller group of 4 to 6 wells.

Another consequence of selecting QA sites in this manner was that the QA sampling locations for each network were widely distributed over the study area. This also helped ensure that QA sampling was distributed in time and over a variety of field conditions encountered during the sampling of each network.

Table V.--Quality-Assurance Measures and Criteria Used in Screening Data Among Samples in a Network

Network ionic balance errors (NIBE) are calculated for all samples and checked for biases in the sum of cations or anions and in relation to the time of sampling (22). The errors in NIBE values are acceptable if the data show no appreciable bias and individual sample values lie between the following limits: [$(1.5 \times IQR) + NIBE_{75th\%}$] < $NIBE_i$ < [$NIBE_{25th\%}$ - $(1.5 \times IQR)$], where IQR is the interquartile range based on the difference between the 75th and 25th cumulative percentile NIBE values. If NIBE data are normally distributed, there is only about a 1% chance that a NIBE value falling outside these limits is a random event. Sample NIBE values which fall outside this limit are assumed to reflect a significant error in the water-quality data.

Test, and field blanks consist of water initially free of organic compounds (Table III). A test blank, including any sample preservatives, is shipped as a sample to the NWQL just prior to each major sampling period to check for contaminants in reagents or due to shipping, handling, and laboratory procedures. A field blank is water that is passed through sampling equipment after wash procedures used between sampling sites. Field blanks are taken at the start, midway through, and near the end of each major sampling period, or when one of the following occurs: (1) a change in field sampling equipment, (2) unusual field conditions (such as areal spraying near a sampling site), or (3) after a contamination incident. A blank analysis is acceptable if no organic compounds are reported.

Modified ruggedness tests, adapted from Taylor (23), are used to isolate the source of contaminants introduced during or after sampling. An example of such a test designed to isolate contaminants X, Y, and Z found in a field blank appears below. A positive ruggedness test isolates the contaminants found in a field blank.

Table V.--Quality-Assurance Measures and Criteria Used in Screening Data Among Samples in a Network--Continued

Sample	Sample make-up	Compounds found
A	Blank, Water Lot 1#	X
A'	Blank, Water Lot 1#, Preservative 1#	X, Y
A''	Portion of A;, run through sampling pump	X, Y, Z
B	Blank, Water Lot 2#	X
B'	Blank, Water Lot 2#, Preservative 2#	X
B''	Portion of B', run through sampling pump	X, Z

Conclusions from test: Contaminant X is from the sample bottles (a test blank showed shipping and laboratory were not the source), Y is from Preservative 1#, and Z is from the sampling operation or equipment. Recommend discarding Preservative 1#, checking sampling pump wash procedure, and submitting subsequent test and field blanks prior to any sampling. Also need to check water samples collected since last clean field blank for X, Y, or Z.

Verification tests are initiated by project and done by the NWQL to verify (disavow) the presence of organic compounds which are found in several or more samples at concentrations near the reporting level. This test can require sample concentration and purification steps, and use of an alternate, but recognized, analytical method for determining the identity and concentration of the organic compound in question. A positive verification test confirms the identity and concentration of the organic compound in question.

Table VI.--An Example Illustrating the Classification of Inorganic Water-Quality Data for Assessments Using Areal Network Samples

A partial listing of QA measures and results considered in classifying a sample

Sample Number	QA class	Temperature	Specific conductance	pH	Replicate alkalinity titrations	Ionic balance error
1	A	Stable	Stable	Stable	Satisfactory	+ 2.1
2	B	Stable	Stable	Stable	**Unsatisfactory**	**- 5.4**
3	C	Stable	Stable	Stable	Satisfactory	**- 12.4**
.	. .	......	......	......	....	......
.	.	......	......	......	....	......
69	D	**Unstable**	Stable	Stable	**Unsatisfactory**	**+ 30.1**

A summary of the QA classification of inorganic water-quality data

QA class	Frequency by class
A	27
B	32
C	3
D	1
X	6

Remarks

Balance errors for A and B samples are normally distributed with a mean near zero (-0.01) and inner quartile range of -7.00 (25th percentile) to +6.91 (75th percentile). This suggests errors in measurements are random. Groundwater with elevated iron and alkalinity concentrations appeared to be the most difficult to analyze. Inorganic data for samples classified as A and B are suitable for assessments. Exercise caution using data from samples classified as either C, D, or for which complete data is missing (X). Recommend resampling at sites with data classified as C, D, or X.

A partial listing of statistics on the variability of chemical measurements for replicate samples

Chemical Constituent	Relative standard deviation (%)	Upper bound of 95% confidence interval
Calcium	1.2	0.22 mg/L
Magnesium	2.1	0.28 mg/L
Sulfate	6.7	0.79 mg/L
.....	...	...

Table VII.-- Quality-Assurance-Sampling Objectives and Design Based on the Delmarva National Water Quality Assessment Program Network Design

Network	Network-based objective	Site selection	Analyte selection
Areal	Estimate regional variability in hydrochemical constituents found in the surficial aquifer, using 35 sites, and sampling a shallow and a deep well at each site.	Eight sites, all shallow wells, spatially distributed among uplands and lowlands and in different land uses.	Replicate samples for all areal network analyses except for organic compounds, which will be fortified replicates.
Peninsula-wide transects	Estimate subregional variability in hydrochemical constituents in the surficial aquifer using 7 hydrogeomorphic subregions, 5 transects, and sampling 6 to 12 wells along a transect.	Seven sites, all shallow wells spatially distributed over five transects and among major hydrogeomorphic subregions.	Similar to areal network, but using fortified triplicate, not replicate, samples for organic compounds.
Local watershed	Estimate local variability in hydrochemical constituents in the surficial aquifer, along shallow-flow paths and under varying land use and hydrogeomorphic subregions, using 7 small watersheds, and seasonally sampling 4 to 8 wells in each watershed.	Seven sites, mainly shallow wells with one in each watershed being sampled over time (seasonally if the budget allows).	Replicate samples, except for organic compounds, which will be fortified replicates or triplicates for selected pesticides for selected suspected or known use.
Confined aquifer	Estimate the variability in measuring the hydrochemistry of the confined-aquifer system, using 11 aquifers, and sampling at least 2 to 3 wells along regional flowpaths in each aquifer.	Seven sites, wells at different depths along the gradient in each of three aquifers that differ in areal extent and location.	Replicate samples, except for organic compounds, which will be fortified triplicate samples for all upgradient wells and selected downgradient wells.

Because samples are sent to the NWQL directly from the field, QA samples also were distributed over a range in shipping and laboratory conditions.

The QA samples were analyzed at NWQL at about the same time as the corresponding network samples. Differences in the concentrations of a given chemical constituent between replicate samples for all such sample pairs in the network are used to estimate the variability in measuring a chemical constituent due to sampling design.

For synthetic organic compounds, two or three replicate samples were collected with each network sample at each QA site (Table VII). The QA samples were each fortified at the site with known organic compounds to yield concentrations similar to those found in groundwater in the study area. Differences in the concentration between the fortified replicates are used to estimate the variability in measuring selected pesticides and volatile organic compounds due to sampling design.

Results from the areal network can be used to illustrate the integrated approach developed for QA sampling. In the areal network, QA site selection among network wells was done by choosing sites in upland and lowland areas that represented a variety of land uses. As a result, the range in hydrochemical composition of the QA samples was fairly representative of the range in hydrochemical composition of all the network samples (Figure 7A). Only sodium bicarbonate-type waters (lower portion of diagram) were not well represented by initial QA sampling. Subsequent QA sampling was used to correct this deficiency.

It is also noteworthy that the range in hydrochemistry of the QA samples was similar to that of those network samples in which pesticides were found (Figure 7B). This result ensures that estimates of both pesticide recovery and the variability in measuring pesticides generated by the QA program are applicable to those groundwaters where pesticides are found.

Assessing the Variability in Analytical Measurements. Two statistical measures of the variability in the concentration of a constituent due to differences in sampling design used by the Delmarva NAWQA project are (Table VI): (1) the relative standard deviation (RSD) and (2) the upper bound of the 95th percentile confidence interval (UCI95%). The RSD estimate for a given constituent can be compared to statistics generated from national QA programs (24). These comparisons can be used to indicate similarities and differences in quality assurance between a NAWQA study unit and larger regional or national assessment programs.

Within the project, RSD estimates among the constituents can be compared to determine which constituent varied the least to the most because of differences in sampling design. Because there is variability in RSD estimates among the constituents, groundwater-quality assessments in the Delmarva NAWQA project are done using groups of water-quality measures or constituents. For example, the impact

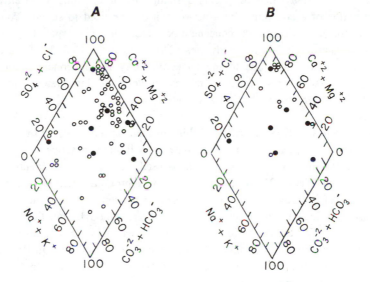

○ Areal network sample

● Quality assurance sample

Percentages in milliequivalents per Liter

Figure 7. Comparison of sample-matrix chemistry for quality-assurance samples and (A) areal-network samples or (B) areal-network samples containing pesticides

of agriculture on water quality is being assessed not only by investigating differences in the concentration of nitrate-N and the occurrence of pesticides, both of which can be difficult to measure at low levels in groundwater, but also by investigating differences in other chemical constituents which are less difficult to measure.

The UCI95% for an individual constituent also can be used in water-quality assessments. It is a rough measure of the minimum difference in the concentration needed to be relatively (95%) sure that two sample concentrations are different. When the UCI95% value represents a small part of the total variability in the concentration of a constituent in a network, it can be used to support conclusions that differences in network concentrations represent real spatial or temporal trends in water-quality in the study area. However, if the UCI95% for a constituent is similar in magnitude to the total variability in the concentration of that constituent, assessments based on this constituent are interpreted with caution and only in conjunction with other water-quality data.

Limitations in Quality-Assurance. While the Delmarva NAWQA QA program has helped detect samples with erroneous data, it is limited in scope. The field and analytical measures and screening methods used will not always identify samples with erroneous data. Nor does the current QA program attempt to identify the major source(s) of the variability in measuring a chemical constituent, except in the case of selected organic compounds. Finally, although statistics such as the RSD and UCI95% are useful in assessing differences in water-quality in the study area, their validity is determined by the extent to which the quality-assurance sampling program is able to represent the range in sampling design features that affect water-quality measurements.

Summary

A ground-water-quality sampling network for the Delmarva Peninsula project was designed to address national, regional, and local water-quality issues. As part of the National Water Quality Assessment (NAWQA) Program, the network design and sampling strategy of the Delmarva project provide nationally consistent information for regional comparisons of water quality. The design and sampling strategy also were developed to assess the effects of natural factors (such as geology and hydrology) and human activities (such as agriculture) on water-quality patterns and seasonal trends within the study area.

The different objectives of the Delmarva NAWQA project and the differences in hydrogeologic and geomorphic settings in the study area required a multi-network sampling design. The resulting design consists of four regional and targeted sampling networks. These networks provide spatial coverage of the major, freshwater parts of the surficial and confined aquifers in the study area. They also provide an indication of the quality of water currently being used in the region. In addition, the targeted networks focus on specific features that can affect water quality at local, subregional, and regional scales.

A comprehensive program of quality-assurance sampling was undertaken by both the Delmarva project and the national NAWQA program to help ensure that interpretations of water quality in the study area are based on accurate and representative data. In the Delmarva project, quality-assurance (QA) samples were integrated into the network sampling design by distributing their collection among the four sampling networks and within each network over the period of sample collection, a range of field conditions, and in relation to differences in the hydrochemistry of groundwaters in the study area.

The integration of QA into the network design and sampling strategy provides the means to monitor, detect, and sometimes correct or eliminate erroneous data. The results of the integrated QA program also make it possible to assess whether differences in water quality result from natural features and human activities in the study area or reflect variations in sampling and analytical procedures and methods.

Literature Cited

1. Hirsch, R.M.; Alley, W.M.; Wilber, W.G. *U.S. Geol. Sur. Circular 1021*, **1988**, 42 p.
2. Wilber, W.G; Alley, W.A. *U.S. Geol. Sur. OFR 88-175*, **1988**, 2 p.
3. Hardy, M.A.; Leahy, P.A.; Alley, W.M. *U.S. Geol. Sur. OFR 89-396*, **1988**, 36 p.
4. Hamilton, P.A.; Shedlock, R.J.; Phillips, P.J. *U.S. Geol. Sur. OFR 89-34*, **1989**, 71 p.
5. Cushing, E.M.; Kantrowitz, I.H.; Taylor, K.R. *U.S. Geol. Sur. PP 822*, **1973**, 58 p.
6. Phillips, P.J.; Shedlock, R.J. Abs. In: *Proc. Chapman Conf. on hydrogeochemical Responses of Forested Watersheds*, **1989**, 1 p.
7. Denver, J.M., In: *U.S. Geol. Sur. Nat'l. Sym. on Water Qual.*: Abs. of Tech Sess., U.S. Geol. Sur. OFR 89-409, **1989**, p.17.
8. Harsh, J.F.; Laczniak, R.J. *U.S. Geol. Sur. PP 1404-F*, **1986**, 126 P.
9. Leahy, P.P.; Martin, M.; Meisler, H.; In *Amer. Water Res. Mon. 9*, Vechioli, J., Johnson, A.I., eds., **1988**, p. 7-24.
10. Vroblesky, D.A.; Fleck, W.B. *U.S. Geol. Sur. PP 1404-E*, **1988**, 123 p.
11. *Bureau of Census*, U.S. Dept. of Commerce, Series P-26, No. 84-S-SC, 117 p.
12. Perkins, H.F.; Parker, M.B.; Walker, M.M. *Georgia Agr. Exp. Sta. Bull. No. 123*, Univ. of Georgia, Athens, Georgia, **1964**, 22 p.
13. Ritter, W.E.; Humenick, F.J.; Skag, F.W.; Chirnside, *Jour. Irrig. and Drain. Eng.* ACSE 115(5), **1989**, p. 807-821.

14. Kruchten, T. *Maryland Pesticide Statistics*, Maryland Dept. Agr., USDA, and Maryland Agr. Stat. Serv., MDA No. 227-87, **1987**, 37 p.

15. Pait, A.S.; Farrow, D.R.G.; Lowe, J.A.; Pacheco, P.A. *Agricultural Pesticide use in Estuarine Drainage Areas*, Nat'l. Coast. Poll. Dis. Inv., U.S. Dept. of Commerce, Nat. Ocean. and Atmos. Admn., **1989**, 134 p.

16. Alexander, W.J.; Liddle, S.K.; Mason, R.E.; Yeager, W.B.; *Ground-water Vulnerability Assessment in Support of the First Stage of the National Pesticide Survey*: Research Triangle Institute. EPA No. 68-01-6646, Off. Pest. Prog., USEPA, Washington D.C., **1986**, 22 p.

17. Bachman, L.J. *U.S. Geol. Sur. WRIR 84-4332*, **1984**, 51 p.

18. Denver, J.M. *Delaware Geol. Sur. RI No. 45*, **1986**, 66 p.

19. Ritter, W.F.; Chirnside, A.E.M.; Scarborough, R.W. *Proc. Irrig. and Drain. Div. Conf. ACSE*, Lincoln, Nebraska, **1988**, p. 468-477.

20. Mattraw, Jr, H.C.; Wilber, W.G.; Alley, W.M. *U.S. Geol. Sur. OFR 88-726*, **1989**, 21 p.

21. Jones, B.E. *U.S. Geol. Sur. OFR 87-457*, **1987**, 36 p.

22. Lucey, K.J.; Peart, D.B. *U.S. Geol. Sur. WRIR 89-4049*, **1989**, 72 p.

23. Taylor, J.K. *Quality assurance of chemical measurements*, Lewis Publishers, Inc. Chelsea, Mich. 4th Printing., 1988, Chapts. 4,13.

24. Lucey, K.J. *U.S. Geol. Sur. OFR 90-162*, **1990**, 53 p.

RECEIVED November 20, 1990

Chapter 7

Groundwater-Sampling Network To Study Agrochemical Effects on Water Quality in the Unconfined Aquifer

Southeastern Delaware

Judith M. Denver

Water Resources Division, U.S. Geological Survey, Room 1201 Federal Building, 300 South New Street, Dover, DE 19901

Understanding local and regional groundwater-flow patterns was necessary to design a sampling network to study the movement and distribution of agrochemicals in the unconfined aquifer in southeastern Delaware. Clusters of wells completed at various depths were installed in the expected direction of local groundwater flow along a transect from the center of a 100-ha cultivated field toward a nearby stream. Contrary to expectations, groundwater flow in the study area is almost parallel to the stream, in the direction of regional flow. Consequently, agrochemicals from the site migrate along flow paths from source (recharge) areas to distant regional discharge areas and do not significantly influence the water quality in the stream. The sampling network was expanded upgradient and downgradient from the original site during a second phase of the study. The expanded network provided better understanding of agrochemical distribution relative to regional groundwater-flow patterns.

Distribution of agrochemicals in the unconfined aquifer is affected by such factors as differences in fertilizer application rates (for corn and soybean crops), recharge timing and magnitude, soil and aquifer properties, upgradient land use (present and historical), and groundwater withdrawal. The degree of agrochemical influence on water quality varies widely, both areally and with depth in the aquifer, because the interrelations among these factors are complex. A three-dimensional sampling network is needed to understand the groundwater-flow system and to interpret groundwater quality in relation to the above factors.

This paper presents the design of a groundwater-sampling network at a research site used for two studies of water quality in an agricultural area

(Figure 1) (1,2). The importance of understanding both local and regional groundwater-flow systems when assessing agrochemical effects is emphasized. Examples of both problems and successes with this network are given.

The original groundwater-sampling network was installed in an approximately 100-ha field with an irrigation well at the center (Figure 2). A major project objective was to study groundwater withdrawal effects on the distribution and movement of agrochemicals in the aquifer. A second objective was to assess agrochemical effects on water quality downgradient from the field and in an adjacent stream.

Based on results of the first study, the sampling network was expanded in a subsequent project to study the distribution of agrochemicals in relation to regional groundwater flow (Figure 3). The expanded network extends from the regional recharge area, which is upgradient from significant agricultural land use, to the regional discharge area, which is downgradient from the predominantly agricultural area.

Study Area

The study area is in eastern Sussex County, Delaware, which is part of the Atlantic Coastal Plain Province (Figure 1). Land uses are predominantly corn and soybean production and forest. Soils are generally well-drained sandy loams. The aquifer, which is approximately 30 m thick, consists mainly of permeable sand and gravel; it is susceptible to contamination by NO_3-N and other chemical constituents associated with agricultural practices.

The site is one of several local watersheds being investigated in conjunction with the U.S. Geological Survey's National Water Quality Assessment (NAWQA) project on the Delmarva Peninsula (Koterba, M.T., Shedlock, R.J., Bachman, L.J., Phillips, P.J., in this volume).

Original Well Network

Groundwater-flow directions were estimated using a published water-table map of the area to select sites for well installation (3). A network was designed that included five clusters of wells. Individual wells within each cluster were screened at different depths in the aquifer to monitor vertical hydraulic gradients and water quality at approximately 6 m intervals from near the water table to the base of the unconfined aquifer, about 30 m below the land surface. Three of the clusters were installed in a transect between the irrigation well and the stream in the expected direction of groundwater flow (Figure 3). Several shallow wells also were installed around the field perimeter for additional water-table control. Vertical hydraulic gradients and water quality were related to irrigation pumping and natural groundwater-flow patterns.

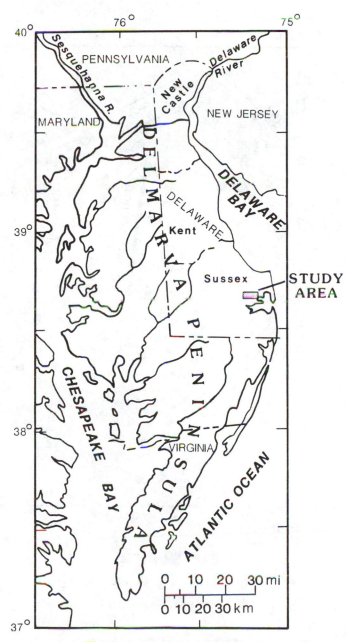

Figure 1. Location of Study Area

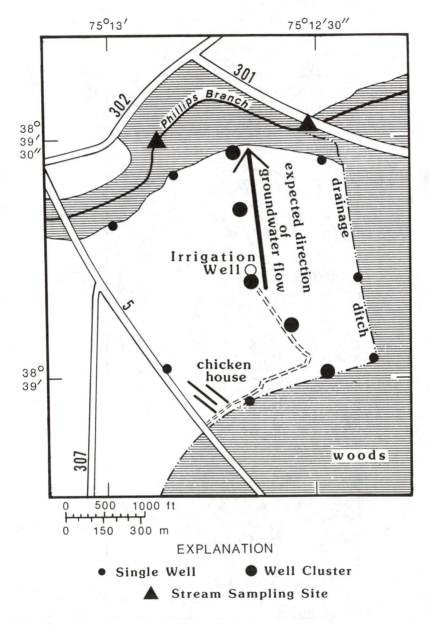

Figure 2.   Configuration of Original Sampling Network

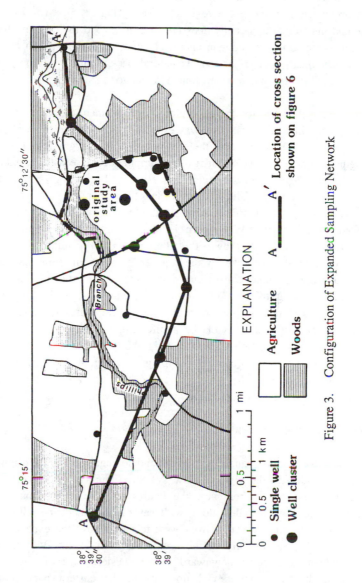

Figure 3. Configuration of Expanded Sampling Network

Groundwater Flow and Water Quality

Initial water-table measurement indicated that groundwater flow at the site is predominantly parallel to the stream, or perpendicular to the expected direction (Figure 4). Streamflow is maintained by discharge from a local shallow groundwater system which lies above the regional system (Figure 5). The local flow system around the stream apparently does not extend beyond the adjacent wooded area. Stream water does not contain chemical concentrations that can be attributed to agrochemicals, but rather reflects natural background conditions. Agrochemicals in the aquifer move from source areas to distant regional discharge areas.

Groundwater withdrawal for irrigation promotes movement of water into deeper parts of the unconfined aquifer. This was especially apparent in the well cluster adjacent to the irrigation well where the downward hydraulic gradient was almost 1 m during pumping. The effects of irrigation pumping decrease with distance from the well: Water level measurements from the well clusters on the field's perimeter had downward hydraulic gradients of less then 2 cm (1).

Water recharging the aquifer contains dissolved ions from fertilizers and lime applied to the crops. When corn is planted, nitrogen is applied to the field. Soybeans, generally grown on alternate years, require no nitrogen fertilizer. As a result, recharge to the aquifer contains different amounts of nitrate and other agrochemical constituents depending on the crop and water in the aquifer is chemically stratified. This stratification was most obvious in samples from the well cluster nearest the irrigation well where water is pulled rapidly downward by pumping. In parts of the aquifer not significantly influenced by groundwater pumping, upgradient land use is the principal control on water chemistry. The effects of groundwater pumping on water quality generally were difficult to distinguish from the effects of agrochemicals carried along regional flow paths, except immediately adjacent to the irrigation well.

Expanded Monitoring Network

The sampling network was expanded upgradient and downgradient of the original site to study water quality in the regional flow system (Figure 3). Six well clusters screened at different depths from near the land surface to approximately 30 m below land surface were installed along with several independent shallow wells. Groundwater flow is predominantly from west to east across the expanded study area. The network successfully defines regional groundwater flow: Vertical hydraulic gradient is directed downward at the western part of the network, indicating groundwater recharge, and up-ward at the eastern part, indicating groundwater discharge (Figure 6).

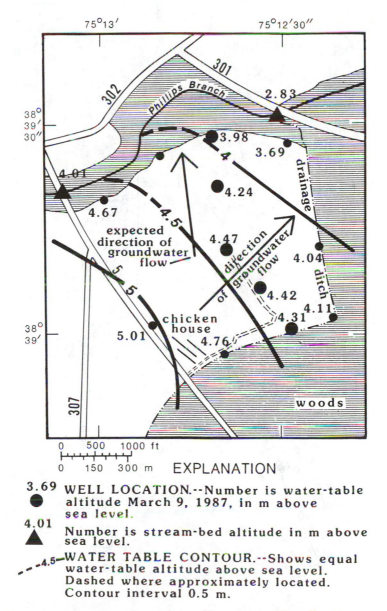

Figure 4.   Direction of Groundwater Flow, Original Sampling Network

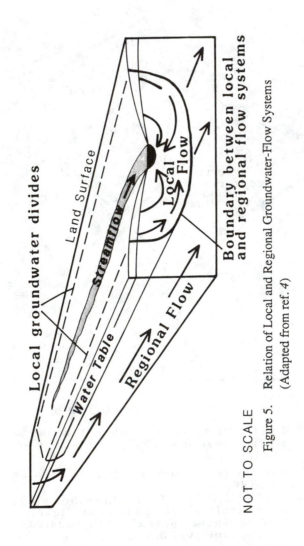

NOT TO SCALE

Figure 5.   Relation of Local and Regional Groundwater-Flow Systems
            (Adapted from ref. 4)

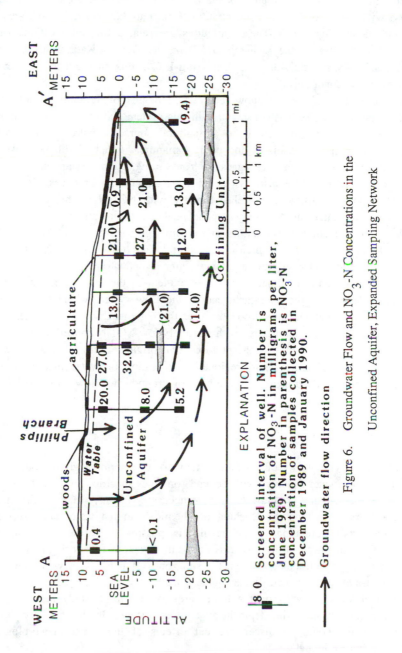

EXPLANATION

8.0   Screened interval of well. Number is
concentration of NO$_3$-N in milligrams per liter,
June 1989. Number in parenthesis is NO$_3$-N
concentration of samples collected in
December 1989 and January 1990.

Groundwater flow direction

Figure 6.   Groundwater Flow and NO$_3$-N Concentrations in the
Unconfined Aquifer, Expanded Sampling Network

Relation of Water Chemistry to Land Use

Water samples from wells aligned in the direction of regional groundwater flow show a wide range of NO_3-N concentrations (<0.01 to 32.0 mg L^{-1}) both areally and with depth Figure 6). These variations are related to well locations with respect to land use in the recharge areas for the wells. Natural background levels of NO_3-N generally are less than 1 mg L^{-1} in this area and the aquifer is under oxidizing conditions (1).

Nitrate-N was selected to show the relative degree of agricultural influence on groundwater because it is conservative in this system and concentrations above background levels are attributable to agricultural sources. Nitrogen sources at this site are inorganic fertilizers and poultry manure. Other dissolved constituents, which include calcium and magnesium from liming, and potassium and chloride from potash fertilizer, are also associated with agricultural practices. Concentrations of these constituents correlate directly with NO_3-N in this area, therefore, they also could be used to study agrochemical effects on water quality.

Water in the aquifer that shows significant agrochemical influence can be related to recharge areas of intense agricultural land use (Figure 6). In contrast, water that recharged the aquifer through wooded areas does not show significant agrochemical influence. This is evident in the low concentrations of NO_3-N at depth near the upgradient end of the flow system where the regional recharge area is predominantly wooded. Near the regional discharge area, shallow water from a wooded recharge area with relatively low NO_3-N concentrations overlies water with higher NO_3-N concentrations that originated in an upgradient area of agricultural land use. Thus, regional stratification of water quality in the aquifer can be explained using the definition of groundwater flow provided by the expanded well network.

Discussion

The extent and complexity of a sampling network to determine groundwater-flow directions and water quality will depend on the specific objectives of the project. Understanding groundwater flow is necessary for any water quality study that considers more than the leaching of chemicals to the water table surface; otherwise, misinterpretation of the results is possible. For example, understanding the groundwater-flow system is not as important as monitoring recharge to study agrochemicals leaching through the soil zone. A network of wells screened near the water table would be adequate for this type of project. A three-dimensional array of wells over a larger area is needed to study agrochemical distribution areally and with depth in an aquifer system. In this case long-term regional trends would be discernible but effects of individual meteorological events would be minimal.

The original well network did not entirely satisfy the design objectives because the transect of wells from the agricultural area toward the stream did not intercept downgradient flow. The network did, however, describe the groundwater-flow system. If the well network had not adequately defined groundwater flow, and expected flow paths had been assumed correct, interpretation of agrochemical effects on water quality would have been quite different. The water-table gradient would have appeared to be toward the stream (except for dry periods when the water table is lower than the stream bed) and base-flow water quality in the stream could have been assumed to represent groundwater discharge from the agricultural area drained by the stream. Because water quality in the stream reflects essentially natural background conditions in the aquifer, it could have been concluded that nutrients and other agrochemicals were not entering the surface-water system during base flow. In reality, as shown with results from the expanded network, agrochemicals are being carried in the regional groundwater-flow system to distant regional surface-water discharge areas, by-passing the local groundwater-flow system near the stream.

Understanding the position of a groundwater-sampling network in relation to regional as well as local flow systems is important to any study. A complete interpretation of groundwater flow is especially important if the results of a study are to be considered representative of a broader area.

Literature Cited

1. Denver, J. M., *Effects of Agricultural Practices and Septic-System Effluent on the Quality of Water in the Unconfined Aquifer in Parts of Eastern Sussex County, Delaware:* Delaware Geol. Sur. Rpt. of Inv. No. 45; 1989, 66 pp.
2. Denver, J. M., In *U.S. Geological Survey 2nd Natl. Symposium on Water Quality: Abstracts of the Technical Sessions:* U.S. Geol. Sur. OFR 89-409: Reston, VA, 1989, p 17.
3. Boggess, D. H.; Adams, J. K.; Davis, C. F., Water-Table, Surface Drainage and Engineering Soils Map of the Rehoboth Beach Area, Delaware: U.S. Geol. Sur. Hydro. Inv. Atlas HA-109; 1964; Scale 1:24,000.
4. Prince, K. R., *Preliminary Investigation of a Shallow Ground-Water Flow System Associated with Connetquot Brook, Long Island, New York, 1980:* U.S. Geol. Sur. WRI 80-47, 1980, p. 5.

RECEIVED September 28, 1990

Chapter 8

Study Design To Investigate and Simulate Agrochemical Movement and Fate in Groundwater Recharge

L. E. Asmussen[1] and C. N. Smith[2]

[1]Southeast Watershed Research Laboratory, Agricultural Research Service, U.S. Department of Agriculture, Tifton, GA 31793
[2]Environmental Research Laboratory, U.S. Environmental Protection Agency, Athens, GA 30613

The vulnerability of aquifers to contamination by agrochemicals is relatively high in the southeastern Coastal Plain. Transport and fate of agrochemicals in either the root, unsaturated, or saturated zones can be simulated by existing mathematical models. However, a linked mathematical model is needed to simulate the movement and degradation from the point of application through the unsaturated zone, and into groundwater. The United States Geological Survey and Agricultural Research Service initiated a cooperative investigation in 1986. In 1988, the United States Environmental Protection Agency joined the research investigation. These agencies are sharing technical expertise and resources to develop an understanding of physical, chemical, and biological processes and to evaluate their spatial and temporal variability; and to develop and validate linked model(s) that would describe chemical transport and fate. Study sites have been selected in the Fall Line Hills district of the Coastal Plain province. The Claiborne aquifer recharge area is located in this district near Plains, Georgia. Instrumentation to measure water and chemical transport has been installed.

The potential for ground-water contamination by chemicals used in agricultural production systems is relatively high in many regions of the United States, including the Coastal Plain of the southeast. In the Coastal Plain of the southeast long-growing seasons, multi-cropping, sandy soils, active ground-water recharge, heavy pest pressure, and low organic residue along with 125 cm of rainfall and high infiltration rates make ground-water systems susceptible to agrochemical contamination.

In 1986, the U.S. Department of Agriculture, Agricultural Research Service (USDA-ARS), and the U.S. Geological Survey (USGS-WRD) initiated a cooperative, interdisciplinary research investigation to address this need and to develop and test linked models (Hicks, D.W., et al., *1990, U.S. Geological Survey Open File Report,* in press). In 1987, University of Georgia scientists became aware of ongoing research and showed interest in developing faculty/ student participation in peripheral research activities. Later in 1987, copies of a comprehensive work plan, drafted by the USGS-WRD and USDA-ARS, were provided to the U.S. Environmental Protection Agency (USEPA) for review. Subsequent to their review, scientists from the USEPA, Athens Environmental Research Laboratory (AERL), Athens, Georgia, became interested and began participating in the project.

The major objectives of this project are to: (1) improve understanding of the processes that affect the movement and fate of certain pesticides and nitrogen fertilizers in the soil-root, the unsaturated, and the saturated zones; (2) develop and test linked mathematical models for applying the process-oriented findings; and (3) use these models to evaluate the impact of agricultural management practices on the chemical quality of groundwater. It is anticipated that during the course of this project secondary objectives may develop as a result of ongoing research (Hicks, et al., *1990).

The cooperating research groups (USGS-WRD, USDA-ARS, USEPA, and University of Georgia) are in general agreement with the project objectives. However, each agency may also develop supporting research that is unique to that agency. For example, the USDA-ARS will continue development of the GLEAMS model. This model will be linked to a vadose zone model, which is presently under development. USEPA's major objectives are to: (1) develop a data base for use in calibration and testing of the VADOFT and SAFTMOD components of the RusTiC (Risk of Unsaturated-Saturated Transport and Transformation Interactions for Chemical Concentrations) model in the unsaturated and saturated zones; and (2) investigate sampling techniques and strategies for both the unsaturated and saturated zones (Hicks, et. al.; *1&2*). These agency sub-objectives will be closely tied to the major objectives, but each agency will have opportunities to expand the research to fit their mission and expertise. However, the data bases developed to focus on the objectives of one research group can obviously be shared.

USDA-ARS and the USGS-WRD propose to develop and test two models: (1) a one-dimensional, soil-root/unsaturated zone flow and transport model; and (2) a two-dimensional, unsaturated/saturated zone solute-transport simulator. The first model will simulate the transport and fate of nutrients and pesticides on a field or regional scale. This model will be based on partial differential equations derived from the conservation of mass principle, and simulate transport in unsaturated porous media. The GLEAMS (Ground Water Loading Effects of Agricultural Management Systems) model, *(3&4)*, will be coupled with the one-dimensional transport model. This root-zone model will define the upper boundary condition for the unsaturated-zone model.

The second modeling approach will quantify the multidimensional-flow component by using mathematical simulation. An existing two-dimensional unsaturated-zone model will be coupled with a saturated-zone, solute-transport model. The output from this model will be used as an upper boundary condition for the ground-water or saturated-zone model. Modifications will be made to the unsaturated-zone model to include parameters critical to the simulation of chemicals in the root zone. Existing saturated zone, solute transport model may be used, or a new model developed. This unsaturated/saturated-zone model could provide predictions over a regional scale, as well as at the plot scale.

Study Site

The initial study area was located in the Ty Ty Creek watershed in central Georgia. The area is approximately 2.67 km² (1.03 mi²) and is typical of watersheds in this district. It is in the Fall Line Hills district of the Coastal Plain physiographic province of southwestern Georgia in Sumter County, (Figure 1). Initially, about 85 reconnaissance wells were installed in the general area of the watershed to define the geohydrology and to develop an understanding of the ground-water and surface-water relationship.

Geohydrologic data indicates that rainfall recharges the Claiborne aquifer, which is the uppermost saturated zone, relatively rapidly. Field observations suggest that within a few days after heavy rainfall, a portion of the recharge water may be returned as surface flow. The general direction of ground-water flow in the Claiborne aquifer is southward. This aquifer was selected because of its recharge characteristics, outcrop at the surface, and regional definition. The configuration of the water table approximates the topography.

Analysis of ground-water samples indicate that the aquifer is low in dissolved minerals, specific conductance ranges from 18 to 70 (uS/cm) and NO_3-N concentration that ranges from less than 0.1 to 4.9 (mg/L). The water is acidic (pH of 4.5 to 5.4), and could affect the fate of some pesticides.

Data collected from monitor wells (37) and soil cores installed in and adjacent to the study plot indicates that the geohydrologic units of importance to this study are, in ascending order, the Tuscahoma Formation, the Tallahatta Formation, and the undifferentiated residuum and alluvium (Figure 2). The Tuscahoma Formation consists of homogeneous, well-sorted, glauconitic, very fine-to-fine, argillaceous quartz sand, and the Tallahatta Formation is composed of fine-to-coarse quartz sand. The undifferentiated residuum and alluvium consists of alternating and intermittent layers of sand, clayey-sand, and clay. The unsaturated zone includes the undifferentiated residuum and alluvium and the upper part of the Tallahatta Formation, and ranges in thickness from a few meters in the toe-slope areas to about 13 meters in parts of the interfluve areas. The saturated zone, or Claiborne

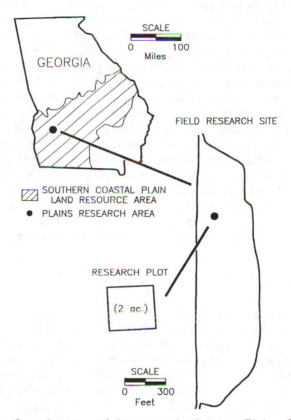

Figure 1. Location map of the research plot near Plains, Georgia.

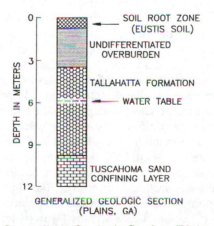

Figure 2. Generalized Geologic Section (Plains, Georgia).

aquifer, is restricted to the lower part of the Tallahatta Formation and ranges in thickness from about 2.5 to 14 m. It is confined below by the Tuscahoma Formation and generally unconfined above (Hicks, et. al.).

After visiting sites throughout the Coastal Plain of Georgia, this watershed was selected, because the size, location, and geohydrology of the watershed are ideal for conducting field research studies. The most critical part of any study is the selection of a site that data on processes selected to study can be measured. Soil spatial variability, aquifer characteristics, and controlled agricultural management that meets the project objects is difficult to find in field conditions. However, an initial site evaluation revealed that the areal extent of the study area and the spatial variability of the hydraulic properties, were not ideal for model validation and testing. For this reason, adjacent farmland, within the Ty Ty Creek watershed, was leased to establish plot-sized research areas (Figure 1). In addition, smaller areas were better suited for study of chemical and biological processes. Time/scale factors must be matched to the project objectives and a realistic assessment of the project resources made.

Data Collection

The test plot (Research Plot Figure 1) has been instrumented to provide data on water and chemical transport in the soil, vadose (unsaturated zone above the water table and below root-zone), and ground-water zones (saturated zone). Data includes runoff measurements, chemical application practices, farm practices (corn/wheat rotation), root-zone properties, unsaturated-zone properties (vadose), saturated-zone properties (aquifer), and chemical movement in the 3 study zones (Table I). The transport and fate of atrazine [6-chloro-N-ethyl-N'-(1-methylethyl)-1,3,5-triazine-2,4-diamine]; carbofuran [2,3-dihydro-2,2-dimethyl-7-benzofuranyl methylcarbamate]; and alachlor [2-chloro-N-(2,6-diethyl-phenyl)-N-(methoxymethyl)acetamide] will be investigated. In addition, the transport and denitrification of nitrate fertilizers will be observed. The movement of bromide and chloride incorporated in commercial fertilizer, will also be monitored (applied summer, 1989). Soil, vadose, and saturated aquifer material are sampled at selected times after application by a drill rig or soil augers. Ground-water samples are taken from wells and surface runoff is sampled for chemical transport. In addition, water from suction lysimeters, located at selected depths, will be sampled.

Permanent monitoring stations were randomly located at 12 sites in the test plot (Figure 3). The monitoring equipment at these sites consists of (1) 4.25 cm diameter, threaded-joint PVC wells tapping the Claiborne aquifer at three-depth intervals, top, middle, and bottom, (at four of the 12 sites, a fully penetrating well was installed, as well as the three zoned wells); (2) stainless steel, vacuum lysimeters (soil-water samplers) installed at seven depth intervals; and (3) soil-moisture sensors (Water Mark sensors) installed at 12 depths, seven of which correspond to the lysimeter depths, and soil-

Table I. Physical, chemical, hydrologic, and
associated data bases necessary for process
research and model development and testing

(PHYSICAL SYSTEM CHARACTERIZATION)

Compartment/Processes	Products
Topography	Maps (local & regional)
Soil Properties	⎡ Classification/morphology (maps, pedon description) – Soil pedon characteristics (chemical/physical) Infiltration Characteristics ⎣ Soil Moisture Release Curve
Stratigraphy **(local/regional)**	⎡ Monitor Wells Geophysical logs [Neutron, Resistivity, Gamma] – Ground penetrating radar ⎣ Geologic cores (undisturbed)
Unsaturated zone properties (vadose)	⎡ Permeability – Composition (size, carbon, pH, etc) Mineralogy (sand,silt, & clay) ⎣
Saturated zone properties	⎡ Aquifer tests – Laboratory tests ⎣ (physical-undisturbed)

(HYDROLOGIC/METEOROLOGIC OBSERVATIONS)

Compartment/Processes	Products
Precipitation	Long & short-term
Radiation	Net
Evaporation	Pan
Temperature	⎡ Ambient (maximum) – ⎣ Soil (by depth)

Continued on next page

Table I. *(continued)*

Wind

Surface Runoff   – ⌈ Watershed
                             ⌊ Plot

Ground-Water Table   – ⌈ Instantaneous
(local/regional)   ⌊ Seasonal/long term

Soil/Unsaturated   – ⌈ Instantaneous
Zone (vadose)   Neutron Probe
                               Gravimetric
                             ⌊ Ground-Penetrating
                                 Radar

(PESTICIDE FATE AND TRANSPORT)

Compartment/Processes   Products

Field Application   Atrazine, carbofuran,
Rate   alachlor

Soil Root Zone   Concentration, time
(Depths)

Unsaturated Zone   – ⌈ Solution-lysimeters
(Depths)   (concentration/time)
                               Continuous Cores
                             ⌊   (concentration/time)

Saturated Zone   Solution by depth
                                  (concentration, time)

Methodology and quality – ⌈ Laboratory
assurance   ⌊ Sampling Methods

Table I. *(continued)*

(BROMIDE TRACER TRANSPORT)

Compartment/Processes	Products
Field Application Rate	Bromide, chloride
Soil and root zone	Concentration/time
Unsaturated Zone (vadose)	⎡ Solution-lysimeters (concentration/time) – Continuous cores (concentration/time) ⎦
Saturated zone	Solution by depth (concentration time)
Methodology/quality assurance	

(NITROGEN TRANSPORT)

Compartment/Processes	Products
Precipitation Input	
Soil and root zone	Concentration/time
Unsaturated zone (vadose)	⎡ Solution lysimeters (concentration/time) – Continuous cores (concentration/time) ⎦
Saturated zone	Solution by depth (concentration/time)
Quality assurance	

Continued on next page

Table I. *(continued)*

(AGRONOMICS)

Compartment/Processes   Products

Cultural Practices   – ⎡ Tillage,planting,harvest
                                    ⎣ Chemical Application

Crop Development

Crop Yields

Chemical Uptake and   ⎡ Nitrogen
** Removal (Crop)**   – ⎢ Pesticides
                                  ⎣ Bromide

Irrigation   Rate and Time, Duration

(RELATED PROCESS STUDIES)

Compartment/Processes   Products

Nitrogen Transformation   Nitrification/
** (By Depth)**   denitrification

Pesticide Transformation   Degradation/transform-
** (By Depth)**   ation (rates/time)

Short/Long-Term Affects   Concentration/time
** on regional ground water**
** (chemical)**

Soil and Pesticide Grouping   Hydrologic/chemical
                                    for model input

Aquifer Vulnerability   Maps (GSI)

Models Development/Testing   Linked Models
**   Development**   (short/long-term)
**   1-D**
**   2-D**
**   Validation**
**   1-D**
**   2-D**

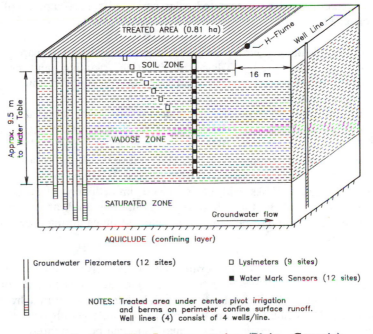

Figure 3. Study Plot Instrumentation (Plains, Georgia).

temperature detectors (thermocouples) at four depths all connected to centrally located data loggers (Figure 3). All instrumentation at the 12 sites was permanently installed within the corn row spacing.

Precipitation data (recorded at 5 min. intervals) are being collected at four sites, two of which adjoin the plot. Pan evaporation, wind speed and direction, ambient air temperature, and rainfall (long-term) are being collected at the University of Georgia Plains experiment station (5 km distance), and at a site adjacent to the plot.

Runoff. A soil berm was built around the perimeter of the plot to control the runoff. An H-flume (0.46 m) was installed at the base of the 0.81-hectare test plot to measure sediment and agrochemicals transported in the runoff. A stage recorder and an automated stage-activated runoff-sample collector were installed.

Chemical Application. During the study, standard agricultural management practices are being used. Tillage, fertilization, planting, and chemical application are being conducted in accordance to procedures outlined by the Georgia Extension Service. The study will continue for 5 years.

Liquid atrazine, carbofuran, and alachlor are being applied at label recommended rates annually to the area. Potassium bromide salt, also in a liquid form, was applied in 1989 as a conservative tracer following the pesticide application (application 1st and 3rd-year of the study). The application rates will be determined from analyses of filter disks located on the treated area. Disks will be removed immediately after chemical application and sent to the laboratory. Following application, soil and geologic materials will also be sampled at selected times during the year and after each crop season.

Root-Zone Properties. In situ permeability tests were conducted using a Guelph permeameter to evaluate the infiltration rate of each soil horizon. These permeameter tests can be used to calculate the saturated hydraulic conductivity. Data obtained from continuous borehole geophysical logs will be used to estimate formation permeability, and will be compared to in situ and laboratory developed estimates of hydraulic properties.

Unsaturated-Zone Properties. Continuous soil cores were collected prior to project initiation, at 3 randomly located sites in each quadrant (12 total). A continuous-coring device (CME hollow-stem split-tube) was used to collect "undisturbed" soil cores initially and 2-times a year for selected depths to the saturated zone.

A constant head, controlled gradient, flexible-wall permeameter will be used to determine the saturated and unsaturated hydraulic conductivity on selected cores. Undisturbed soil segments, approximately 1.9 cm in length, are cut from the undisturbed core. These samples are analyzed in the laboratory and hydraulic conductivity will be calculated using a constant-head equation. These hydraulic conductivity values will be compared to those developed in situ using the Guelph permeameter.

Mineralogy, bulk density, pH, organic carbon content, porosity, particle-size distribution and pesticide degradation rates will be determined from the soil cores. Subsamples will be selected for tracer and pesticide analyses.

Saturated-Zone Hydraulic Properties. Rising-head and falling-head aquifer tests will be conducted in the test-plot and adjoining area. These tests will be used to estimate the transmissivity, hydraulic conductivity, diffusivity, and specific yield of the Claiborne aquifer.

Undisturbed cores collected from the saturated zone and from the lower confining unit will be analyzed to determine vertical hydraulic conductivity.

Chemical Compounds in the Unsaturated Zone. Before agricultural management was initiated, soil cores from selected depths, within the unsaturated zone, were collected to determine background chemistry. Sections of core were sealed in metal containers, chilled and transported to the laboratory to be analyzed for residual pesticides, bromide, and chloride.

Undisturbed-continuous cores will be collected from the surface to the saturated zone at 3-sites randomly located in each quadrant 2-times each year. A hollow-stem auger sampler will be used. In addition to the core samples, soil samples (20) will be collected at 12 cm intervals to 3.2 m and analyzed at selected times after chemical treatment (7 or 8 times). A soil auger will be used to collect these samples. The sampling sites (4 per quadrant) will be randomly selected prior to each sampling event.

Sampling event times will be determined by previous chemical analyses and the occurrence of significant rainfall. Immediately following periods of heavy rainfall, lysimeters will be purged and sampled. The water samples will be analyzed for the same chemicals as the soil samples. The analytical results of these samples will be used as a guide to determine the depth interval for subsequent collection of soil samples.

In addition to the random soil sampling sites, eight soil sampling sites next to the lysimeter installations will be sampled when water samples are taken. During each sampling event, these soil samples will be collected at intervals that correspond to the lysimeter depths, (0.6, 1.3, 1.9, 2.5, 3.2, 3.8, 4.5 m) as well as intermediate depths. These samples will be used to determine the gravimetric soil-moisture content at the approximate time that each of the lysimeter samples were collected. From these data, a relationship of soil moisture to unsaturated hydraulic conductivity will be developed.

Soil samples will be pulverized and a volume of deionized water will be added. A representative sample volume will be extracted for bromide and chloride analysis using an ion-Chromatograph. Samples will be divided and part will be analyzed for pesticides. These analyses will be run on all samples taken from the plot.

Chemical Compounds in the Saturated Zone. Subsequent to the collection of soil and water samples from the unsaturated zone, each monitor well in the test plot will be purged and sampled. A minimum of five ground-water sampling events are planned annually. Down-gradient wells will not be sampled until chemical compounds (pesticides, nutrients, bromide and chloride) are detected in wells located within the test plot.

Residue analyses for atrazine, de-alkylated atrazines, and hydroxy-atrazine from selected ground-water samples will be conducted. In addition, determinations for alachlor, alachlor carbinol, ortho-diethylaniline, carbofuran, carbofuran phenol, 3-hydroxy-carbofuran, and 3-ketocarbofuran are being determined.

Ground-Water Monitoring. Monitoring wells, constructed identically to those within the test plot, have been established up- and downgradient from the test plot. Two groups of wells were installed upgradient of the test site. These wells provide background control for the test site and are sampled for ambient ground-water quality. Four well lines have been installed down-gradient from the plot at approximately 15 m intervals. Additional lines of wells will be installed as the movement of the chemical compounds is detected in the in-place downgradient monitoring wells.

Two, large diameter, fully penetrating, monitoring wells were installed upgradient, and downgradient. Each of these wells is equipped with a continuous water-level recorder. Continuous water-level data will be collected to estimate rainfall and recharge relationships for the plot area and the water table gradient.

Long-term, seasonal, and instantaneous water-level changes are monitored. The water level in the Claiborne aquifer is monitored continuously at 6-sites in the watershed in addition to the 2-sites at the plot. Ground-water levels are measured monthly in approximately 145 Claiborne wells in the watershed and surrounding area. The water levels at selected sites are obtained from multilevel wells. The elevation differences will be compared to evaluate vertical-head gradients within the saturated zone.

Nitrogen Cycle Studies. Various forms of nitrogen, originating from fertilizer application, have been discovered in ground water in southeastern agricultural areas. Moreover, it is hypothesized that nitrogen cycling may be an important factor associated with pesticide transport. Organic compounds may be coupled with the nitrogen cycle via microbial processes. For example, denitrifying bacteria could utilize the pesticide as a carbon source and NO_3-N as an electron acceptor.

Studies are proposed to (1) establish a 3-year nitrogen budget for the study area (a number of physical-transport mechanisms, microbial reactions and harvesting would be considered); (2) address the denitrification part of the cycle in detail by quantifying in situ rates of denitrification in undisturbed cores of aquifer material and the effects of NO_3-N and pesticide in aquifer microcosms; (3) quantify denitrification on a field scale; and (4) field test the hypothesis that denitrification and other nitrogen cycling reactions result in significant changes in the isotropic composition of ground-water constituents (Hicks, et. al.).

Microbial Studies. It is likely that microbial processes have a significant impact on the transport and degradation of organic compounds. Field and laboratory experiments are proposed to evaluate the role of microbial processes. Undisturbed cores will be collected from the soil, unsaturated, saturated, and the aquiclude zones. Samples will be taken from the interior of each core for incubation. Microbial colonies will be exposed to labeled atrazine, alachlor, and carbofuran to evaluate the impact on the microbes as well as the degradation of the compounds.

Overview of the Research Perspective

One year of test data is available for all parameters (1989). Data collection will continue for 4 more years. Specific conductance and NO_3-N increases have been noted in some wells. The water and chemical transport is progressing as originally expected. However, complete data sets and additional information is necessary before specific conclusions can be determined.

During the course of the investigation, data sets will be assembled to test models which are in various stages of development. It will be incumbent on each researcher and agency to share data and information in a timely manner and use such information respecting agency protocols and individual scientists contributions. Data sets are expected to have wide applicability and be in demand by groups not directly involved with the investigation. At the conclusion of the project, a joint-final report will be prepared that will serve as the data release to individuals and agencies outside the cooperating agencies. In the interim, any data release to those other than on the initial project team should be with mutual knowledge and consent by all cooperating parties. This does not exclude widening of the project with additional cooperators. Such activities will be jointly planned and revised by team members representing the three primary federal agencies and cooperating university.

The USGS, ARS, USEPA, and University of Georgia are sharing technical expertise and resources in a ground-water/agricultural management research project of mutual interest that involves complex, interdisciplinary investigations by scientists from all cooperators. The project success will require careful planning and implementation of research activities, as well as frequent communication among the various scientists.

Conceptually, each research participant will concentrate on testing a hypothesis that relates to specific physical, chemical, or biological processes that affect pesticides or nitrogen in the environment. The data sets assembled as a result of these research activities will be used to test and validate modeling efforts in each component of the hydrologic system. As the studies proceed, data and technical expertise will be exchanged among the cooperating scientists. Information exchange is an essential element of the research project.

Each scientist has a vital interest in the project and to collect data to test specific hypotheses related to processes of interest. In normal pursuit of his or her professional responsibility, each scientist is expected to author relevant publications. Most publications will involve joint authorship by researchers from various combinations of the three federal agencies and their cooperators, such as scientists from the University of Georgia.

Acknowledgments

The authors wish to acknowledge the assistance of D. W. Hicks and J. B. McConnell, cooperators on the project. A great deal of the material used in this publication was extracted from Hicks, D. W., et al., *1990. U.S. Geological Survey Open File Report* (in press.)

Literature Cited

1. Carsel, R.F., Mulkey, L.A., Lorber, M.N., and Baskin, L.B., *Ecol. Modeling. 1985, 30,* 49.
2. Carsel, R.F., Smith, C.N., Mulkey, L.A., Dean, J.D., and Jowise, P., *Users Manual for the Pesticide Root Zone Model (PRZM): Release 1.,* U.S. Environmental Protection Agency, Env. Res. Lab., Athens, GA, EPA-600/3/84-109, 216 p.
3. Leonard, R.A., Knisel, W.G., and Still, D.A. *TRANS. Amer. Soc. Ag. Engr., 1987, 30,* 1403.
4. Knisel, W.G. (Ed.). *USDA, S&E Adm. Cons. Res. Report No. 26, 1980,* 643.

RECEIVED September 5, 1990

Chapter 9

Design of Field Research and Monitoring Programs To Assess Environmental Fate

Russell L. Jones and Frank A. Norris

Rhone-Poulenc Ag Company, P.O. Box 12014, Research Triangle Park, NC 27709

Field research and monitoring study design should depend on study objectives, environmental conditions, chemical properties, and use patterns. Comprehensive groundwater research studies will usually involve sampling both the unsaturated and saturated zone after a carefully controlled application but often monitoring program objectives may be satisfied by collecting only water samples. In comprehensive research studies, timely analysis of samples is essential so that results from previous sampling intervals can be used to guide activities at future sampling dates. Sampling procedures should be tailored to agricultural chemical properties and site characteristics. Regardless of the study design or objectives, sample contamination should always be avoided by using trained and conscientious personnel with cleanliness always being a primary concern.

The discovery of agricultural chemical residues in some drinking water wells during the past decade (1-3) has resulted in development of field and modeling research to better understand and predict agricultural chemical movement and degradation in soil and groundwater. Although field dissipation studies have been conducted for many years, these studies until recently have usually been concerned with residue behavior near the soil surface. Therefore new types of studies (protocols for these studies are still evolving) using different sample collection techniques have been developed during the past decade for evaluating potential impact of an agricultural chemical on groundwater quality. This article will only discuss study design; soil and groundwater sampling techniques are described in companion papers (Norris et al.; Kirkland et al., in this work).

Three study types are discussed in this paper. The first type, comprehensive groundwater research studies, measures degradation rates and extent of movement in both soil and groundwater by sampling both the unsaturated and saturated zones. The second, groundwater monitoring studies, determines magnitude and extent of residues

0097–6156/91/0465–0165$06.00/0

rather than degradation rates. The third, potable well monitoring studies, determines the occurrence and magnitude of residues in drinking water.

Comprehensive Field Research Studies

This category is used to refer to those studies which measure the degradation and movement in both soil and groundwater following an agricultural chemical application. The U. S. Environmental Protection Agency (EPA) refers to this study type as "prospective leaching" studies. This section is intended to update information in a previous paper (*4*).

General Study Design. The typical field research program designed to address the potential for residues to appear in groundwater can be separated into two distinct parts. The first deals with movement and degradation in the unsaturated zone (the soil present above the water table including the root zone) while the second focuses on saturated zone behavior. Unsaturated zone behavior is studied by collecting and analyzing soil samples at regular time intervals after a carefully controlled application. Sampling is usually continued until most residues are no longer present in the unsaturated zone, either as a result of degradation or due to movement into the saturated zone. Residue movement into groundwater is detected by collecting and analyzing water samples from clusters of shallow monitoring wells. Samples from these wells are used also to track any residue plume to determine the rate of movement and degradation. If a consistent residue plume appears during the study, additional monitoring wells may also be installed to provide more detailed information. Sampling is usually continued until residue concentrations fall below established health advisory levels or until saturated zone degradation rates can be determined.

The most basic information usually obtained from a comprehensive field research program are degradation rates in the unsaturated and (if residues reach groundwater) the saturated zones. These rates can then be used in model simulations to predict the extent of movement in a variety of soils and aquifers under a range of weather conditions (*5-6*).

Usually field research studies are conducted using normal agronomic practices, especially for those compounds applied to the soil surface or incorporated into the soil. However, exceptions may be necessary to meet study objectives. For example, the use of multiple, foliar sprays makes determining degradation rates in soil difficult, if not impossible. One solution to conducting studies with foliar applications is to conduct two different studies (*7*). In the first, soil degradation rates are measured using an application sprayed directly onto the soil, while in the second the magnitude of residues appearing in groundwater is monitored under conditions approximating worst case conditions (light soils and shallow water tables) with normal foliar applications, perhaps using a design such as described in the section on groundwater monitoring studies.

Site Selection. Site selection is a critical study phase which can be a time consuming process requiring several months of elapsed time. The first step is to determine the exact study requirements: plot size, soil type, water table depth, irrigation requirements, aquifer characteristics, and other special considerations. Then potential areas can be selected for further investigation using a combination of soil maps, other references, and telephone inquiries to local contacts (government officials, university personnel, and individual farmers). Finally, site visits, probably including limited soil coring by hand to determine the actual soil profile and the water table depth, will be required to evaluate the acceptability of the proposed sites. Pretreatment soil and water analyses should be completed prior to application to confirm that no residues from previous applications are present.

Typically the authors have used a plot size of 1 to 2 ha for conducting most field research studies involving both unsaturated zone and saturated zone monitoring. However, larger plots have been employed if the site is relatively uniform and crop destruction is not required. Sometimes the test plot has consisted of a portion of a large field but the agricultural chemical has been applied only to the test plot. If only unsaturated zone monitoring is being conducted in the study, the plot size can be reduced down to about 0.1 ha.   Usually, field research studies (except those examining residues remaining from previous applications) are not conducted on plots recently treated with the agricultural chemical under study to prevent residues from prior applications being present in either soil or groundwater at the time the application is made (studies which examine residues remaining from previous applications are discussed under groundwater monitoring studies).

The study objectives determine the soil characteristics sought in the site selection process. If the objective is to measure movement and degradation under typical agricultural conditions, then a typical soil should be chosen. However, often studies must be conducted in soils which favor residue movement (typically soils with low organic matter and low water holding capacity). In the United States, sensitive soils used to grow a particular crop in an area can often be identified using existing soil data bases such as the SIRS data base (*8*), DBAPE (*9*), or county soil survey information. Similar information is available in many other countries.

The water table depth is also usually a key factor in site selection. Because the water table depth can vary significantly even within an individual field, definitive information is often hard to obtain without an on-site inspection. Local contacts are an invaluable resource for providing information on ranges commonly found in an area.   The presence of nearby streams, ponds, and marshy areas can serve as indicators to areas with potentially shallow water tables. Usually an area with shallow water tables (less than about 6 m below the soil surface; the proposed EPA guidelines require a water table less than 9 m deep) is desirable for a study which includes saturated zone monitoring.

Irrigation can also influence site selection. Certain studies require that irrigation be used to supplement rainfall or normal irrigation practices. The ability to apply this irrigation to a test area at a carefully prescribed rate may make conducting such a trial difficult in commercial fields. Also, the use of supplemental irrigation requires a nearby residue-free water source such as a pond or well. Often field study costs are substantially higher if supplemental irrigation is required. Also, the irrigation method (for example, furrow versus sprinkler) may significantly effect residue movement in soil (*10*).

Aquifer characteristics may also be important in site selection. The presence of clay lenses may result in the formation of a perched water table or in a relatively impermeable boundary between two aquifers. Placing well screens in clay or silt layers may result in inadequate water flow into these wells during sampling.

Groundwater flow direction in relation to site features is often important. For example, if a test site is located next to a stream into which groundwater underneath the test site is discharging, there will be little time to follow the residue plume to determine saturated zone degradation rates. Therefore, plots should be located to permit following any residue plume for at least 100 m before encountering any nearby streams or ponds. If possible, test plots should be situated so that any residue plume does not move onto adjacent land of other property owners. Test plots should also not be located near housing developments or private or municipal drinking water wells.

Unsaturated Zone Monitoring.   Factors to be considered in unsaturated zone monitoring design include number, location, depth, timing, and analysis of soil cores and potential use of soil-suction lysimeters.

   Number and Location of Soil Cores.   The number and location of soil cores collected during a single sampling interval is greatly influenced by the inherent variability of soil samples as well as the application. Studies (for example, (*7,11*)) indicate coefficients of variation ranging from 100 to 200 percent. Therefore, usually 15 to 20 soil cores must be collected at each sampling interval (*12*). If necessary to reduce analytical costs, these cores can be composited in the laboratory to a smaller number of samples. However, several different samples from the same depth increment must be analyzed so that variability can be estimated. Similar guidelines have been adopted by EPA for field dissipation (*13*) and proposed for prospective leaching studies.

   One scheme used by the authors is to divide the test plot into four subplots. At each sampling date 16 cores are collected, four from each subplot. The location of the four cores per subplot are selected using a random number routine.   If compositing is necessary, then the four samples collected for each depth increment in each subplot are combined, resulting in four composite samples for each sampling date for each depth increment. By assuming a normal distribution, the variability of an individual sample can be estimated by the variability of the composites.   For pretreatment samples where only the presence or absence of residues is being determined, collection of only a single core from each subplot has been sufficient in studies conducted by the authors.   Recently issued EPA guidelines for field dissipation studies (13) require 15-20 cores which may be composited to one sample.

However, the authors recommend that compositing, when necessary, be more limited (for example, only compositing cores in each subplot.

The application method can influence the location of the soil cores in the test plot, especially when the application is not spatially uniform (*14*). The authors' usual practice for row crops (where the agricultural chemical is applied in bands which are relatively closely spaced) is to collect all samples from the middle of the treatment bands. For applications made to the seed furrow, cores are located in the middle of the plant row, while for other types of applications samples may be collected somewhat away from the middle of the plant row.

The previously described procedure cannot provide an adequate mass balance when bands of agricultural chemicals are spaced relatively far apart as is often the case for such crops as citrus or grapes. In these cases, the authors recommend a procedure involving collection of samples spaced approximately at 0.3 m intervals on a transect located perpendicular to the treatment band and spanning one or two treatment bands. Usually four transects, each of 12 to 16 cores, will be required per sampling interval. An example of such a sampling scheme is presented elsewhere (*14*). Depending on the study objectives, the use of the transect procedure may not be necessary for all sampling intervals. For example, residue movement could be monitored using 16 cores collected in the middle of treatment bands during earlier sampling intervals and the transect procedure could be used in later sampling increments to provide more precise estimates of the remaining mass for calculating degradation rates.

Sampling Intervals. Soil sampling intervals should be tailored to the expected degradation rates of the agricultural chemical under study. For field studies designed to address groundwater concerns, these intervals may be significantly different from those normally used in field dissipation studies. For example, precisely determining quite rapid degradation rates is not usually necessary, since the potential for significant groundwater residues will usually be quite low. A typical sampling schedule would be to collect soil samples prior to and immediately after application, and 0.25, 0.5, 1, and 2 months and at two month intervals thereafter until the unsaturated zone monitoring is terminated. The earlier intervals may be eliminated when residues have relatively long soil degradation rates. The authors' goal is to place the first sampling interval when about 3/4 of the applied material is present.

The need to collect samples immediately after application is questioned under certain circumstances by some researchers. The amount applied can usually be better determined by weighing the amount used or by calibration of the application equipment. Determining actual rates using soil sampling, especially for banded applications, is less precise due to the variability of residues in soil samples. Variability problems may be compounded in earlier sample intervals when granules are applied rather than liquid formulations (the analytical procedure used must also extract the active ingredient from the granule). In spite of these problems, samples are usually collected immediately after application to demonstrate that the material was actually applied at a rate near the desired amount and to show that the analytical procedure is effective under field conditions.

The duration of the unsaturated zone monitoring depends on the degradation rate and movement observed in the study. Normally little useful information can be obtained after most of the applied amount has degraded or moved into groundwater. One criteria used is to terminate soil sampling after four to six half-lives have elapsed or when less than 5 to 10 % of the applied amount (parent plus major metabolites) remains in the unsaturated zone.

Core Depths.   Sample core depth depends on chemical properties, soil characteristics, and the climatological conditions encountered during the study. Except for samples collected immediately after application, soil samples should be taken to the water table or to a depth sufficient to include all unsaturated zone residues. For field studies directed toward groundwater concerns, depth increments are likely to be significantly different from a typical field dissipation study. Generally, the purpose of soil sampling in groundwater studies is the determine movement through the soil profile and into groundwater, while distinguishing residue distributions in the upper few centimeters is not important. The depth increments generally used by the authors are 0.3 m increments down to 0.6 m and 0.6 m thereafter.   These depth increments are significantly larger than the 0.15 m increments currently required in field dissipation studies (*13*). However, unsaturated zone residue plumes of mobile compounds below about 1 to 2 m are often spread over vertical distances exceeding one meter. Plumes with vertical spreads greater than 5 m have been observed, even within 4 months after application (Hornsby et al., *Water Resources Res.*, in press). Because vertical residue distribution varies from core to core, reducing the number of cores in a composite, rather than shortening depth increments will be more informative.

During a study, core depth is likely to increase as downward residue movement is observed.   Pretreatment cores should be collected to the water table or the maximum residue depth expected (usually at least 3 m). Post-treatment cores should be collected such that all the soil strata containing residues are sampled. In post-treatment sampling intervals and especially at the last sampling interval, at least one residue-free stratum should be present at the bottom of the soil core. Because residue analyses from a sampling interval are often a useful guide to the appropriate sampling depth for the next interval, soil samples should be analyzed in a timely manner.

Cores collected immediately after application require special consideration. Usually only one depth increment per core is recommended for samples collected at this sampling interval. Depending on whether the application was to the soil surface or incorporated, this depth increment (typically 0.1 to 0.3 m) may be less than the initial depth increment collected afterwards Collecting deeper samples immediately after application usually yields no additional information on residue movement (an exception is in a study with multiple applications; then the required core depth after all but the first application will be determined by the location of residues from previous applications). The potential for contaminating subsurface samples is highest just after an application to the soil surface; and the occurrence of erroneous residues due to contamination will, at the least, require additional effort to prove otherwise.

Sample Analyses. Typically pretreatment samples are used to characterize the soil profile with at least mechanical analyses, soil pH, and organic matter determinations. Soil-water and residue concentrations should be measured in all pretreatment and post-treatment samples. As discussed earlier, samples should be analyzed in a timely manner so the results can be used to guide sample collection at the next sampling interval.

Interpretation of Data. The primary use of the soil residue data is to estimate the degradation rate. If the agricultural chemical was applied in bands, the calculation is not straightforward since the soil data represent the concentrations in the bands and not a field average concentration. Also the band width may increase during the test due to lateral dispersion or physical mixing as the result of planting, harvesting, or tillage operations. Therefore, a simple regression of residue concentrations in the middle of the treatment bands as a function of time may overestimate the degradation rate. For row crops where treatment bands are usually relatively closely spaced, one option is to estimate the degradation rate based on the elapsed time and the change in the amount applied and that measured at the last sampling interval where residues were present (*15*). This procedure may tend to underestimate the degradation rate. For crops where treatment bands are more widely spaced such as on citrus or grapes, proper averaging of the sample core data obtained using the transect design can be used to calculate the amount of residues remaining at the last sampling interval (Norris et al.; Jones; unpublished reports). If significant changes in degradation rate occur with depth, then the use of an unsaturated zone model may be necessary to estimate the degradation rates. As mentioned earlier, models can also be used to extrapolate the results to other climatological conditions or soil types.

Soil-Suction Lysimeters. In some field research studies, water samples from lysimeters have been used to supplement or replace analytical data from soil samples. The EPA requires the installation of soil-suction lysimeters in prospective leaching studies. The two main advantages of lysimeters include the ability to monitor regularly the movement of agricultural chemical residues in one location and improved analytical sensitivity. This increased sensitivity is due to two factors. Often the sensitivity of methods for analyzing water are more sensitive than those used for soil. Also agricultural chemical residues below the root zone are usually in the soil-water rather than bound to the soil particles because of reduced organic matter. Since soil-water comprises only 5 to 30 percent by weight of a typical soil sample, then (assuming the same analytical sensitivity for soil and water analyses) the analysis of a soil water sample can detect residue amounts 3 to 20 times less than could be detected in a soil sample.

Several disadvantages are associated with soil-suction lysimeters. For instance, a mass balance, needed for the estimation of unsaturated zone degradation rates, is much more difficult to obtain. This is of particular concern for volatile compounds and those compounds for which concentrations may vary as a function of pore size because concentrations present in lysimeter samples may not reflect average concentrations in soil water.

Lysimeters samples cannot be collected in dry soils. Sometimes this problem can be overcome by coordinating sampling times with irrigation or rainfall. The inability to collect samples is mostly encountered in coarse sand soils, due to their low water holding capacity. However, low water holding capacity also increases potential for downward movement of water and agricultural chemicals, so such soils are the most common soil type used in comprehensive research studies. Also due to the low water holding capacity, the analytical sensitivity advantage of soil-water samples over soil samples is the greatest in these soils.

The absence of residues in a lysimeter sample indicates only that the residue plume is not at the sample point. Residues may have degraded or could be located either above or below the lysimeter. Therefore, lysimeter samples must be taken more often than is necessary for soil samples.

Lysimeters cannot be thoroughly cleaned between samples. Therefore, if low residues are found in a sample collected from a lysimeter in which the previous sample contained residues, residues may be present in the soil water or result from cross contamination with the previous sample. This potential for contamination is also highest in sand soils where sample volumes are usually low.

A disadvantage not usually considered by users of lysimeters is the spatial variability of residues. Since there is no indication that variability of lysimeter samples would be less than currently observed in soil samples, 15 to 20 lysimeters at each depth increment would be required to provide comparable information.

In the authors' opinion, lysimeters cannot be used as a replacement for soil cores at this time. However, there are circumstances where the qualitative results provided by lysimeters could potentially enhance results from field research studies. One such situation would be to monitor residue movement when soil core depth is limited by the sampling procedure or subsoil properties (for example the presence of a stoney layer). Another situation would be to confirm the absence of soil residues below the depth indicated by soil cores when the detection limit in soil is significantly less sensitive that with soil water (this would not be necessary if the water table was sufficiently close to the measured depth of movement since the same function then could be provided by a monitoring well). Another use would be to confirm instances of suspected contamination that sometimes occur in deeper increments of soil cores.

Additional experiments are needed to determine the magnitude of any potential channeling of soil water in the vicinity of installed lysimeters and to demonstrate that regular sample collection does not influence movement of agricultural chemicals near lysimeters. Although several comprehensive field research studies have been conducted recently which include soil-suction lysimeters, the authors are not aware of any situation where soil-suction lysimeters provided significant insights that were not obtained eventually from soil and groundwater sampling. In general, the authors do not believe that the additional information obtained from soil-suction lysimeters is normally significant enough to justify their inclusion in comprehensive field research studies.

Saturated Zone Monitoring. Factors to be considered in saturated zone monitoring design include the number, location, and depth of monitoring wells and the sampling frequency.

Monitoring Well Design. The number and location of monitoring wells depends on site characteristics and on whether residues reach groundwater during the study. One approach that has been used successfully in a number of studies is to install an initial grid consisting of five or six well clusters which monitor the upper 3 m of the saturated zone (*10, 16-18*). A typical well cluster is composed of wells screened at different depths so that the vertical position of any residue plume can be determined. For field studies conducted by the authors where the water table was relatively shallow, the clusters have generally consisted of three wells per cluster with 0.3 m long slotted screens located just below the water table and at 1.5 and 3 m below the water table at the time of installation. Relatively short well screens are used so that the vertical distribution of residues of agricultural chemicals can be more accurately defined. If clay or silt layers are present in shallow groundwater, well placement should consider the different groundwater strata and wells should be properly constructed to prevent them from acting as a channel from one stratum to another. Well design, material, and installation are discussed elsewhere (Kirkland et al. in this work).

A typical layout for the initial five well clusters is illustrated in Figure 1. One well cluster is located immediately upgradient of the treatment area, two well clusters are located in the treatment area, and two well clusters are located 3 to 30 m downgradient (the tops of wells placed in the treatment area must often be located beneath the soil surface to prevent damage from cultivation, planting or harvesting equipment). The direction of groundwater flow can usually be estimated satisfactorily based on surface topography or the location of nearby streams, lakes, or drainage ditches. In some instances, where the direction of groundwater flow within the plot is not known or may be temporally or spatially variable (perhaps due to the presence of irrigation ditches throughout the plot), well clusters may be placed on each side of the plot as illustrated in Figure 2.

If residues are found in shallow groundwater, then this well network is expanded as needed by adding additional well clusters or deeper wells in existing clusters. The direction of groundwater flow (determined using water table elevation measurements from the initial well clusters) and the pattern of residues in the monitoring wells can be used to select the location and depth of additional wells. This approach optimizes well installation and sample collection because wells are only installed and sampled when and where they are needed. The quality of the study is also improved because, in the authors' experience, accurate predictions of where additional monitoring wells will be needed cannot be made at the beginning of a study. However, the use of such

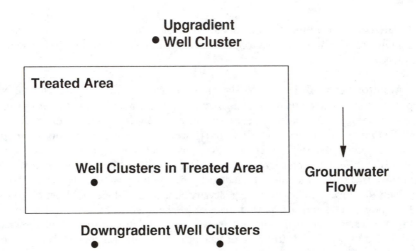

Figure 1. Typical locations for the initial well clusters at a site where the direction of groundwater flow is clearly defined. (Adapted from ref. 4.)

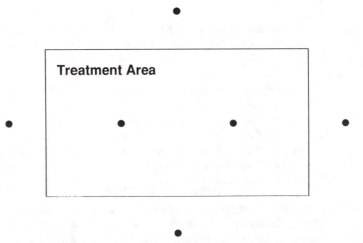

Figure 2. Alternate design for locating initial monitoring wells for use at a site with no clearly defined direction of groundwater flow. (Adapted from ref. 4.)

an approach is contingent on the ability to analyze water samples within several days after collection and the ability to install monitoring wells, if needed, prior to the next sampling date.

As an example of the effect of interim study results on the monitoring well network, if residues are found in one well cluster located in the field but not the other, then additional well clusters may be needed near the cluster containing residues to see if the residues were very localized (perhaps from contamination or channeling). In addition, other well clusters could be installed in other portions of the field to better characterize residues in groundwater beneath the treatment area. An example is shown in Figure 3 where clusters 8, 9, and 10 were installed after residues were detected in cluster 4 but not 2. Cluster 11 was installed after residues were found in clusters 6 and 7 (Jones, unpublished report). If a widespread residue plume is found at a site then many additional well clusters may be required to characterize this plume underneath the treatment area and to track the plume movement. An example is shown in Figure 4 where 58 clusters (total of 174 wells) were installed and sampled over a period of 40 months (*17*). At this site, initially five well clusters were installed. After residues were found in the clusters located in the treated grove, additional clusters were placed in and immediately adjacent to the treatment area to characterize the residue plume. As the plume moved towards the creek, additional clusters were added as needed to track its movement.

Sampling Intervals. Samples should be collected from monitoring wells prior to application and at regular intervals thereafter. Because agricultural chemical residues are applied over a relatively large area while movement of residue plumes in the saturated zone are usually no more than about a meter a day (typically less than 0.2 m/day in studies conducted by the authors), water samples normally do not need to be collected in response to recharge events as is required for lysimeter samples. One common schedule is to sample monthly during the first six to 12 months after application and if necessary every two or three months thereafter. After each sampling date, the potential need to install new wells prior to the next sampling date should be evaluated. If no residues are detected in monitoring wells during a study, the sampling can be terminated a few months after soil sampling has been terminated due to the absence of residues. The study length will depend on the observed degradation rate and mobility, climatological conditions, the distance to the water table, and other site characteristics. If residues are detected in monitoring well samples, sampling is usually continued until the saturated zone degradation rate is determined or residues drop below guideline levels in all wells for two consecutive sampling.

Groundwater Monitoring Studies

Groundwater monitoring studies with objectives different than those of comprehensive research programs are often performed. Such objectives generally include determining whether an agricultural chemical is present in groundwater under normal use conditions, and if so, the magnitude of these residues and perhaps their persistence or movement. However, groundwater monitoring studies usually do not

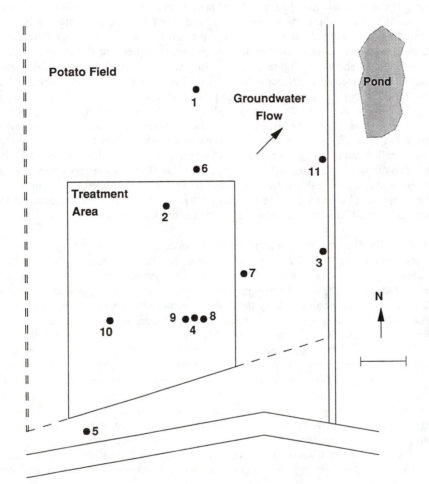

Figure 3. Monitoring well network used at an experimental site near Savannah, New York. (Adapted from ref. 4.)

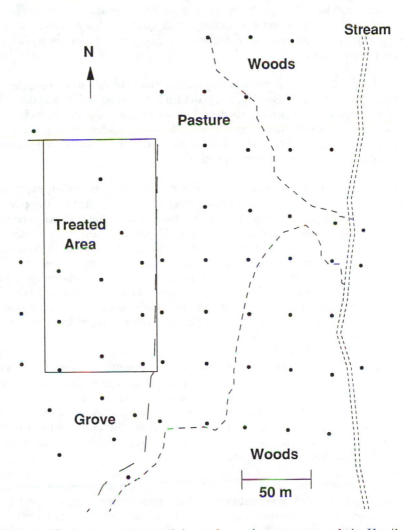

Figure 4. Monitoring well network located at a citrus grove near Lake Hamilton, Florida. (Adapted from ref. 17.)

have as an objective determining degradation rates in both soil and groundwater. Groundwater monitoring protocols need to be tailored to the study objectives, chemical properties, regional agricultural practices, and site characteristics. Although collecting soil samples may not be always necessary, groundwater sampling will be a necessary component in all studies of this type. If the study focuses on the potential for persistent residues in groundwater beneath and downgradient of treated fields, then rarely will sampling more frequently than quarterly be necessary to meet scientific objectives, although more frequent sampling may be necessary to meet regulatory requirements.

The design of groundwater monitoring studies should be developed on a case-by-case basis. Specific protocols will not be provided here, but objectives and designs of four different groundwater studies conducted by the authors will be used to illustrate the development of appropriate study designs. Much of the information presented for comprehensive field research studies will be applicable to groundwater monitoring studies (especially the section on site selection).

One study type occasionally used by the authors is to gather exploratory information on the environmental fate of an agricultural chemical in a region (for example, several counties in a state). In these studies two to four soil cores down to the water table are gathered from at least four fields throughout the area which have been previously treated with the agricultural chemical under study. If the water table is relatively shallow, then water samples may be collected from the bottom of the soil core holes. Analyses of these soil cores will indicate the depth of residue movement. Also sometimes a crude soil half-life can be estimated if cores are collected in the fall for applications made in the spring or winter. Usually such a program will be conducted to assess the need for more detailed research and most likely will not be suitable to satisfy regulatory requirements.

A second study (7) involved the measurement of potential residues under actual use conditions. Because the agricultural chemical was applied to a crop canopy with multiple foliar sprays, determination of accurate soil degradation rates was essentially impossible. In the study, two in-field well clusters were installed at two different worst-case sites and water samples were collected at regular intervals for 4 months following the multiple applications. Soil samples were collected to confirm the absence of soil residues after 2 months.

Another study's objective (17) was to determine the adequacy of restrictions preventing applications of an agricultural chemical near shallow drinking water wells. In 40 locations, a single shallow well (with a 1.5 m well screen) was placed 90 m downgradient of a previously treated area. Wells at each site were monitored quarterly for 2 to 4 years and additional well clusters (with 0.3 m screens) were installed at sites where residues were detected in the initial well. Because the desire was to maximize the number of sites examined (previous research had already provided more comprehensive data at several sites), the resources were most efficiently allocated by installing one well in forty sites, rather than six wells at six or seven sites, with increased resources then going to those few sites showing horizontal residue movement of at least 90 m.

The fourth groundwater monitoring study was conducted to determine residues present in soil and groundwater from previous applications in vulnerable fields, one field located in each of five different use areas (Norris, unpublished report). At each of the fields, two in-field well clusters were installed and sampled and four soil cores down to the water table were collected and analyzed. No residues were found in any soil below 0.15 m or in any groundwater sample. Low levels of residues were found in some surface soil samples at two of the sites. After this evaluation was completed, then comprehensive field research studies were conducted at each site to determine degradation rates in soil and to monitor any potential movement into groundwater.

Potable Well Monitoring

Because protecting drinking water supplies is the impetus for most groundwater research conducted with agricultural chemicals, some studies focus on potable well sampling. There are two basic types of potable well programs, programs that are designed to statistically estimate exposure to agricultural chemicals and programs designed to assess the potential presence and magnitude of residues in drinking water wells located in a region.

Statistically based exposure studies determine the exposure of a population to residues potentially present in drinking water supplies. Because the studies are designed to statistically evaluate results from carefully selected wells to draw conclusions about an entire population, a considerable portion of the study effort is directed toward the statistical design and selection of the wells. The authors are aware of only five studies of this type. The design of such studies has been discussed elsewhere (*19-20*).

In non-statistical potable well monitoring programs, several to many wells are sampled in an area to determine presence/magnitude of residues. In non-statistical monitoring, most of the labor is directed toward sample collection and analysis. This study type, especially when the sampled wells are generally the most vulnerable in a region, is a powerful tool for assessing the potential presence of an agricultural chemical in drinking water. Such monitoring programs have successfully identified areas where wells may sometimes contain residues, often with the analysis of less than 25 samples.

One simple design frequently used by the authors is to collect 200 drinking water samples from a use area (generally composed of a state or portions of several adjoining states). Sales information is used to determine the number of samples to be collected in each county. The specified number of samples is then collected in each county from potable wells located within 150 or 300 m of fields that have been treated with the agricultural chemical under study (sometimes the lack of wells near treated fields may prevent the collection of the specified number of samples). If significant residues (instances and/or concentrations) are found, then additional sampling may be necessary to identify other potable wells containing residues. Also, comprehensive research studies or groundwater monitoring studies may be needed to better understand the situation, and/or to develop appropriate management practices. However, if no significant residues are found, then additional monitoring would normally not be necessary.

General Recommendations

The design of groundwater studies should be tailored to the study objectives, the environmental conditions at the test sites, and the specific properties of the agricultural chemical under study. Although the development of generalized guidelines may be useful, regulatory requirements that dictate strict adherence to inflexible, previously established protocols may result not only in unnecessary effort but may compromise the quality of the study results.

The authors' experience with groundwater research and monitoring programs indicate that anticipating study events is often difficult. Therefore, timely analyses of both soil and groundwater samples is essential so that results from previous sampling intervals can be used to guide activities at future sampling dates. Examples of such actions are installation of additional wells to better define any groundwater residue plume, collection of deeper soil samples, or investigations of potential sample contamination.

One problem often encountered in these studies is ascertaining whether trace residues in a soil or water sample are the result of residues actually present in the media being sampled, or the measurements are caused by analytical interferences or contamination introduced during sampling or analyses. However, the challenging task of collecting uncontaminated samples (often in circumstances where dust or surface soils may contain residues three to four orders of magnitude higher than the sensitivity of the analytical method) requires that samples always be collected by conscientious and trained personnel using appropriate sampling techniques, with cleanliness always being a primary concern.

Literature Cited

1. Zaki, M. H.; Moran, D.; Harris, D. *Am. J. Public Health* 1982, *72*, 1391-1395.
2. Cohen, D. B. In *Evaluation of Pesticides in Ground Water*; Garner, W. Y.; Honeycutt, R. C.; Nigg, H. N., Eds.; ACS Symp. Ser. 315; American Chemical Society: Washington, DC, 1986; pp. 499-529.
3. Cohen, S. Z. In *Book of Abstracts, Seventh International Congress of Pesticide Chemistry, Hamburg, August 5-10, 1990*; Frehse, H.; Kesseler-Schmitz, E.; Conway, S., Eds.; IUPAC, ISBN-Nr. 3-924763-25-9; vol. III, p. 424.
4. Jones, R. L. In *Environmental Fate of Pesticides*; Hutson, D. H.; Roberts, T. R., Eds.; Progress in Pesticides in Biochemistry and Toxicology, vol. 7; John Wiley and Sons: Chichester, U. K., 1990; pp. 27-46.

5. Carsel, R. F.; Mulkey, L. A.; Lorber, M. N.; Baskin, L. B. *Ecol. Modeling* 1985, *30*, 49-69.

6. Jones, R. L. In *Hazardous Waste Containment and Treatment*; Cheremisinoff, P. N., Ed.; Encyclopedia of Environmental Technology, vol. 4; Gulf Publishing: Houston, TX, 1990; pp. 355-376.

7. Jones, R. L.; Hunt, T. W.; Norris, F. A.; Harden, C. F. *J. Contam. Hydrol.* 1989, *4*, 359-371.

8. Thompson, P. J.; Young, K.; Goran, W. D.; Moy, A. *An Interactive Soils Information System User's Manual*; CERL TR-N-87/18; U. S. Army Construction and Engineering Research Laboratory: Champaign, IL, 1987.

9. Imhoff, J. C.; Carsel, R. F.; Kittle, J. L.; Hummel, P. R. *Data Base Analyzer and Parameter Estimator (DBAPE) Interactive Computer Program User's Manual*; EPA 600/3-89/083, U. S. EPA Environmental Research Laboratory: Athens, GA, 1989.

10. Jones, R. L. *J. Contam. Hydrol.* 1989, *1*, 287-298.

11. Jones, R. L.; Rourke, R. V.; Hansen, J. L. *Environ. Toxicol. Chem.* 1986, *5*, 167-173.

12. Hormann, W. D. *Proc. Eur. Weed Res. Coun. Symp.* 1973, 129-140.

13. Fletcher C.: Hong, S.; Eiden, C.; Barrett, M. *Environmental Fate and Effects Division Standard Evaluation Procedure Terrestrial Field Dissipation Studies*; EPA 540-09/90-073; U. S. EPA Office of Pesticide Programs: Washington, DC, 1989.

14. Rao, P. S. C.; Edvardsson, K. S. V.; Ou, L. T.; Jessup, R. E.; Nkedi-Kizza, P.; Hornsby, A. G. In *Evaluation of Pesticides in Ground Water*; Garner, W. Y.; Honeycutt, R. C.; Nigg, H. N., Eds.; ACS Symposium Series No 315, American Chemical Society: Washington DC, 1986; pp. 100-115.

15. Porter, K. S.; Wagenet, R. J.; Jones, R. L.; Marquardt, T. E. *Environ. Toxicol. Chem.* 1990, *9*, 279-287.

16. Hornsby, A. G; Rao, P. S. C.; Wheeler, W. B.; Nkedi-kizza, P.; Jones, R. L. In *Proceedings of the Conference on Characterization and Monitoring of the Vadose (Unsaturated) Zone, Las Vegas, NV Dec 8-10, 1983*; Nielson, D.; Curl, M.; Eds.; National Water Well Association: Dublin, OH, 1984; pp. 936-958.

17. Jones, R. L.; Hornsby, A. G.; Rao, P. S. C.; Anderson, M. P. *J. Contam. Hydrol.* 1987, *1*, 265-285.

18. Jones, R. L.; Kirkland, S. D.; Chancey, E. L. *Appl. Agric. Res.* 1987, *2*, 177-182.

19. Liddle, S. K.; Whitmore, R. W.; Mason, R. E.; Alexander, W. J.; Holden, L. R. *Ground Water Monit. Rev.* 1990, *10*(1), 142-146.

20. Mason, R. E.; Benrud, C. H.; Iannacchione, V. G. *Stratification Proposed for the National Pesticide Survey*; Report RTI/3030/03-06F; Research Triangle Institute: Research Triangle Park, NC, 1986.

RECEIVED October 9, 1990

Chapter 10

Soil Map Units

Basis for Agrochemical-Residue Sampling

D. L. Karlen[1] and T. E. Fenton[2]

[1]National Soil Tilth Laboratory, Agricultural Research Service,
U.S. Department of Agriculture, 2150 Pammel Drive, Ames, IA 50011
[2]Department of Agronomy, Iowa State University, Ames, IA 50011

Representative sample collection is the most critical step in any program designed to determine how soil and crop management affects the presence of agrochemical residues such as nitrate nitrogen or pesticides. Soil map units within the soil classification system, can be used to develop sampling plans with comparable soil bodies despite natural soil diversity. Soil map unit data can be analyzed statistically and used to provide information for geographic information systems (GIS). Use of soil map units for selecting sampling sites to evaluate current and alternate management practices on agrochemical residues is recommended.

Analytical measurement of agrochemical residues such as nitrate nitrogen (NO_3-N) or pesticides in soil samples must meet QAQC (quality assurance quality control) standards, but for useful and valid interpretation of the analyses each sample must represent an individual and specific soil phase. This is important because soils have different biological, chemical, and physical properties. The ultimate fate of many agrochemicals will be determined by interactions controlled by soil and agrochemical properties. Developing a sampling plan that provides representative samples is a very critical process. It is important to understand landscape variability in relation to the soil patterns, and to have elementary knowledge of current soil characterization and classification concepts. Our objective is to describe why soils vary and to demonstrate how soil map units can be used to locate particular soil bodies from which representative samples can be collected to quantify agrochemical residue concentrations.

0097–6156/91/0465–0182$06.00/0
© 1991 American Chemical Society

Drainage Systems

Drainage systems have been defined as open or closed basins. Using an analogy of thermodynamic systems (*1*), an open system was theorized to have matter and energy imported and exported across boundaries and energy transformed uniformly to maintain a steady state. A drainage basin of any size is a natural open system that is confined at its head (origin) and along its sides by the perimeter (outermost) divide, but it is open at its mouth. Surface water from rainfall or melt water is collected by the drainage net that forms the basin and is discharged through the outlet or mouth (*2*). In a natural closed system, there is an interior basin and the drainage net descends in a centripetal pattern. Surface water originating from atmospheric sources descends the drainageways and collects in the basin. Loss of water is only through evaporation, transpiration by plants, and subsurface percolation (*2*).

Drainage patterns are important because movement of many agrochemical residues is directly or indirectly related to the path of water and sediment movement. Identifying the drainage basin type and boundaries must be given high priority when selecting sites and developing agrochemical sampling plans. Although the same soil map units can occur within either an open or closed drainage system, fate of agrochemicals applied to soils will differ. Materials that move with runoff or drainage water will accumulate in a closed basin, but continue to move away from application sites in an open basin. If samples representing a single map unit are collected from different types of drainage basins, differences in the amount of residues that are measured may erroneously be attributed to variability among soil samples rather than to differences in drainage patterns.

Landscape Position

The idea that soils are landscapes as well as profiles has been generally accepted for 40 years (*3*). Each soil occupies space, is defined in three dimensions, can be evaluated relative to evolution of elements within the landscape, and can be mapped. Recognition of this idea increased awareness of soil-geomorphic relationships and resulted in development of several models to explain soil-landscape relationships (*4-7*). Each model can make important contributions toward understanding landscape position, but the most important part is recognizing that all soils on the landscape are not the same age (*8*).

Soil landscapes have been described as the geographic distribution of soils on landscapes (*2*). When this simple classification system is combined with a model of landscape evolution, both soil materials and relative soil water relations can be accurately predicted. These factors are important for understanding processes and especially in helping understand soil variability within a local landscape (*8*). This knowledge is important for developing representative agrochemical sampling plans because the fate of many chemicals will be determined by interactions that occur because of identifiable soil and/or chemical properties.

The current landscape terminology originated after previously suggested (9) segments of a "fully developed hillslope" were modified and renamed (10). A summit, shoulder, backslope, footslope, and toeslope are the five elements currently recognized. This slope-profile terminology can be applied to the geomorphic components of headslope, sideslope, and toeslope (Figure 1). Relative stability and water movement for these slope-profile elements in humid regions was discussed previously (11), but are summarized here because of their importance for understanding factors that influence soil variability and their reaction with various agrochemicals.

Summit. This position is considered to be the most stable element of the landscape. There is little runoff where the summit is at least 30 m wide. Water movement is predominantly vertical except near the transition to the shoulder or on summit undulations. In these areas some lateral water movement and accompanying surface and deep percolation occur.

Shoulder. This position has slopes that are usually convex. Surface runoff is maximized in this element resulting in a highly erosional and relatively unstable surface. Probability of lateral subsurface flow is high. Solum thickness and organic matter content are usually a minimum on this element.

Backslope. Dominant processes on this position include both surface and subsurface transportation of material and water. Slopes are nearly linear and steepest on this landscape position. Surface transport of material may be in the form of flow, slump, surface wash, or creep. This position is considered to be relatively unstable.

Footslope. Concavity is characteristic of this landscape position. The concavity results in deposition from upslope of particulate material as well as material carried in solution. The position is dominantly constructional and relatively unstable. Seepage zones are common and the water content is usually much higher than on shoulder or backslope positions. Cumulic soils are associated with this position in the Midwest.

Toeslope. This position is constructional and relatively unstable. Alluvial material in the toeslope position is derived from up valley and from upslope elements.

The slope shape and position classification system (2) is a very useful tool for field investigations, because it helps predict the soil composition, relationships to surrounding landscape features, and soil water regimes (8). However, for evaluating soil productivity, it has been suggested that the backslope should be divided into linear, nose and head slope positions (8). This recommendation was made because different water flow regimes among these positions can create large differences in soil-plant environments. A similar argument can be made for evaluating agrochemical movement.

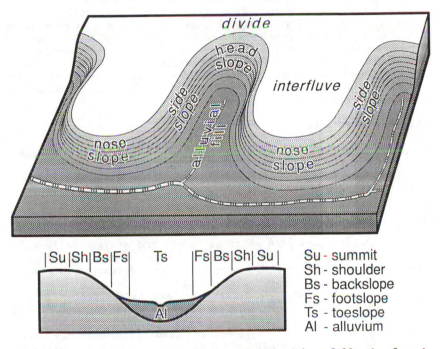

Figure 1.   Landscape elements associated with a fully developed
hillslope. (Redrawn with permission from ref. 5. Copyright 1969
Iowa State Univ. Press).

What is a Soil?

Soil is defined as "the unconsolidated mineral matter on the surface of the earth that has been subjected to and influenced by genetic and environmental factors of: *parent material, climate* (including moisture and temperature effects), *macro-* and *microorganisms,* and *topography,* all acting over a period of *time* and producing a product- -soil--that differs from the material from which it is derived in many physical, chemical, biological and morphological properties and characteristics (*12*).

The basic soil model is a function of the five soil forming factors, but occasionally the model has been modified to include man as a sixth soil forming factor. Unique combinations resulting from overlapping of these factors gives rise to different soils (*13*).

An alternate model views soil formation or genesis as consisting of two steps that in some cases can be overlapping (*14*). The two steps are parent material accumulation and horizon differentiation. Additions, removals, transfers, and transformations that occur over time cause horizon differentiation. Various process combinations are operative in all soils, and the process balance determines the nature of the soil formed.

Soil Forming Factors

Parent material (mineral content, particle size, etc.) influences the inherent fertility, chemical reaction, and soil texture. The deposition method (residual, or transported by ice, water, or wind) primarily affects soil texture and landscape topography. Formation time, in conjunction with intensity (ie temperature, rainfall, etc.), determines the degree of progress in soil development.

Topography influences soil water, temperature, and erosion. Soils on sloping land lose water as runoff and are generally drier than non-sloping soils in the same area. Depressional soils usually have higher water content than sloping soils. Sloping topographies are subject to more erosion than flatter land under similar land cover. Slope orientation and elevation are topographical factors that influence local microclimate and thus influence the soil forming processes.

Climate influences soil type because precipitation, temperature, and the amount of erosive wind and water action determine weathering rates of parent materials. Water and temperature influence biological and chemical reaction rates including solution, hydration, and leaching. Climate also determines the kind and quantity of vegetation found throughout various landscapes and the amount of organic material added each year.

The biota type and quantity determine the kind and amount of organic materials that are returned to the soil. Biota influence spatial organic material deposition, ie., trees deposit most organic matter on the surface, while grasses distribute organic matter via their root systems throughout large volumes of soil. Vegetation influences many other biological processes by providing energy sources for microbial processes including nitrogen mineralization, fixation, and immobilization, as well as organic matter and crop residue decomposition processes that influence nutrient cycling.

Soil Variability

Recognition of spatial variability in soils is important in designing a sampling scheme. Two broad categories of spatial variability, systematic and random, are commonly recognized (*15*). Systematic variability is a change in soil properties as a function of landform, geomorphic component, or a soil forming factor. One example from northwest Iowa (Figure 2) shows the relationship between glacial till and ground surface elevation in a loess over glacial till landscape. If surface elevation is known, the elevation at which glacial till will be encountered can be predicted quite accurately.

Systematic change has been documented for hillslope sediments (*16*) which are formed by processes including slope wash, faunal activity, creep, and frost heave. Water movement sorts the materials downslope creating changes in soil texture that are predictable. For an Iowa landscape (*16*) formed in glacial till, sorting resulted in a systematic change in particle size from the summit to a midslope point where a sand lens was encountered. However, this change could be described by equation 1, where L is the distance from the summit. Below the intersection with the sand lens, mean sediment diameter increased abruptly and the relationship was described by a logarithmic function presented in equation 2.

$$Y = 139.73 + 1.95L - 3.67L^2 \tag{1}$$

$$Y = 325.25 - 228.12 \log L \tag{2}$$

Another hillslope sediment study (*4*) showed that particle size, organic carbon, cation exchange capacity, and extractable Ca and Mg were directly related to the upslope source. Being aware of these soil changes is important when developing agrochemical sampling plans because of organic matter, pH, and other interactions between soil and agrochemical properties, and because most sloping landscapes have large areas of sediment reworked by one or more processes (*8*). Even in virgin forest (*17*), and throughout the Southern Piedmont, about 50% of the landscape is covered by material other than residuum (*18-19*). The most difficult impediment to obtaining this information is that only highly detailed soil morphology normally recognizes hillslope sediments (*8*).

If changes in soil properties can not be related to a known cause, the changes are classified as random or chance variations. The categorization of variability, however, is often dependent on observation spacing. Variability originally considered random may in some cases be shown to be systematic if the sampling intensity is increased, or if controlling mechanisms are identified.

Anticipated variability among soils can be predicted based on selected parameters. By studying morphologically matched pedons (three-dimensional bodies of soil having an area ranging from 1 to 10 m² and identifiable horizon shapes and relations), the following

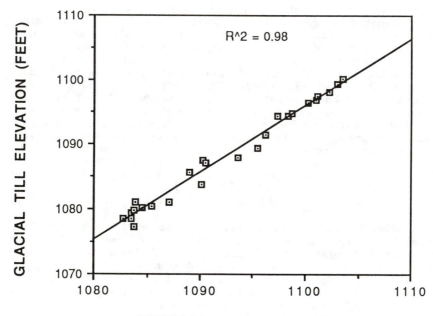

Figure 2.   Systematic variability as shown by soil surface and glacial till elevations in northwest Iowa.

generalized order for variability in physical properties has been reported as a function of parent material: loess < glacial drift < alluvium = residium (20). Similarly, the following generalized array for spatial variability in physical, chemical and elemental soil properties was suggested (21):

> Parent material-- Loess < glacial till < glacial outwash
> = glacial lacustrine = alluvium;
>
> Elements---   K = Ti < Zr < Fe < Ca;
>
> Horizons---   No consistent trend among A, B, and C
> horizons.

In an overall sampling project, the magnitude of data variability associated with various sources was probably greatest to least in the following order (21):

> Landscape body >>> choice of pedon >> pedon sampling > laboratory analyses.

Soil Map Units

Natural soil variation because of differences in drainage, landscape position, soil forming factors, or possibly long-term management practices may appear to make it impossible to collect a representative sample. However, by using soil map units the natural variation can be grouped into identifiable sampling units that can be analyzed statistically.

A soil map unit is a collection of areas within a landscape that can be defined and named in terms the same as their soil components (22). Each map unit identified on a soil survey represents an individual pedon, a collection of very similar pedons (polypedon), or polypedon parts that consist of contiguous similar pedons and thus represent a "specific soil". Map units may consist of one or more components that are identified in the name of the map unit. Minor components that are not identified in a map unit name are considered inclusions. All components, whether dominant or inclusions, considered to be important for interpretation and use or understanding of a soil map unit are included in the soil map unit description. With regard to agrochemical sampling, it is important to know that different soil map units may respond differently to various agricultural chemicals because of inherent differences in properties such as pH or organic matter content.

Each map unit differs in some respect from all others within a survey area, is bounded on all sides by pedons of unlike character, and can be uniquely identified as a delineation on a soil map. An important aspect, however, is that soil boundaries can seldom be shown with complete accuracy on soils maps because many boundaries are gradational in character. Therefore, parts and pieces of adjacent polypedons are sometimes inadvertently included (inclusions)

or excluded (exclusions) from each soil map unit. The purity and
kinds of map units depend primarily on the scale and purpose for
which a soil survey map was developed and the pattern of soils and
miscellaneous areas within the landscape.

When developing agrochemical sampling plans, it is important to
recognize differences in scale because soil survey maps have been
prepared at ratios of 1:12000, 1:15840, 1:20000, or 1:24000. These
scales correspond to 8.33, 6.31, 5.00, and 4.17 cm km^{-1} (5.28, 4.00,
3.17, and 2.64 inches mile^{-1}). Map scale determines the number of
inclusions in each soil map unit, because each section of land (640
acres or 259 hectares) is drawn on 27.9, 16.0, 10.0, or 7.0 inches^2
(180, 103, 65, or 45 cm^2). Examples of map unit detail associated
with scales of 1:1200 and 1:15840 are shown in Figure 3 and Figure
4, respectively. Figure 3 is drawn for an area of approximately 20
acres (8 hectares) and provides much greater detail than Figure 4
which shows two adjacent 40 acre (16 ha) areas.

Soil map units provide an excellent basis for developing an
agrochemical sampling scheme for several reasons. First, taxonomic
classes provide the basic sets of soil properties that define soil
map units. The soil taxonomic classes provide predefined sets of
soil properties that have been tested for genetic relationships and
for interpretive value. Taxa provide stable and consistent criteria
for recognizing the components and most probable characteristics of
potential map units in an unfamiliar area. Established taxa also
make it much easier to identify similar soils for each statistical
class designation. Soil map units thus summarize an immense amount
of research and experience related to the significance of individual
and combinations of soil properties (23).

Soil map units also provide a basis for sampling and grouping
the natural variation caused by landscape position (24). To support
increasing interest in soil map units, more research is occurring to
quantify morphological map unit differences or "purity". Presently,
no more than 25% of a map unit should be comprised of dissimilar
soils and no more than 10 to 15% should have characteristics more
limiting than the named soil(s) (22).

Sampling Procedure

To develop an agrochemical sampling scheme using soil map units, a
soil survey map must be obtained from the Soil Conservation Service
or other agencies in the National Cooperative Soil Survey. If a
different mapping scale is needed, specialized maps must be developed
for the site by trained soil scientists. After obtaining a map with
appropriate detail (Figure 3 and Figure 4) samples can be collected
randomly from within a soil map unit, or along transects with random
or fixed spacings.

After establishing transects, samples are collected and handled
to prevent contamination, coordinates of each sampling site are
determined, soil map units are identified from digitized or hard-
copy maps, and data are analyzed statistically using soil map units
as class variables for statistical programs such as the SAS General
Linear Model (25). This technique was recently used to demonstra-

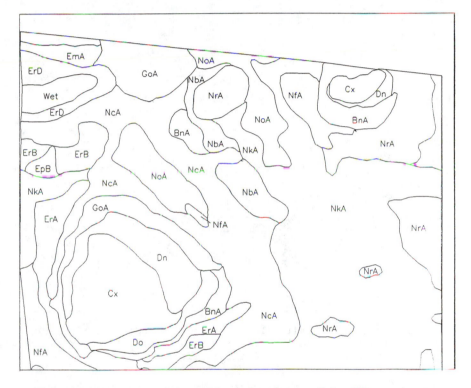

Figure 3. Soil map unit delineation for an 8 ha (20 acre) area associated with a mapping scale of 1:1200 which is equivalent to 83.3 cm km^{-1} or 52.8 inches mile^{-1}.

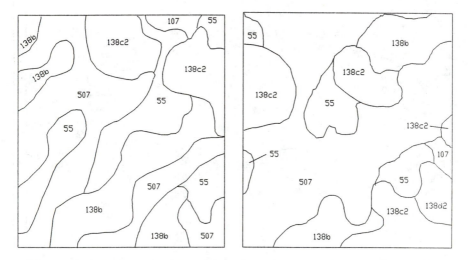

Figure 4.   Soil map unit delineation for two adjacent 16 ha (40 acre) areas associated with a mapping scale of 1:15840 which is equivalent to 6.3 cm km^{-1} or 4 inches mile^{-1}.

te how soil map units could be used to quantify crop yield variation
in a Coastal Plain field (*26*). Identifying the exact soil map unit
at each sampling site is currently the most tedious process, but as
global positioning devices and digitized soil maps become more
available, this task will be simplified.

Information Transfer

Use of soil map units as a basis for sampling for agrochemical
residues will facilitate information transfer among experiments
conducted at different geographical scales. Results obtained from
plot- or field-scale experiments can be compared with those measured
for farm-, watershed- or basin-scale studies that represent areas of
10, 100, or 1000 ha by identifying common soil map units at each
scale of experimentation.

Geographical information systems (GIS) can transfer information
collected for individual map units across spatial or temporal scales.
Techniques for using GIS and small-scale digital soil maps to study
natural resource problems have recently been reported (*27*). By using
currently available data bases such as the Soil Survey Geographic
Data Base (SSURGO), the State Soil Geographic Data Base (STATSGO),
or the National Soil Geographic Data Base (NATSGO), interpretive maps
can be made by overlaying soil data with other spatial resource data
(*28*).

Conclusion

As soil survey maps throughout the U.S. and in several countries
around the world are completed and subsequently digitized, soil map
units should be used to develop sampling schemes to measure agrochemical residues in various soil matrices throughout all agroecological
zones.

Literature Cited

1. Strahler, A. N. *Geol. Soc. Am. Bull.* **1952**, *63*, pp. 923-938.
2. Ruhe, R. V. *Geomorphology*; Houghton Mifflin Co.: Boston, MA.
   1975. pp. 246.
3. Soil Survey Staff. *Soil Survey Manual*, USDA Handbook 18. U.S.
   Gov. Print. Office, Washington, DC. 1951. pp. 503
4. Ruhe, R.V.; Daniels, R.B.; Cady, J.G. *Landscape Evolution and
   Soil Formation in Southwesterm Iowa.* USDA Tech. Bull. 1349.
   1967. pp. 242.
5. Ruhe, R. V. *Quaternary Landscapes in Iowa.* Iowa State Univ.
   Press., Ames, IA. 1969. pp. 255.
6. Hack, J.T. *Am. J. Sci.* **1960**, *258A*, pp. 80-97.
7. Conacher, A.J.; Dalrymple, J.B. *Geoderma.* **1977**, *18*, pp. 1-154.
8. Daniels, R.B.; Bubenzer, G.D. In *Proceedings of soil erosion and
   productivity workshop.* Larson, W.E.; Foster, G.R.; Allmaras,
   R.R.; Smith, C.M., Eds.; Univ. Minnesota: St. Paul, MN 1990.
   pp. 1-22.
9. Wood, A. *Geol. Assoc. Proc.* **1942**, 53, pp. 128-138.

10. Ruhe, R. V. *Transactions 7th Intl. Congr. Soil Sci, Madison, WI.* 1960; pp. 165-170.

11. Hall, G. F. In *Pedogenesis and Soil Taxonomy I. Concepts and Interactions*; Wilding, L.P.; Smeck, N.E.; Hall, G.F. Eds.; Elsevier Publishing Co., Amsterdam. 1983. pp. 303.

12. Soil Science Society of America. *Glossary of soil science terms*; Soil Sci. Soc. Am.: Madison, WI, 1979.

13. Arnold, R.W. In *Pedogenesis and Soil Taxonomy I. Concepts and interactions*; Wilding, L.P; Smeck, N.E.; Hall, G.F. Eds.; Elsevier Publishing Co., Amsterdam. 1983. pp. 303.

14. Simonson, R.W. *Soil Sci. Soc. Am. Proc.* **1959**. *53*, pp. 152-156.

15. Wilding, L.P.; Drees, L.R. In *Pedogenesis and Soil Taxonomy I. Concepts and Interactions*; Wilding, L.P.; Smeck, N.E.; Hall, G.F. Eds.; Elsevier Publishing Co., Amsterdam. 1983. pp. 303.

16. Kleiss, H.J. *Soil Sci. Soc. Amer. Proc.* **1970** *34*, pp. 287-290.

17. McCracken, R.J.; Daniels, R.B.; Fulcher, W.E. *Soil Sci. Soc. Am. J.* **1989**, *53*, pp. 1146-1152.

18. Eargle, D.H. *Science* **1940**, *91*, pp. 337-388.

19. Whittecar, G.R. *S.E. Geol.* **1985**. *26*, pp. 117-129.

20. Mausbach, M.J.; Brasher, B.R.; Yeck, R.D.; Nettleton, W.D. *Soil Sci. Sci. Am. J.*, **1980**, 44 pp. 358-363.

21. Drees, L.R.; Wilding, L.P. *Soil Sci. Soc. Am. Proc.*, **1973**. *37* pp. 82-87.

22. Soil Survey Staff. *National Soils Handbook*. U.S. Gov. Print. Office, Washington, DC. 1983. pp. 618.

23. Hudson, B.D. *Soil Survey Horizons*, **1990**, 31, pp. 63-72.

24. Gilliam, J.W.; Cassel, D.K.; Daniels, R.B.; Stone, J.R. In *Erosion and Soil Productivity*; Proceedings ASAE National Symp. on Erosion and Soil Productivity, New Orleans, LA; Am. Soc. Agr. Eng., St. Joseph, MI. 1985. pp. 75-82.

25. SAS Institute Inc. *SAS User's Guide: Statistics*. Version 5 Edition. Cary, NC. 1985. pp. 956

26 Karlen, D.L.; Sadler, E.J.; Busscher, W.J. *Soil Sci. Soc. Am. J.* **1990**, *54*, pp. 859-865.

27. Bliss, N.B.; Reybold, W.U. *J. Soil Water Conserv.* **1989**, 44, pp. 30-34.

28. Reybold, W.U.; TeSelle, G.W. J. Soil Water Conserv. **1989**, 44 pp. 28-29.

RECEIVED November 5, 1990

Chapter 11

System Design for Evaluation and Control of Agrochemical Movement in Soils Above Shallow Water Tables

System Design for Water Table Management

Guye H. Willis, James L. Fouss, James S. Rogers, Cade E. Carter, and Lloyd M. Southwick

Soil and Water Research Unit, Agricultural Research Service, U.S. Department of Agriculture, P.O. Box 25071, Baton Rouge, LA 70894–5071

The rate of pesticide transport through soils may be significantly affected by various soil-water/watertable management methods. Bordered plots (16 each, 35 by 61 m, each surrounded by a subsurface 2-m vertical plastic film barrier) will be equipped with 102-mm diameter slotted plastic tubing 1.0 m below the soil surface and with appropriate sumps and pumps for microprocessor controlled subdrainage/subirrigation. Each plot will also be equipped with watertable measuring pipes with depth sensors, soil moisture (matric potential) sensors, soil-water pressure sensors, tensiometer-pressure transducers with ceramic cups, watertable sampling tubes, piezometers, and soil temperature sensors (current generating diode type), all placed at various depths in the root/vadose/watertable zones and at appropriate distances from the center drainline in each plot. Initial treatments will include four replications each of (i) surface drainage only, (ii) conventional subsurface drainage at a 1-m depth, (iii) controlled watertable at 45 ± 5 cm depth and (iv) controlled watertable at 75 ± 5 cm depth.

Pesticide and other organic chemical contamination of groundwater has become a national concern that needs timely and rational solution. There are major economic reasons for the continued use of pesticides for the foreseeable future in U.S. agriculture, and there is potential for groundwater contamination from continued, long-term pesticide use. Since groundwater provides drinking water for about half of the U.S. population (1), prudence suggests that steps be taken to rectify this potential problem.

About 25%, i.e., 40 million hectares, of the total U.S. cropland needs drainage (2). Much of this land is usually flat, highly fertile, and has no serious erosion problems. These potentially productive wet soils are primarily located in the prairie and level uplands of the Midwest, the bottom lands of the Mississippi Valley, the bottom lands in the Piedmont and hill areas of the South, the coastal plains of the East and South, and irrigated areas of the

West (3). During most or part of the year these soils have shallow watertables that are potential sinks for pesticides that may leach below the root zone.

Pesticides and fertilizers are used extensively in the lower Mississippi River valley (LMV), the agriculturally important Mississippi River flood plain in Arkansas, Mississippi, and Louisiana (Fig. 1). Although large quantities of water flow down the Mississippi River, most fresh water supplies for domestic and agricultural use come from the Mississippi River alluvial aquifer (MRA) which underlies the LMV. In south Louisiana most water supplies come from shallow wells and surface waters. Pesticide contamination of groundwaters in the LMV has been reported (4-8).

The Mississippi River alluvium is generally less than 70 m thick and grades downward from silt and clay at the surface to coarse sand and gravel at the base (8-10). The thickness of the overlying silt and clay is generally less than 12 m. Rainfall, ranging from 1150 to 1500 mm annually, is the major source of recharge for the aquifer (8,11). The amount of recharge depends not only on the amount and rate of precipitation, but also on the permeability and thickness of the overlying silt and clay. These deposits are relatively permeable compared to typical clay because of their high content of organic material and because they have not been fully consolidated by heavy overburden (8). Water levels in the MRA generally are less than 9 m below land surface, and are much closer to the surface (0 to 2 m) in southern areas (8,11). These shallow watertables fluctuate considerably and respond mainly to rainfall.

Conditions are present in the LMV for surface water and groundwater pollution including (a) shallow watertables, (b) high pesticide use, and (c) high rainfall.

Concepts of Watertable Management

The "optimal" management of soil-water for agricultural cropland in humid areas of the U.S. via control of watertable depth in the soil profile involves complex daily operational/management decisions because of the erratic spatial and temporal distribution of rainfall. The farm management decisions are even more complex because soil-water management must be integrated with improved fertilizer and pesticide application practices. Integrated methodology is needed to manage soil, water, ground cover, pesticide applications, and fertilizer applications in such a way that pesticides and fertilizers are contained within their "action zones", thus reducing the risk of surface and groundwater pollution. Improved soil-water management technology, e.g., watertable control, may reduce the amount of pesticides and fertilizer used, thus increasing crop production efficiency and farmer profitability, while reducing pollution. Periods of excess and deficit soil-water conditions in the active root-zone often occur within the same growing season. Thus, controlling watertable depth within a desired range relative to the root-zone requires facilities for regulating both subsurface drainage flow from and irrigation into the soil profile. A popular field-scale watertable management system uses a subsurface draintube system for both controlled-drainage and subirrigation. Controlled-drainage permits retention and temporary storage of infiltrated rainfall in the soil profile at an elevation above the drainline depth. Conventional, or "free", subsurface drainage to the full depth of the draintubes is often needed during periods of extended or heavy rainfall to control the watertable rise and reduce the duration of excess water in the active root-zone. During subirrigation, the water level at the drain outlet is maintained at an elevated position by pumping from an external source (e.g., a well) to force water back through the draintube system and into the soil profile to manage the watertable at the desired elevation above the drain outlet for proper plant growth.

The primary purposes of watertable control are to minimize the time of excess or deficit soil-water conditions in the root-zone, and to maximize the utilization of natural rainfall, thus minimizing the amount of subirrigation water required from external sources. Watertable management technology has also begun to be used to improve water quality. Gilliam and associates developed controlled drainage practices for reducing nitrogen and phosphorus levels in surface/subsurface effluent from agricultural lands (12-14). These practices are being used in eastern and southern coastal plains soils. Watertable management has a high potential for achieving maximum crop production, water use efficiency, and improved water quality if properly controlled to compensate for changes in weather conditions. Determining when changes are needed in controlled-drainage and subirrigation to manage the watertable depth optimally is a major problem for farmers, especially in coastal areas with fine textured soils. In the Mississippi Delta frequent rainfall events can cause large variations in watertable depth because of the small, 3 to 8%, drainable soil porosity. Rainfall probability information included in daily forecasts issued by the U.S. National Weather Service can be used to aid the farmer in making management decisions in anticipation of predicted weather changes (15,16).

For level and low-lying topography where subsurface drainage by gravity flow outlets is not feasible, a sump-type structure can be used for controlling the water level at the drainage system outlet (Fig. 2). Water is pumped out of the sumps for subsurface drainage and into the sumps for subirrigation. The controlled-subirrigation mode of watertable management is illustrated in Fig. 3. The monitored watertable depth (WTD) midway between the subsurface conduits is the controlling performance parameter. For conventional subsurface drainage the water level in the sump is maintained (by pumping) below the drainline depth. Where gravity flow drainage outlets are feasible, a float-activated control valve on the outlet pipe can be used to regulate drainage effluent (17).

Research Objectives

The research objectives of this study are to:
1. Identify and characterize chemical and physical factors and processes that affect the rate and mode of pesticide and plant nutrient transport in surface runoff and in the root, vadose, and saturated zones of shallow watertable soils.
2. Characterize and quantify the effects of water-soluble organic matter on pesticide transport in soil.

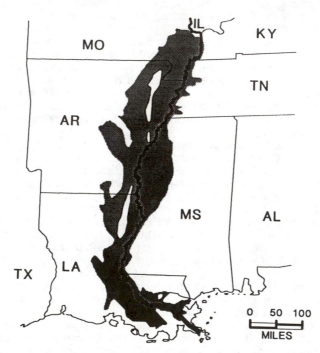

Figure 1. Map of the lower Mississippi River Valley.

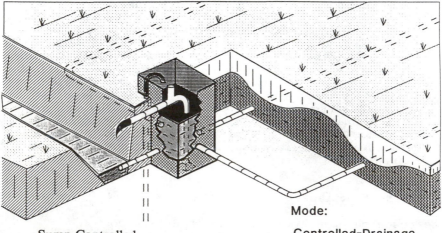

Figure 2. Schematic drawing of sump-controlled watertable
management system in the controlled-drainage mode of operation.

3. Determine the effects of watertable management on losses of pesticides and plant nutrients via surface runoff, subsurface drainage outflow, and leaching to groundwater.
4. Develop models needed to devise watertable management strategies that will avoid and/or alleviate groundwater contamination by pesticides and fertilizers.

General Plot Layout and Site Characterization

The study will be conducted on 16 bordered, 0.21-ha plots (located on the Louisiana Agricultural Experiment Station's Ben Hur farm near Baton Rouge, LA) instrumented for automatic, microprocessor-controlled measurement and sampling of surface runoff and subsurface drain outflow, and watertable management (Fig. 4). The cropping system will be conventionally-tilled corn with common rye grass as a winter cover crop. Previous research (18) has shown that 0.21-ha plots are large enough to minimize "border effects". The plots are on a Commerce silt loam soil (fine-silty, mixed, nonacid, thermic, Aeric Fluvaquents) (19), which consists of layers of silt and clay mingled with sand lenses that were deposited by past Mississippi River overflows. The top 45 cm of the soil profile is relatively high in clay (= < 34%). Consequently, the hydraulic conductivity is relatively low and the soil is easily compacted by wheel traffic (20). At depths from 45 to 90 cm the clay content decreases to about 22% while the silt and sand contents range up to = <44 and = <47%, respectively. Saturated hydraulic conductivity values determined from auger holes on the same soil type near the plots were 15 ± 12 mm/h for 60-cm deep holes, 23 ± 16 mm/h for 90-cm deep holes, 38 ± 28 mm/h for 120-cm deep holes, 44 ± 22 mm/h for 150-cm deep holes, and 30 ± 10 mm/h for hole depths between 180 and 240 cm (21). Watertable depths averaged between 30 and 50 cm below the soil surface. Hydraulic conductivities of the soil profile below 240 cm have not been reported. However, the soil clay content below 150 cm increases somewhat and the hydraulic conductivity should decrease accordingly.

Each plot, 35 by 61 m, will have a 15-cm high dike at the outer edge of each border, a 0.15 mm polyethylene subsurface barrier installed 30 cm below the soil surface and extending down 2 m, three 102-mm diameter subsurface drain lines installed 15 m apart and 1.0 m below the soil surface, a 1.2- by 1.2- by 3.0-m steel sump to control drainline outlet water levels, and an H-flume at the surface runoff outlet (Fig. 5). Each plot is precision-graded to a 0.2% slope with 0.2% cross-slope. The drainlines next to the longitudinal borders will control border effects between adjacent plots. The area centered over the middle drainline (15 by 61 m) is assumed to be representative of an area in a larger field with the same drain spacing. Surface runoff and subsurface drain outflow will be measured, sampled, and directed, via a 300-mm diameter PVC subsurface "main" (Fig. 6), to a collector sump for diversion into a surface drainage ditch. The risers shown in Figure 5 are for access and cleanout of the connecting mains.

A National Weather Service Class-A automated weather station is located 250 m from the plots. Meteorological data (e.g., rainfall, relative humidity, pan evaporation, wind speed and direction, total

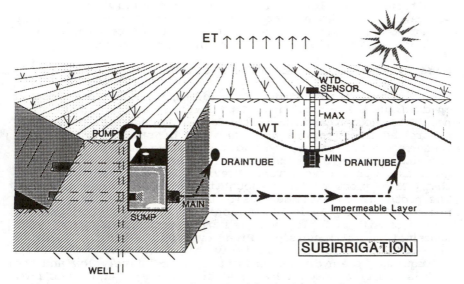

Figure 3. Schematic drawing of subirrigation mode for
sump–controlled watertable management system with well–supplied
irrigation water.

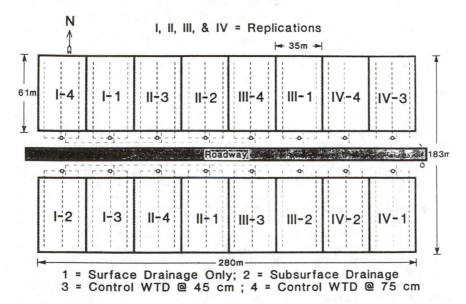

1 = Surface Drainage Only; 2 = Subsurface Drainage
3 = Control WTD @ 45 cm ; 4 = Control WTD @ 75 cm

Figure 4. Field-plot layout.

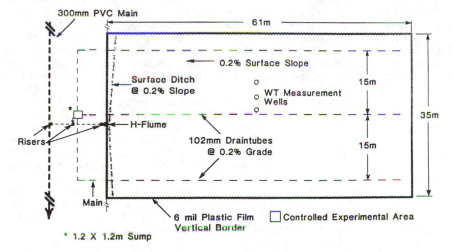

Figure 5. Schematic diagram (top view) of a controlled-watertable research plot with surrounding plastic-film subsurface barriers.

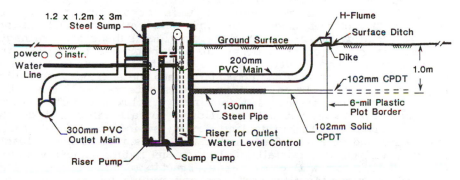

Cross-Section Elevation-View of Outlet Sump

Figure 6. Schematic drawing (longitudinal cross-section) of the outlet-sump structure, H-flume, and subsurface PVC main for each plot.

radiation, air temperature) from the weather station's automatic data-logger will be used in conjunction with the study. Evaporation will be estimated from pan evaporation and the modified Penman equation (22).

During excavations for installing drains, borders, and sampling/sensing devices, soil samples will be taken for soil physical and chemical property determination. Bulk density, hydraulic conductivity, porosity, and water retention characteristics will be determined from soil core samples. Core samples will be taken by pressing, rather than hammering, a coring device into the soil. Rogers and Carter (23) have shown that pressing is superior to hammering when taking core samples. Bulk samples will be taken for the determination of particle size distribution, water retention characteristics, nutrient status, pesticide residues, organic carbon, and cation exchange capacity. In general, procedures for soil analysis as outlined in Methods of Soil Analysis (24) will be used for making the above measurements. The auger hole method has previously been used to determine hydraulic conductivity in the area just outside the borders of most plots. Several transects with different auger hole depths and watertable depths were used to assess spatial variability and variability with soil depth (25).

The treatments, imposed in a randomized complete block design, will consist of four replications each of (i) surface drainage only (the subsurface drain lines will be plugged), (ii) conventional subsurface drainage (watertable kept at the level of the drainline or below), (iii) controlled watertable at 45 ± 5 cm depth, and (iv) controlled watertable at 75 ± 5 cm depth.

Treatment 1 is the control treatment and will be used to characterize pesticide and nutrient transport in surface runoff and soil leachate in the absence of subsurface drainage and watertable control. Chemical movement in this treatment should be representative of that under farming practices typical to millions of hectares of alluvial soil. Treatment 2 provides subsurface drainage but no watertable control: pesticide and nutrient transport in surface runoff should decrease (26,27), but the amounts leached below the plant root zone and potentially into the shallow watertable may increase (12). In this treatment the watertable will not be allowed to rise closer than 1 m (nominally) below the soil surface (during wet periods), but will be allowed to fall below that depth through normal watertable decline during periods of low rainfall. The year-round presence of an unsaturated soil surface zone at least 1 m deep will enhance rainfall infiltration and reduce runoff volume. Greater infiltration will cause larger fractions of surface-applied chemicals to enter the plant root zone and reduce losses in surface runoff. Depending on rainfall amount and distribution patterns, and the chemical's persistence and water solubility characteristics, there will be a greater chance for the chemical to move into the shallow watertable zone. Treatments 3 and 4 provide constant water-tables 45 cm and 75 cm below the soil surface. Those depths will provide less storage for infiltrated rainfall than treatment 2, but will greatly reduce the potential for chemical leaching below the 1-m-deep drainlines. These treatments will encourage rainfall infil-tration (treatments 4 more so than treatment 3) and concomitantly reduce chemical losses in surface runoff. The continuous presence of a watertable above the drainlines will prevent downward chemical movement except for those times when water is pumped from the drainlines to lower the watertable to the prescribed elevation. Thus, except during controlled drainage periods, the infiltrated chemicals will be kept in the soil profile above the drainlines where they will be subjected to extended periods during which utilization and/or modification/degradation can occur. Plant growth and yield data and pesticide/fertilizer use efficiency for the 45- and 75-cm watertable depths will be compared as the initial step in developing a management program of controlled variable-depth watertables designed to provide and optimized combination of profitable crop production, efficient pesticide and fertilizer use, and decreased chemical transport in surface runoff and leaching to groundwater.

Subsurface Drainage/Subirrigation Systems

The dual purpose controlled-drainage and subirrigation system will be installed in each plot as shown in the plan view of Fig. 5. The drainline spacing was selected so that the watertable depth could be accurately controlled in the experimental area of each plot for all three modes of system operation: subsurface drainage, controlled-drainage, and subirrigation.

The 15-m drain spacing selected was based upon computer simulation results for various system designs and operational modes. The water management model DRAINMOD (*28*) was used to conduct the simulations and predict the performance of the various system designs over a ten-year period of hourly climatological record (1979-1989) at the experimental site. All soil parameters at the research site required as inputs to DRAINMOD were previously reported by Fouss, et al. (*29*). Additionally, a short-term simulation model based upon the more comprehensive Boussinesq-Equation of subsurface flow (*30,31*) was used to evaluate the performance of selected system designs for extreme "dry" and "wet" years, e.g., 1986 and 1987, respectively. The performance of the various modes of watertable control were evaluated in terms of the following parameters predicted by the simulation models: (1) daily fluctuation in watertable depth, (2) average and standard deviation of the field watertable depth during the growing season, (3) excess soil-water within 30 cm of the soil surface (expressed as SEW-30 in cm-days), and (4) the number of dry days during the growing season when potential evapotranspiration demand could not be met by upward flux from the watertable. For a watertable management system to meet accepted performance requirements for a silt loam soil, drainage should control excess soil-water such that SEW-30 is maintained in the range from 100 to 150 cm-days on a 5-year recurrence interval (R.I.), and subirrigation should provide sufficient soil-water so that no more than 8 to 10 dry days occur during the growing season on a 5-year R.I. (*32*).

The results of the simulations indicated that a drain spacing of 18 m would provide adequate control of the watertable depth in the controlled-drainage and subirrigation modes of operation, with only occasional excess or deficit soil-water conditions (*33*). Since this is an experimental system where the emphasis is on watertable control rather than economical drain installation, a drain spacing of 15 m was selected. Simulated results for the 15-m spacing indicated that even with the water level at the drain outlet held constant at a

60-cm depth during the growing season, the excess soil-water in the root-zone (30 cm depth, expressed as SEW-30) was less than 150 cm-days, which is an acceptable level of performance.

Subsurface Conduit Materials

Corrugated, perforated, polyethylene tubing with a smooth inside wall (102 mm i.d.) was selected for the subsurface drainlines. The smooth inside tube walls provide significantly less hydraulic roughness to conduit flow, and thus flow velocities will be greater for the same gradient. Consequently, the potential for pesticide-sediment entrapment problems that may occur in regular corrugated-wall tubing may be greatly reduced at the low flow volumes in the relatively short (61 m) and flat grade (0.2% slope) drainlines. As added protection against sedimentation of the draintube in the fine-textured, silt loam soil, a polyester woven fabric material (Bean Sock) (Trade and company names are included in this paper for the benefit of the reader and do not imply endorsement or preferential treatment of the listed product be USDA.) was selected as a synthetic envelope (often referred to as "drain filter sock") to surround the corrugated-smooth tubing.

Sump Outlet Structure. Because of the relatively flat, low-lying land at the experimental site and the large amounts of rainfall that often fill the surface drainage ditches in the area with runoff, sump outlets for the subsurface drains are necessary. A relatively large sump (1.2- by 1.2- by 3-m) was selected for each plot. All drainage effluent is pumped from the sump into a 300-mm diameter main buried close to each sump. Irrigation water is supplied from a pressurized water line into each sump. This sump is large enough to house outlet water level control, flow measuring, and flow sampling equipment, and provide room for a technician to enter the sump for servicing the instrumentation. The sumps were constructed with 6.35-mm thick sheet steel on four sides and 9.5-mm thick sheet steel on the 1.8- by 1.8-m base. The base was made larger than the cross sectional area of the sump to provide a flange onto which concrete could be poured (during construction) to hold the sump in place against the buoyancy force when the watertable is high and the sump is empty.

Watertable Control System

Each drainline will outlet into the sump separately, and the water levels at the draintube outlets will be controlled in 300-mm diameter plastic pipe risers mounted inside the sump (Figs. 6 and 7). The cross-sectional area of the 300-mm diameter risers for controlling the 15- x 61-m land area is proportional to that previously used in a sump-controlled system on a full field-scale experimental system (*34*). The water levels in the risers will be controlled by pumping out the risers with small electric pumps (for drainage or controlled-drainage), or adding water from a pressurized water supply line (for subirrigation). The drainage pump and irrigation water supply capacities are sized such that they also are proportional to

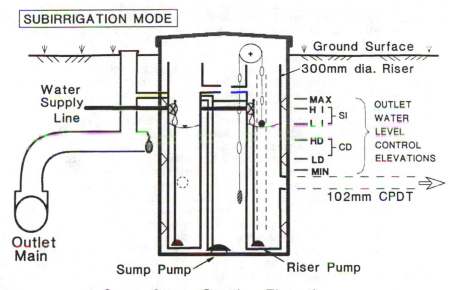

Figure 7.   Schematic drawing of the automated control for a single research plot (middle and buffer drainlines controlled by right and left risers, respectively).

those used for drainage and irrigation in the earlier field-scale experiment. The water level in the riser pipe for the experimental (center) drainline in each sump will be automatically controlled with an electronic data-logger/controller system which operates the small electric drainage pump or the electrical solenoid valve on the irrigation supply line. The water level control in the other riser pipe for the buffer drainlines will be "slaved" to the experimental drain riser pipe in the same sump.

All water pumped from the risers will be discharged into a 300-mm diameter plastic gravity flow drainage conduit (outlet main) buried nearby (Fig. 6). Each sump structure will also be equipped with an electric sump-pump to remove any effluent from the emergency overflow pipes on the risers, that collects in the bottom of the sump, and to discharge it into the outlet main.

The plot surface slope is parallel to the grade on the subsurface draintubes to provide "good" surface drainage. Crop rows will be planted parallel to the drainlines. Surface runoff will be collected in a shallow ditch along the sump end of each plot and routed through an H-flume flow measurement device (Fig. 6). The average elevations, at the centroid of each graded plot, are the means of the elevations of the crown of the crop row and the furrow between rows. These average elevations will serve as datum for measurement of watertable depths midway between drainlines in each plot area.

The operation of the experiment will be continuously monitored and controlled by four microprocessor-based, data-logger/controller systems (Campbell Scientific, Inc., Model No. CR7X). Each data-logger/controller unit will monitor and operate all the plots in one experimental replication (Fig. 4). The microprocessor in the data-logger will be programmed to record (every 5 seconds) the water level in the riser pipe for the experimental drainline in each sump and the watertable depth (midway between drainlines) in each field plot, and to independently control the drainage pumps and irrigation solenoid valves on the research site.

The water level in the outlet risers (OWL) for each experimental drainline can be measured in a perforated 102-mm diameter plastic stand-pipe "stilling well" mounted inside the 300-mm diameter riser; the water level will be measured (for example) with a potentiometer transducer coupled to a large- diameter wheel and float mechanism. The field watertable depth (WTD) will be continuously monitored with a linear-resistor type water level sensor (Metritape, Inc., Type LA) housed within a 50-mm diameter, perforated, plastic pipe, installed to a depth of 1.5 m. The recorded riser water levels and field watertable depth data will be summarized (averages and standard deviations) by the data-logger on an hourly and a daily basis. The field watertable depth will also be recorded (as a backup) with a standard FW-1 water stage recorder and float in a 150-mm diameter WT measurement well (35).

For the automatically controlled plot treatments, a cross-sectional schematic of the experimental drain outlet riser pipe and the riser water-level-evaluation sensor mechanism is shown for the subirrigation mode in Fig. 7; for the controlled-drainage mode of operation (not shown) the water level in the riser would typically be maintained at a lower elevation than for subirrigation. The watertable management system operation will be automatically switched as needed from the controlled-drainage to subirrigation, and vice versa, to maintain the OWL between the maximum (MAX) and minimum (MIN) water elevations specified. Feedback of the measured field WTD midway between drains is an optional control parameter (Fig. 3).

Control With and Without Feedback. with no feedback control, the water level in the riser (OWL) will be maintained between preset high (HD) and low (LD) elevations (about 10 cm apart) for controlled drainage and between preset high (HI) and low (LI) elevations (also about 10 cm apart) for subirrigation (Fig. 7). The preset high and low elevations will be stored in microprocessor memory. With feedback control, the measured field WTD (Fig. 3) will be used to automatically adjust the water level control threshold elevations up or down as a function of the deviation of the measured WTD from preselected standard elevations.

A more detailed description of the data-logger/controller instrumentation, microprocessor program, optional automated control modes, and remote telecommunications control features possible to override the on-site system controller when significant rainfall is predicted, etc., are presented by Fouss et al. (*36*).

Experimental Field Operation. For the conventional subsurface drainage treatment (No. 2) the drainage effluent collected in the outlet riser pipe will be pumped out so that the OWL always remains below the drainpipe outlet into the riser. For the automated watertable management treatments (nos. 3 and 4) the systems will typically be operated in a conventional subsurface drainage mode from November through March; in the controlled-drainage mode with the OWL maintained about 10 to 30 cm above the outlets during April and again from mid-September through October; and in the subirrigation mode from May through mid-September with the feedback option activated, except as overridden by the rainfall amount/time threshold, or by high watertable conditions which are governed by cumulative rainfall amounts. The system control may occasionally be switched from the subirrigation to controlled-drainage mode, via remote computer communications, in advance of predicted significant rainfall (*36*), or to adjust the watertable depth prior to application of fertilizer or pesticide.

Soil-Water and Temperature

Soil Water. A neutron scattering device will be used to measure soil water content. Three access tubes will be installed near the center of each plot at 1/2, 1/4, and 1/8 of the distance between drains (located with respect to the center drain in each plot). The neutron device will be calibrated from neutron readings and known volume gravimetric samples taken outside of the plot area. This will reduce disturbance of the test area. Readings will be taken at least weekly and more often during intense study periods where knowledge of water content will be important to assessing the status of applied chemicals and crop response.

Soil Matric Potential. Soil matric potential sensors of the current generating diode type (Agwatronics Model AGWA-II Sensors) will be located near the center of each plot at depths of 30, 60, and 90 cm and at distances from the drain of 1/2, 1/4, and 1/8 the drain spacing. Soil matric potential data will be used to characterize soil water gradients, i.e., direction of water movement, to correlate with pesticide movement in soil. The sensors will be controlled by the data-logger system and will be read hourly during normal operations. When the soil is near saturation these sensors can be read as often as every 5 minutes. The higher frequency of reading will be used to determine when an upward-moving wetting front approaches the sensor location.

Soil Temperature. Soil temperatures will be obtained with resistance thermometers (Campbell Scientific,Inc. Model No. 108B) at depths of 10, 20, and 30 cm in each plot. Resistance sensors will be read every 10 sec and an hourly average temperature will be recorded by the data-logger.

Subsurface Water Sampling. Two sets of porous-cup soil water samplers will be installed midway between drains in each plot. A set will contain 4 cups located at depths of 30, 60, 90, and 120 cm. These cups will be used to sample water for both unsaturated and saturated conditions. In addition 2.5 cm diameter piezometer wells will be installed to depths of 150, 180, and 210 cm for collecting water samples under saturated conditions.

Surface Runoff

A 0.46 m (1.5 ft) H-flume with a FW-1 chart recorder will be used to measure surface runoff rate and volume. A potentiometric transducer on the FW-1 float will be used to permit electronic recording of flow data by the data-logger. The runoff measuring system is designed to handle a 25-year frequency, 24-hour duration rain storm (254 mm).

Sampling Techniques and Frequency

Surface Runoff. An ISCO 2710 pump sampler, modified to hold 34 L, was selected to collect a discharge-weighted composite sample for each runoff event. The sampler is programmed to take a 130-ml sample for each 1.0 mm (2.135 m^3) of runoff. After collection the samples will be taken to the laboratory for analysis.

Drainline Effluent. Drainline effluent samples for the analysis of pesticide and nutrient content will be collected as water is pumped from the outlet water level control riser pipes for each experimental drainline. A composite effluent sample will be collected for each storm event; a minimum 6-hour period with no more than 2.5 mm of rainfall will be used to define the start of a new storm event. A turbine or paddlewheel type of flow meter that generates an electrical pulse, or switch closure, for incremental amounts of effluent pumped through it will be used to provide a direct input to the data-logger instrumentation for recording cumulative flow versus time. An orfice-type sampling device will be used on the discharge pipe of the riser pump to collect an effluent sample that represents about 0.5% of the total flow volume. The data-logger/controller system can be programmed to automatically advance the sample collection containers between storm events occurring on the same day, if desired. The data-logger/controller programmed functions can also permit the scientist to activate an alternative orifice for flow sampling (via commands communicated electronically from remote PC and modem equipment) so that the sample volume collected can be decreased in advance of predicted large storm events (e.g., 0.25% for a hurricane).

An effluent sampling subroutine will be incorporated into the Boussinesq-based model so that additional simulations can be conducted to evaluate the effluent sample sizes typically obtained for historical "dry" and "wet" weather periods at the experimental site. This information will permit defining a site-specific value for the percent of flow collected for the effluent sample. It is desired to collect at least one liter for small flow events and no more than 35 liters for large storms, with the samples collected daily, if needed.

Water from Unsaturated and Saturated Soil. Specific sampling frequencies will depend on rainfall and pesticide persistence. In general, water will be collected from the porous-cup soil water samplers and the piezometer wells weekly. Sampling may be more frequent following rains and for several weeks following pesticide application to the plots. Conversely, sampling may be less frequent during periods of little rain or several months after pesticide application.

Soil. The top 2.5-cm soil layer will be sampled by collecting all the soil within a 2.5-cm deep by 10-cm diameter metal ring pressed into the soil. Ten "ring" samples will be composited for each plot for each sampling. The surface soil will be sampled before pesticide application, 0 and 2 days, 1 and 2 weeks, and 1, 2, 4, 6, 9, and 12 months after application. Soil samples will be collected in 15-cm increments with a 7-cm diameter auger down to the watertable. Following sample collection, all augured holes will be back-filled with soil to prevent serious disruption of normal water flow paths in the soil profile. Five samples will be composited at each depth on each plot for each sampling. Samples will be taken before pesticide application and at 2 and 6 weeks and 3, 6, 9, and 12 months after application. All soils will be air-dried at ambient temperature in the laboratory, ground to pass a 2-mm sieve, and frozen until extraction and analysis.

Summary and Conclusions

The paper discusses a controlled-watertable experiment designed to (i) help characterize pesticide and nitrate leaching in soil and (ii) devise a management strategy for minimizing pesticide and nutrient leaching in soil. The description of a system of surface- and subsurface-drained, bordered plots equipped for microprocessor control of watertable depths follows brief reviews of potential groundwater contamination in the shallow-watertable soils of the lower Mississippi River valley and the concepts of watertable control. Sampling frequencies and techniques for surface runoff, subsurface water, and soil are also presented.

The described system has the potential for integrating methods for crop production efficiency and pollution reduction. The study should lead to a strategy for managing watertable depth to enhance plant fertilizer use efficiency, thereby reducing fertilizer needs and potential pollution by fertilizers. Further, there should be an opportunity to reduce pollution by pesticides through management of pesticide applications and watertable depths as dictated by prevailing and predicted weather conditions.

Literature Cited

1. Pye, V.I.; Patrick, R.; Quarles, J. "Groundwater Contamination in the United States;" Univ. Pennsylvania Press: Philadelphia, 1983; p. 38.
2. "Farm Drainage in the United States: History, Status and Prospects."   Pavelis, G.A., Ed.   U.S. Dept. Agric., Miscellaneous Pub. No. 1456. 1987.
3. Schwab, G.O.; Frevert, R.K.; Edminster, T.W.; Barnes, K.K. "Soil and Water Conservation Engineering." John Wiley and Sons, New York, NY. 1981.
4. Williams, W.M.; Holden, P.W.; Parsons, D.W.; Lorger M.N. "Pesticides in Groundwater Data Base: 1988 Interim Report." U.S. Environ. Protection Agency, Washington DC. 1988.
5. Calhoun, H.F. "1987 Survey of Louisiana" Groundwater for Pesticides." Louisiana Dept. Agric. Forest., Baton Rouge, LA. 1988.
6. Cavalier, T.C.; Lavey, T.L. Ark. Farm Res. 1987. 36, 11.
7. Acrement, G.J.; Dantin, L.J.; Garrison, C.R.; Stuart, C.G. "Water Resources Data for Louisiana, Water Year 1988".   U.S. Geologic Survey, Baton Rouge, LA. Rept. No. USGS/WRD/HD-89/262. 1989.
8. Whitfield, Jr., M.S. "Geohydrology and Water Quality of the Mississippi River Alluvial Aquifer, Northeastern Louisiana." Louisiana Dept. Public Works, Baton Rouge, LA.   Water Res. Tech. Rept. No. 10. 1975.
9. Morgan, C.O. "Groundwater Conditions in the Baton Rouge Area, 1954-59".   Louisiana Geological Survey, and Louisiana Dept. Public Works, Baton Rouge, LA. Water Res. Bull. No. 2. 1961.
10. Poole, J.L. "Groundwater Resources of East Carroll and West Carroll Parishes, Louisiana." Louisiana Dept. Public Works, Baton Rouge, LA. 1961.
11. Dial, D.C.; Kilburn C. "Groundwater Resources of the Gramercy Area, Louisiana." Louisiana Dept. Transpor. Develop., Water Res. Tech. Rept. No. 24. 1980.
12. Gilliam, J.W.; Skaggs, R.W.; Weed, S.B. J. Environ. Qual. 1979, 8, 137-142.
13. Gilliam, J.W.; Skaggs, R.W.; In "Development and Management Aspects of Irrigation and Drainage Systems." Amer. Soc. Civil Engrs. 1985, pp 352-362.
14. Deal, S.C.; Gilliam, J.W.; Skaggs, R.W.; Konyha, K.D. Agric. Ecosys. Environ. 1986, 18, 37-51.
15. Fouss, J.L.; Carter, C.E.; Rogers, J.S. Trans. Amer. Soc. Agric. Engrs. 1986, 29, 988-994.
16. Fouss, J.L.; Cooper, J.R. Trans. Amer. Soc. Agric. Engrs. 1988, 31, 161-167.
17. Fouss, J.L.; Skaggs, R.W.; Rogers, J.S. Trans. Amer. Soc. Agric. Engrs. 1987, 30, 1713-1719.
18. Schwab, G.O.; Thiel, T.J.; Taylor, G.S.; Fouss, J.L. "Tile and

Surface Drainage of Clay Soils. 1. Hydrologic Performance with Grass Cover." Res. Bull. 935. Ohio Agric. Exp. Sta. 1963.
19. "Soil Survey of East Baton Rouge Parish, Louisiana." U.S. Department of Agriculture, Soil Conservation Service. 1968.
20. Lund, Z.F.; Loftin, L.L. "Physical Characteristics of Some Representative Louisiana Soils". U.S. Dept. Agric., Agric. Res. Ser., ARS 41-333. 1960, 83p.
21. Rogers, J.S.; McDaniel, V.; Carter, C.E. Trans. Amer. Soc. Agric. Engrs. 1985, 28, 1141,-1144.
22. Jensen, M.E. "Consumptive Use of Water and Irrigation Water Requirements". Irrigation and Drainage Division of Amer. Soc. Civil Engrs. 345 East 47th St., New York, NY 10017. 1973.
23. Rogers, J.S.; Carter, C.E. Soil Sci. Soc. Amer. J. 1987, 51, 1393-1394.
24. "Methods of Soil Analysis." Part I, 2nd ed. Klute, A. ed. American Society of Agronomy, Madison, WI. 1986.
25. Rogers, J.S.; Selim, H.M.; Carter, C.E.; Fouss, J.L. ASAE Paper No. 90-2084. ASAE, St. Joseph, MI 49085. 1990.
26. Southwick, L.M.; Willis, G.H.; Bengtson, R.L.; Lormand, T.J. Bull. Environ. Contam. Toxicol. Accepted for Pub., Oct. 1989.
27. Bengtson, R.L.; Carter, C.E., Morris, H.F., Bartkiewicz, S.A. Trans. Amer. Soc. Agric. Engrs. 1988, 31, 729-733.
28. Skaggs, R.W. "Drainmod-Reference Report; Methods for Design and Evaluation of Drainage-Water Management Systems for Soils with High Water Tables". USDA-SCS, National Technical Center, Ft. Worth, TX. 1980. 182 p.
29. Fouss, J.L.; Bengtson, R.L.; Carter, C.E. Trans. Amer. Soc. Agric. Engrs. 1987, 30, 1679-1688.
30. Smith, M.C. " Subirrigation System Control for Water Use Efficiency." M.S. Thesis. North Carolina State Univ., Raleigh, NC. 1983, 182 p.
31. Fouss, J.L.; Rogers, J.S. Dynamic Simulation Model to Optimize Management/Control of Dual Purpose Subdrainage-Subirrigation Systems. ASCE Irrigation and Drainage Div. Speciality Conf., Planning Now for Irrigation and Drainage in the 21st Century. Lincoln, NE, July 18-21, 1988.
32. Skaggs, R.W. Third National Drainage Symposium Proceedings, Amer. Soc. Agric. Engrs. Publication 1-77, ASAE, St. Joseph, MI 49085. pp.61-68. 1977.
33. Fouss, J.L.; Rogers, J.S.; Carter, C.E. ASAE Paper No. 88-2105, ASAE, St. Joseph, MI 49085. 1988.
34. Fouss, J.L; Rogers, J.S.; Carter, C.E. Trans. Amer. Soc. Agric. Engrs. 1989, 32, 1303-1308.
35. Bengtson R.L.; Carter, C.E.; Morris, H.F.; Kowalczuk, J.G. Trans. Amer. Soc. Agric. Engrs. 1983, 26, 423-425.
36. Fouss, J.L.; Carter, C.E.; Rogers, J.S. Proc. Third International Workshop on Land Drainage, Workshop Group A, Drainage Models, Ohio State Univ., Columbus, OH, 1987. pp. A-55 to A-65.

RECEIVED September 5, 1990

GROUNDWATER SAMPLING

Chapter 12

Well Installation and Sampling Procedures for Monitoring Groundwater Beneath Agricultural Fields

S. Dwight Kirkland, Russell L. Jones, and Frank A. Norris

Rhone-Poulenc Ag Company, P.O. Box 12014, Research Triangle Park, NC 27709

The installation and sampling of monitoring wells are important components of most studies of agricultural chemicals in groundwater. For many agricultural chemicals, requirements for well materials and sampling techniques can be simplified compared to those often used in other types of groundwater monitoring programs. These simplified techniques allow for quicker reaction to events occurring in a study and installation of wells in areas inaccessible to drilling equipment, while reducing unnecessary expenses.

In recent years, new types of field research studies using different sample collection techniques have been developed for use in evaluating the potential impact of agricultural chemicals on groundwater quality. These techniques include the ability to determine the location and concentrations of agricultural chemicals in the saturated zone. Because groundwater studies are relatively new, protocols for their conduct are still evolving; however, these studies generally involve the sampling of shallow groundwater. This paper will describe some of the procedures that are being used to install and sample shallow groundwater monitoring wells and is intended to be an update of the information in a previous summary (1). This paper is not intended to be a discussion of the design of studies which use groundwater sampling; this is the topic of a companion paper (Jones and Norris in this work).

Well Design

Monitoring wells usually consist of a well screen which is attached to a casing. In order to obtain information on concentrations of agricultural chemicals as a function of depth, often several wells with different screen depths are installed nearby (such a grouping has often been referred to as a well cluster). Another approach to obtaining information on concentration as a function of depth has been to use multi-level samplers (2-4) or packers (5) (potential problems with packers are discussed in ref. 6)

0097–6156/91/0465–0214$06.00/0

in single bore holes. The scope of this paper will be restricted to single wells or clusters of wells.

The objectives of the field research and site characteristics are factors which determine the appropriate length of the well screen used in a monitoring well. For example, if the objective is to obtain a representative sample of an aquifer, then the well screen should be long enough to penetrate the entire aquifer. Usually the objective is to determine the concentration of an agricultural chemical as a function of both depth and spatial position. Under these circumstances, clusters of wells with relatively narrow screen depths are more appropriate. For example, the authors' typical design in studies where the water table is less than about 8 m uses clusters of three wells. Each well in the cluster has a screen 0.3 m in length, with the three screens in each cluster placed just below the water table and about 1.5 and 3.0 m below the water table. Generally in those areas with deeper water tables, screens at least 1.5 m in length are used, with often only one or two wells comprising a well cluster.

The diameter of a monitoring well is dependent on the objectives of the study, the nature of the site, and the procedures used to sample the well. If the objective for the monitoring well is to determine the concentration of an agricultural chemical at a specific point rather than obtain a representative sample of the aquifer, then the well diameter should be as small as practical to minimize the amount of water that must be evacuated prior to sampling. For wells located where the water table is less than about 8 m deep (allowing the use of sampling pumps located above the ground surface), a pipe diameter of 38 mm (1.5 inches) has proven to be satisfactory in studies performed by the authors. Nevertheless, even with the use of this pipe diameter, repeated sampling of wells has apparently resulted in drawing residues deeper into the saturated zone in some studies. Since the volume of water evacuated is usually proportional to the volume of water standing in the well, this problem is most severe for larger diameter wells or for wells located deeper below the water table in areas where lateral groundwater movement is quite slow and the same part of the residue plume is subjected to repeated sampling (the effect of diameter is more pronounced than length since the volume is proportional to the length but proportional to the square of the diameter). For areas with water tables deeper than about 8 m, pipe diameters of at least 50 mm traditionally have been used to allow the use of submersible pumps. Recently, the development of inertial pumps (7) appears to have made possible the sampling of monitoring wells with diameters as small as 19 mm down to water table depths of 40 to 50 m below ground surface.

In some studies, the tops of wells located within agricultural fields must be below ground surface at least part of the time to allow planting, application of agricultural chemicals, cultivation, and harvesting. This can be accommodated by a threaded fitting and cap at the top of the well at the appropriate depth. A cap should be used rather than a plug to help reduce any potential for soil being introduced into the well when the top of the well is being removed. For those times when the well does not have to be located below ground an extension pipe can be added using the threaded fitting. When the well top is located below ground, it is desirable to place metal on top of the well in addition to having relative accurate distance measurements so that the well can be located in the future with a combination of a measuring tape and metal detector. The presence of soil or any ponded water around the fittings can also be reduced by placing a protective cover such as a plastic bag or another piece of larger diameter pipe around the top of the well. Of course, the top of the well and all fittings located below ground should be adequately sealed to prevent entry of any ponded water into the well.

Well Materials

The materials used to construct monitoring wells depend on the properties of the agricultural chemical under study and the analytical techniques. Most agricultural chemicals for which field research is being conducted to address groundwater concerns are relatively non-volatile and somewhat water soluble. Therefore, adsorption to well screens, casings, and sampling equipment are usually not of concern. If significant adsorption of an agricultural chemical occurred during the short residence time in the well during sampling, then normally adsorption to soil would be high enough to prevent significant mobility. Usually the use of PVC pipe (including use of glued joints) is acceptable. This is in sharp contrast to the materials which may be required for other monitoring applications, especially those involving relatively volatile, chlorinated hydrocarbons. Nevertheless, suitability of all materials used in well construction and sampling should be verified by appropriate laboratory experiments. These experiments should show that no significant loss of agricultural chemical residues occurs or that no compounds which interfere with the chemical analysis are introduced.

Well Installation

A number of techniques can be used to install monitoring wells. These techniques can usually be divided into those requiring or those not requiring commercial drilling equipment. Regardless of the procedure used, introducing residues from one boring into another and clogging the well screen with sealing materials such as bentonite must be avoided. After wells are installed, a surveyor's transit is normally used to determine the relative elevations of the tops of the well casings. These measurements, plus the depth to the water table, as measured from the top of the well casings, can be used to determine groundwater flow gradients.

Monitoring wells are installed using commercial drilling equipment by drilling a bore hole down to the desired depth, inserting the well screen and casing, placing sand into the bore hole around the well screen, and then sealing the well by filling the remainder of the hole with layers of bentonite and grout. A variety of drilling methods may be used to drill and install monitoring wells. Advantages and disadvantages have been discussed in a recent review (8-9). Methods commonly used in field research studies with agricultural chemicals, where rarely is it necessary to drill through rock formations, include hollow stem augering and mud rotary drilling. Hollow stem augering is generally preferred since drilling mud is not introduced into the borehole. A discussion of monitoring well installation using hollow stem augering is presented elsewhere (*10-11*).

Advantages of the commercial drilling techniques include the ability to install wells to any depth in a variety of geological settings. Also the ability to install a sand pack around a well screen helps increase the water flow around the well. A disadvantage is that usually several hours are required to install each well since the equipment must be moved and cleaned between each hole. Bringing a drilling rig into an agricultural field may also result in compaction of surface soils, potentially reducing movement in the unsaturated zone. Also, drilling rigs may not be brought into certain locations (for example, into heavily wooded areas, marshy areas near streams, or the middle of the test plot during the growing season).

In some field research studies, monitoring wells have been installed manually without the use of commercial drilling equipment. One procedure (*12*) uses a bucket auger to bore a hole down to the water table. The well point and casing (often schedule 80 PVC) is then inserted into the hole and driven downward to the desired depth using a sledge hammer. Short sections can be added to the top of the casing in approximately 0.8 m lengths as needed. The well is sealed by filling the borehole with bentonite. This procedure has been used to install 38 mm (1.5 inches) diameter wells as deep as 6 m below the water table in sites with water tables less than about 7.5 m.

Another manual installation procedure is similar except that after the bore hole to the water table has been drilled, a temporary casing is placed in the well. Sediment is bailed from the bottom of the casing which drops downward under its own weight. After the casing has dropped to the proper depth, the well is placed inside the temporary casing. Then a filter pack is placed around the well screen and the well is sealed with bentonite during the process of removing the temporary casing. This procedure can be used to install 40 mm or smaller diameter wells as deep as 7 m below ground surface.

Advantages of the manual installation procedures include rapid and economical installation of shallow wells with a minimum of disruption to the surrounding environment and less disturbance of the geological strata near the well point. Also wells can be installed in areas that are not accessible to commercial drilling rigs. However, this technique cannot be used to install wells when rock layers must be penetrated. The first manual technique described also has the disadvantages of not being able to penetrate hardpan and clay layers as well as the inability to install a sand pack around the well (which may result in insufficient flow of water in areas where silts and clays are present in the saturated zone). In areas where the saturated zone is composed primarily of sands, usually the first manual installation procedure is preferred because it provides equivalent results with less effort.

In most research studies with agricultural chemicals, which are usually conducted in areas where the water table is shallow and subsoils are predominantly sand, wells can normally be installed either manually or by the use of drilling equipment. The authors' preference is to use the manual installation procedures whenever possible because of the decreased cost and minimal disruption to the surrounding environment. Because installing wells manually requires less advance planning, additional wells can be more easily added during the course of the study. Also, more options usually exist for the placement of these wells. For example, in the study described in ref. 12 commercial drilling rigs could not have been used to install wells in the wooded and swampy areas and installation of wells in the citrus grove using drilling rigs required more trimming of the trees than when wells were installed manually. Of course in areas with deeper water tables, when rock layers are present, or when larger diameter wells are necessary, monitoring wells must be installed using commercial drilling equipment.

Sampling Procedures

Samples are usually collected from monitoring wells after purging water standing in the well. The distance from the top of the well casing to the water in the well is usually measured for use in developing groundwater elevation contour maps and (depending on the sampling technique) calculating the quantity of water to be purged prior to sampling. Also during sample collection the temperature, pH, and conductivity of the groundwater is usually measured.

The choice of sampling technique and sampling equipment will depend on the properties of the agricultural chemical being studied. Techniques which may not be acceptable for general monitoring applications such as those involving landfills may be acceptable for many agricultural chemical studies. Most agricultural chemicals for which field research is being conducted to address groundwater concerns are usually relatively non-volatile and at least somewhat water soluble. Prior to the selection of a sampling technique or equipment, laboratory studies need to be conducted with all materials such as sample containers, tubing, and pumps or bailers that will come in contact with the samples to demonstrate that no significant loss of agricultural chemical residues occurs or that no compounds which will interfere with the analysis are introduced.

Two procedures are commonly used to determine the amount of water to be evacuated prior to sample collection. One is to purge until the pH and conductivity remain constant, while the other procedure specifies a certain number of well volumes (usually 3 to 6) to be evacuated (a recent study (13) indicates that only minimal purging may be necessary if the sample intake is placed near the bottom of the well screen). The first procedure is probably the most appropriate when a representative sample of an entire section of an aquifer is desired. However, in many studies with agricultural chemicals, the constant well volume approach is preferred because the concentration of residues at a specific depth and location is desired. In the latter case, excessive pumping may result in the sample containing water which was located at a different depth than the well screen prior to the onset of pumping. Excessive pumping may also draw agricultural chemicals deeper into the saturated zone.

A variety of sampling devices can be used to remove water from monitoring wells. Some of the most commonly used are bailers, pumps which are inserted into the well (such submersible or bladder pumps), and pumps which are located above the ground surface (such peristaltic pumps).

Sample collection using pumps inserted into the bottom of a well is probably the best approach when samples are being analyzed for volatile compounds although inertial pumps have also been shown to perform satisfactorily (*14*). Advantages of these approaches include the ability to collect samples regardless of the depth of the water table and keeping the pressure in the sampling equipment above atmospheric pressure. Because loss of volatile compounds during sampling is minimized by this approach, the use of submersible or bladder pumps are often preferred by regulatory agencies for all groundwater sampling. One disadvantage of submersible or bladder pumps is that usually the well casing must be at least 5.1 cm (2 inches) in diameter and often twice as large for many submersible pumps. Therefore, the use of submersible or bladder pumps is not compatible with the 3.8 cm (1.5 inches) diameter wells usually installed with the manual installation procedures previously described. Also due to the larger well diameter, more water must be evacuated during purging, increasing the possibility for drawing agricultural chemicals deeper into the saturated zone.

Since most agricultural chemicals for which field research is being conducted to address groundwater concerns are relatively non-volatile, other sampling procedures may be equally acceptable (or even preferred due to the ability to use smaller diameter wells). Pumps located above the ground are often used for collection of samples from wells where the water table is less than about 8 m. For 3.8 cm diameter wells, relatively high capacity (about 1 L/minute) peristaltic pumps have been widely used.

The choice of sampling device may also be influenced by the need to prevent contamination of samples in an environment where mg/kg concentrations may be present in dust and surface soils. One way to help prevent contamination under such conditions is to minimize the introduction of equipment into the well. However, access to the inside of the well must be maintained in order to make water table measurements. For sites where surface pumps are used, one option is to permanently place a relatively small diameter rigid sampling tube inside the well. The cost of foot valves used in inertial pumps is low enough to make a similar approach practical. The cost of submersible or bladder pumps may make their dedication to a single well infeasible, especially at sites with a large number of wells (such as the site in ref. *12* where 174 wells were located). In such situations, equipment placed inside a well must be carefully cleaned before insertion into another well.

Bailers are another device commonly used to remove water from monitoring wells. Advantages of this approach include that any size diameter well can be sampled as long as an appropriately sized bailer is constructed, the equipment is relatively cheap, and no power supply is needed. Disadvantages are that this procedure can be quite labor intensive when used to purge wells which are screened more than a few meters below the water table and that the continual insertion and removal of any device increased potential for contamination in an environment where residues of agricultural chemicals may be present in dust or surface soils.

One common use of a bailer is in conjunction with other sampling techniques. For example, a pump may be used to purge the well and then the actual sample is collected using a bailer to eliminate any concern about materials used in the construction of the pump. However, for a bailer to collect a valid sample, all of the water originally in the well must be removed during the purging process (the studies in ref. *13*, indicate that the water above the sample intake is stagnant and not removed during purging so that either the sample intake must be located at the top of the water standing in the well or the pump capacity must be sufficient to completely evacuate all of the water in the well). To minimize problems associated with purging or potentially introducing residues into the well, the authors' opinion is that whenever possible the sample should be collected at the end of the purging process using the same equipment.

Regardless of the type of equipment or the sampling procedure used, careful attention must always be given to cleanliness. Care should be taken not to introduce soil into the well during sampling. This may be especially difficult for wells in which the top of the casing is located below ground surface. All equipment should be carefully washed between each well and the wash water should not be discarded in the test plot. Sample containers should be triple-rinsed before sample collection. Hands should be kept as clean as possible. All sampling equipment and bottles should be kept off the ground and away from dust which might contain residues of agricultural chemicals. All sample containers (before and after sample collection) should not be transported in vehicles used to transport agricultural chemicals.

Conclusions

The selection of the techniques used for installation and sampling of monitoring wells in a field research study of an agricultural chemical should consider both the properties of the agricultural chemical and site characteristics. For studies with many agricultural chemicals, requirements for well materials and sampling techniques can be simplified compared to those often used in other types of groundwater monitoring programs. These simplified techniques allow for quicker reaction to events occurring in the study, installation of wells in areas inaccessible to drilling equipment, and reducing unnecessary expenses.

During the collection of groundwater samples, care should be taken to avoid contamination of samples. Therefore, samples should be collected by trained personnel using appropriate techniques, with cleanliness always being a primary concern.

Literature Cited

1. Jones, R. L. In *Environmental Fate of Pesticides*; Hutson, D. II.; Roberts, T. R, Eds.; Progress in Pesticides in Biochemistry and Toxicology, vol. 7; John Wiley and Sons: Chichester, U.K., 1990; pp 27-46.
2. Pickens, J. F.; Cherry, J. A.; Grisak, G. E.; Merritt, W. F.; Risto, B. A. *Ground Water* 1978, *16*, 322-327.
3. Harkin, J. M.; Jones, F. A.; Fathulla, R. N.; Dzantor, E. K.; Kroll, D. G. In *Evaluation of Pesticides in Ground Water*, Garner, W. Y.; Honeycutt, R. C.; Nigg, H. N., Eds; ACS Symposium Series 315, American Chemical Society: Washington, DC, 1986; pp 219-255.
4. Ronen, D.; Magaritz, M.; Levy, I. *Ground Water Monit. Rev.* 1987, *7*(4), 69-73.
5. Welch, S. J.; Lee, D. R. *Ground Water Monit. Rev.* 1987, *7*(3), 83-87.
6. Taylor, K.; Hess, J., Mazzella, A.; Hayworth, J. *Ground Water Monit. Rev.* 1990, *10*(1), 91-100.
7. Rannie, E. H.; Nadon, R. L. *Ground Water Monit. Rev.* 1988, *8*(4), 100-107.
8. Keely, J. F; Boateng, K. *Ground Water* 1987, *25*, 300-313.
9. Keely, J. F; Boateng, K. *Ground Water* 1987, *25*, 427-439.
10. Hackett, G. *Ground Water Monit. Rev.* 1987, *7*(4), 51-62.
11. Hackett, G. *Ground Water Monit. Rev.* 1988, *8*(1), 60-68.
12. Jones, R. L.; Hornsby, A. G.; Rao, P. S. C.; Anderson, M. P. *J. Contam. Hyrol.* 1987, *1*, 265-285.
13. Robin, M. J. L.; Gilham, R. W. *Ground Water Monit. Rev.* 1987, *7*(4), 85-93.
14. Barker, J. F.; Dickhout, R. *Ground Water Monit. Rev.* 1988, *8*(4), 112-120.

RECEIVED October 9, 1990

Chapter 13

Sampling Groundwater in a Northeastern U.S. Watershed

H. B. Pionke, J. B. Urban, W. J. Gburek, A. S. Rogowski, and R. R. Schnabel

Northeast Watershed Research Center, Agricultual Research Service, U.S. Department of Agriculture, 110 Research Building A, University Park, PA 16802

The sampling of groundwater, particularly for nitrates, is examined in a flow system and watershed context. A groundwater flow dominated watershed located in east-central Pennsylvania provides an example and basis for this analysis. Groundwater sampling is also viewed from a groundwater recharge (percolate) and discharge (streamflow) perspective. Some spatial and timing controls are described and examined in terms of where and when to sample.

Nitrate-N (NO_3-N) is the primary agricultural chemical of concern in northeastern U.S. watersheds. It approaches or exceeds the health advisory level (10 mg L^{-1} NO_3-N) in groundwaters draining agricultural areas far more frequently than does any other agricultural chemical (1-3). Also, the well documented decline of the Chesapeake Bay estuary has been blamed in part on excess nitrogen (N) input (4). The major N source is NO_3-N in stream inflows, largely originating from agriculturally impacted groundwaters (4). In this humid climate, where groundwater recharge rates are characteristically high and streamflow consists mostly of groundwater discharge, N transfer from farm field to aquifer to estuary is controlled by NO_3-N recharge and NO_3-N transport through the groundwater system. Thus, to sample these systems properly, we must identify the critical NO_3-N and flow contributing land use-soil combinations, estimate the controlling flow pathways and rates through the subsurface system, and properly place this groundwater system in the watershed-streamflow context.

The objectives of the paper are to: 1) identify the basic sampling issues for estimating NO_3-N flux from soil to groundwater to streamflow for an agricultural watershed located in east-central Pennsylvania, and 2) determine where, when, and how to sample this flux relative to the critical source areas, aquifers and recharge times that exist for this watershed. Where, when and how refers to strategies and approaches, not sampling mechanics. The text consists mainly of three sections following a groundwater recharge-transport-discharge sequence or specifically soil-vadose zone, groundwater system, and streamflow. NO_3-N flux provides the interface and ties the three sections together. Research done on the Mahantango Creek watershed provides the context, data and examples used throughout.

Basic Sampling Strategies. A sampling program designed in the context of the watershed system greatly increases the probability of a proper and representative sampling. Initially, specific goals and targets must be established. Without doing this, the when, where, and how of sampling cannot be decided. The decision to estimate the NO_3-N load annually exported from a watershed requires much different sampling positions and frequencies than does the decision to link either that load or NO_3-N concentrations in local recharge to a land use/management activity on the watershed.

Basically, our approach is to establish or hypothesize the NO_3-N sources and sinks which are then positioned within the flow system of the watershed. This requires a concept of the existing flow system, and the primary zones of NO_3-N generation and denitrification. When viewed this way, a relatively small part of the watershed can control the NO_3-N recharge or loss which greatly simplifies the sampling program. Incorporated into compatible simulation techniques, this concept of flow and position can provide good sampling insights and strategies.

Study Area Description. The Mahantango Creek Watershed is located in east-central Pennsylvania and drains into the Susquehanna River (Figure 1). It contains the two research watersheds (GK-27, WE-38) used to discuss the recharge, groundwater and streamflow systems.

Land use for the 7.4 km^2 WE-38 watershed is: 57% cropland, 35% forest, 8% permanent pasture, 0% urban and 0% industry. Deciduous mature forest dominates the ridges in the north with cropland dominating the valleys in the center and south. Major farming activities are livestock and cash cropping with most manuring and fertilizer application to corn in various corn-small grain-hay rotations.

Geology, topography, and soils of the WE-38 watershed are typical of the unglaciated, intensely folded, and faulted Valley and Ridge Province of the Appalachian Highlands in Pennsylvania. The Trimmers Rock Formation (late Devonian) crops out as shale with a nearly horizontal dip at the watershed outlet (5). To the north, dip increases to an estimated 22°. The Catskill Formation (late Devonian-early Mississippian) crops out as siltstone in mid-watershed, and as a relatively pure quartz sandstone-conglomerate at the northern watershed divide. The dip increases from 22° to 30° at the north watershed divide. There are no limestone outcrops. Bedrock is overlain by a 2-10 m thick blanket of periglacial talus on mountain slopes. Beneath the soil, a 3-15 m layer of weathered, highly fractured rock exists throughout the watershed (6). During winter through spring, water tables exist within this layer. Subsurface water flows south with most flow resurfacing to discharge at the watershed outlet. Most soils are residual, from 1-2 m deep,

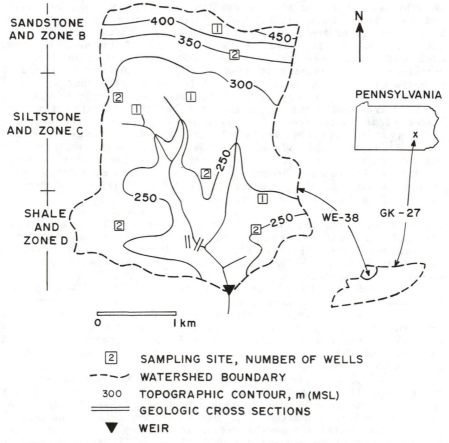

Figure 1.   Location of the Mahantango Creek Watersheds (GK–27 and WE–38).

and generally grade from being shallow and coarse-textured on ridge tops to deeper and finer-textured in the valley. Because soils on the ridge tops are so highly permeable, nearly all rainfall infiltrates. Soils adjacent to stream channels characteristically have high water tables or fragipans that restrict internal drainage and often function as groundwater discharge zones during the winter-spring period. The 100 km^2 GK-27 subwatershed is similar to WE-38.

The climate is temperate and humid with the average precipitation (P) from 1983-7 being 1,156 mm y^{-1}. Streamflow (Q) averaged 544 mm y^{-1} at the WE-38 weir, of which 76-86% was subsurface return flow. Evapotranspiration (ET) loss accounts for the remaining 612 mm y^{-1}. Most groundwater recharge occurs during the late fall, winter, and spring months from rainfall.

Sampling NO$_3$-N Flux from Soils

In most agricultural watersheds, the primary source of groundwater NO$_3$-N is the soil layer. The NO$_3$-N flux potential combines the computed, long-term percolate quantity with the potential NO$_3$-N content of the soil. Thus, soils with low water holding capacities and large excesses of applied N would have the greatest N leaching potential. Although soils can operate as N sinks temporarily by NO$_3$-N transformation to organic matter or permanently through denitrification, the N source from manuring, fertilization and legumes is considerably larger than the sink term for most intensively farmed agricultural soils. Percolate draining from the soil is used here as a surrogate for groundwater recharge because travel times through soil are generally much less and sampling access to soil is much easier. This assumes that storage changes in N and water below the soil are minimal and most flux reaches perched or regional groundwater. Our focus is on the soil layer instead of the root zone because soil maps with supporting data are published and properties of the soil layer are basically stable, whereas rooting depth is a time-space variable within the soil-land use complex.

Spatial and Temporal Delineation of NO$_3$-N Contributing Soils. At the watershed scale, spatial and temporal distributions of percolate and its NO$_3$-N potential vary. This variability depends on land use, hydrologic, chemical and physical properties of soil and overburden, climatic conditions, and season.

Spatial Consideration. The two most important considerations are land use and soil type. On cultivated watersheds, soil N and NO$_3$-N contents depend mostly on the type and intensity of land use. Land use is directly and easily detectable for purposes of sampling by remote sensing. However, the intensity of land use varies by operator and crop. Usually, cornland will generate much more NO$_3$-N excess than will small grain, forest stands or unfertilized pasture. Thus, N loss potential for corn will be higher, but also can vary greatly, depending on the N mass balance for a particular field. High NO$_3$-N concentrations remaining in soil at harvest translate into high loss potentials during fall and winter. The excess N in soils can result from overfertilization or improper timing of N applied as fertilizer, manure or legume plow-down. Pionke and Urban (1) found a 22 kg ha^{-1} yr^{-1} excess for cropland in the WE-38 watershed.

The potential percolation loss also varies over the watershed and is related to the texture and thickness of the soil, and differential ET losses, particularly between forest, grass and croplands. Macropore flow could affect both water and NO_3-N loss, but data is not yet available.

Temporal Considerations. Soils, percolate, and ground-water recharge occur mostly during winter through spring in the northeastern U.S. (Figure 2), provided the ground is not frozen. Exceptions sometimes occur in summer during major storm periods. The surplus (S) becomes percolate where surface runoff is small. This figure can be constructed where available-water capacity of soil, monthly mean temperature, precipitation and PET (8) are known or can be calculated.

Fertilizer-N applications and NO_3-N contents in soil also vary seasonally. Here, manures and most N fertilizers are applied in spring to cornland, with legume plow-down being done then or in the preceding fall. Thus, the NO_3-N concentrations in soil are often highest over summer when leaching is minimal, but can also be high in late fall when groundwater recharge normally starts. Atypical situations such as major droughts or floods can alter the seasonal pattern.

Estimating Spatial and Temporal Distribution of NO_3-N Flux. If sufficient data are available or can be generated, there are useful simulation models available such as CREAMS (9) and EPIC (10). However, some very simple approaches can provide sampling insights.

Leaching Index (LI). Williams and Kissel (Chapter 4, Managing Nitrogen for Groundwater Quality and Farm Profitability, in press) developed a simple NO_3-N leaching index based on the percolation potential for a given site. The LI combines a Percolation Index (PI) which incorporates gross soil hydrologic properties and annual precipitation, and a Seasonal Index (SI) which emphasizes precipitation received from fall to spring. Computation of PI is similar to the SCS runoff curve number equation (11). Soils are hydrologically classified as groups A (high infiltration rate, little runoff, high percolation potential), D (low infiltration rate, low percolation potential) or B and C (intermediate). The LI was computed for soils located in the GK-27 watershed to identify those areas with the highest and lowest percolation potential. Given this spatial distribution, and by knowing the N-excess balance or land use, the critical NO_3-N contributing areas can be located for further sampling. The soil series distributions needed to apply the LI concept are usually available from county soil surveys while precipitation can be estimated from available records. The LI approach has the same limitations as soil surveys and does not incorporate variations of ET due to major vegetation differences (12).

Travel Time. When and where to sample the soil-geologic column can be a major issue if travel times are unknown.

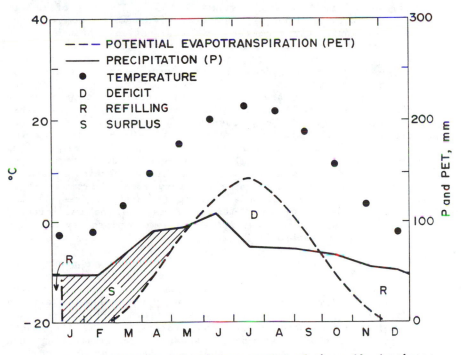

Figure 2. Climatic and moisture regime of the soils in the humid U.S. adapted from (7).

Assumptions of unit hydraulic gradient and constant hydraulic conductivity utilized by Davidson et al. (13, Rogowski, A.S., J. of Agric. Management, in press) to compute residence time and percolate flux may also hold between the root zone and water table. If this distance (L), and the total porosity (P_0) are known, and the percolate flux (q) is set equal to the groundwater recharge flux, the travel time to the water table becomes L times P_0 divided by q. By adding in the residence times for the root zone, a time distribution of groundwater recharge flux can be generated at the water table. This can provide insights on when the effect of management changes at the land surface can be sampled at the water table.

Geographic Information Systems (GIS). The use of models requires input data which may be obtained directly from GIS and soil maps or indirectly from land use designations through remote sensing. Formed from these data, composite maps of land use and important soil properties can provide better information on where to sample for NO_3-N than does the LI approach provided that the proper data bases exist.

The parameters that are important in estimating the spatially distributed flow in soils can be evaluated on a field or watershed scale, or abstracted from county soil surveys. Hydraulic conductivity, soil water, and percolated NO_3-N concentration determined at a number of locations throughout GK-27 were kriged and spatial distributions of recharge and NO_3-N loads interpolated. Such interpolated values differ at times from projections based on soil surveys. Because soil properties vary greatly from point to point, not all sites within a soil mapping unit will respond the same way. While accurate, soil maps may not be sufficiently accurate for making recharge estimates and some exploratory measurements are usually advisable (12).

Sampling Implications. To sample NO_3-N recharge to groundwater under agriculture, it is necessary to have some prior knowledge regarding NO_3-N contributing zones based on the properties of the overlying soil, climate, land use and management. Climatic effects determine when recharge is likely to occur. Leaching Index approach distributes potential recharge spatially, while the travel time computations provide timing of recharge, thus indicating when to sample. GIS either when combined with the proper data collection program or when coupled with more sophisticated modeling methods, can delineate sensitive or critical zones where the potential for contamination of groundwater with NO_3-N may be severe.

Sampling NO_3-N Flux in the Groundwater System

Sampling groundwater requires that the groundwater body be viewed as a continuum of paths of groundwater flow vectors--a flow system. The soil sources of N are visualized as contributing to various parts of the flow system, thus linking these percolates to the recharged groundwater and its internal flow paths. To sample within the flow system, we must consider 1) the size and shape of the region of groundwater flow, 2) boundary conditions around that

region with recharge as input and discharge as output, 3) the
spatial distribution of parameters that control flow and transport,
and 4) initial conditions within the region (for time-variant
problems). Much of the information and data base required can be
extracted from geologic-topographic maps and field well tests.

The focus of the section is to describe the controls on areal
groundwater flow and its domain using the WE-38 watershed as an
example (Figure 1). The aquifer behaves as a two-layer, fractured
aquifer system, the shallow layer (<15 m) being much more weathered
than the deeper one (15-100 m). These fracture and flow patterns
have been verified by extensive well drilling, well-based testing,
and seismic and chemical analyses.

Aquifer Hydrology. To define the flow system, the overall
aquifer boundaries, geologic properties and the fracture layer
system are examined.

Defining the Boundaries for the Regional Aquifer. Aquifer
boundaries can be impermeable geologic formations, but more often
are defined by hydrologic criteria such as a groundwater ridge or
divide. The groundwater divides, defined by a groundwater table
map, coincide with major land surface divides in WE-38 thereby
establishing the size and shape of the region of groundwater flow.
The most difficult boundary to determine is often the aquifer
bottom. In the absence of some prominent physical control, the
effective depth may be determined by that well depth where the
water yield determined by a pumping test becomes negligible. In
our watershed, this was at about 80-100 m, where rock fractures are
effectively closed.

**Evaluating the Hydrology of the Geologic Formations Within
the Regional Aquifer.** Aquifer properties are typically determined locally using long-term pumping tests. However, areal
characterization of an aquifer by pumping tests is rarely practical.
Instead, the change in drawdown is determined at available production wells pumped at constant rates over a few hours. The pumping
rate to drawdown ratio is the specific capacity (SC) of the well
and may be related graphically (14) to aquifer transmissivity. The
SC frequency distribution for GK-27 was determined and then interpolated to a smaller study area such as WE-38 in order to establish
the basic hydrogeologic setting. The SC distribution was correlated with geologic mapping units, and then groundwater ridges were
located by superimposing a water table map upon the geologic and
topographic maps. These maps and SC distributions together provided a means to formulate a groundwater flow system based on the
estimated hydrologic boundary conditions and aquifer properties.
Because wells in sandstone and siltstone yielded very little more
water than wells in shale on this watershed, the conclusion was
rock fracturing controlled flow more than did rock type. Thus,
fracturing depth defines the lower boundary of this aquifer system.

Importance of Fracture Layers. Fracture layers can greatly alter the path and timing of recharge. Where they form multilayer aquifers, multidepth sampling schemes are required.

Within WE-38 and the Appalachian region (15-16), a shallow, weathered,fracture layer exists that transmits large quantities of subsurface flow. The question becomes, "What role does this highly permeable layer play in delivering NO_3-N to the stream or deeper less fractured aquifers, and how should this groundwater be sampled?" The shallow fractured layer controls whether percolate becomes deeper groundwater or is routed laterally and quickly within the layer to become streamflow. Near the valley bottoms, the shallow fracture layer acts as a drain for the remergent deeper groundwaters, thus enhancing the discharge of all groundwaters near the stream. Areally, the flow pattern varies. Flow lines diverge in the areas dominated by recharge such as ridges and converge in areas dominated by discharge such as the stream channel. Thus, sampling strategies must be based on a knowledge of these flow patterns.

In WE-38, the extent of the fracture layer was determined by coring, and by seismic refraction which reasonably estimated the fracturing depth observed in cores. Two cross-sections (Figure 1), were cored to characterize the physical and hydrologic controls imposed by the shallow fractured layer near the stream (Gburek, W.J., Urban, J.B., Ground Water, in press). Bedrock fracturing was deepest directly under the channel and most concentrated in the upper 5-10 m (Figure 3). Based on slug tests, the shallow fractured layer was about three times more permeable than the under-laying aquifer. The measured rock matrix and total porosity is extremely low in both layers. Overall, flow within the shallow weathered fracture layer is essentially governed by fracture geometry, with little potential for interaction with the matrix.

Land Use and Geologic Interaction on Groundwater Quality. A groundwater quality sampling scheme was developed for WE-38 in the context of the now-defined flow system and land use distributions. Figure 4 depicts a cross-section of the flow system produced by using the hydrogeologic approach. Then, chemical data were collected from wells positioned to sample the deeper aquifer layer and used to test this flow framework.

Flow is from left to right (Figure 4). Zone B is primarily a recharge zone dominated by forestland, sandstone, high recharge rates and vertical downward flows. Zone C, located in siltstone, combines groundwater recharge mostly from cropland with throughflow from zone B. This groundwater recharge dominates the shallower not deeper groundwater, which flows horizontally from zone B. Zone D located predominantly in shale, is primarily a mixing and discharge zone, where horizontal flows from zone C converge and resurface at the stream. Recharge from cropland to the shallow layer occurs, but is less important due to the strong upward groundwater flow component. The combination of geologic structure (decrease in dip from 30-22° downgradient), stratigraphy (presence of low permeability shale, particularly at the lower X' boundary) and the geometry of the flow system favors groundwater discharge to the stream in zone D.

The effects of both geochemical weathering and the overlying agricultural practices (additions of N and Cl) were sampled over a 10-year period (1) and evaluated by geologic zone (Figure 5). Zone-B data (recharge-forest) are widely dispersed because these

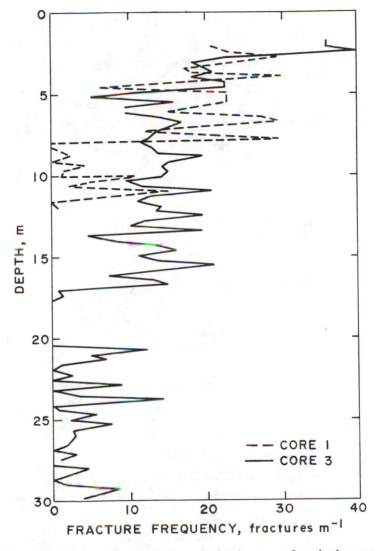

Figure 3. Fracture frequency from bedrock cores of geologic cross-sections. (Reprinted with permission from ref. 17. Copyright 1988 American Society for Testing and Materials.)

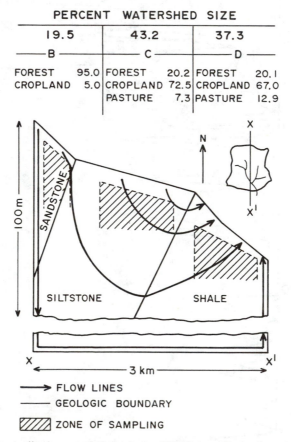

Figure 4. Conceptualized cross-section of the WE-38 watershed adapted from ref. 1. (Copyright 1985 National Water Well Association.)

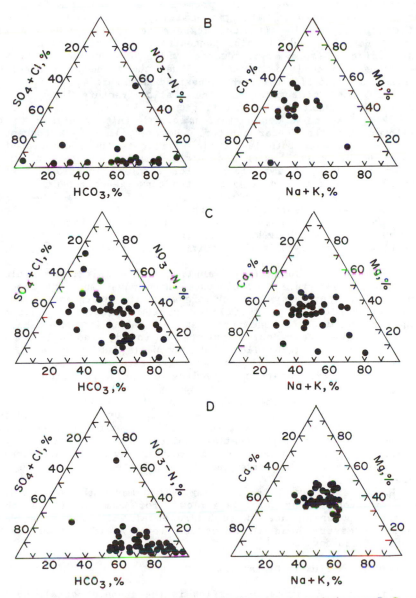

Figure 5. Cationic and anionic composition of groundwater for geologic zones B, C, and D. (Reproduced from ref. 1. Copyright 1985 National Water Well Association.)

waters represent recharge from nonfertilized soils and groundwaters in sandstone. Zone C (lateral flow-cropland) displays a cation shift to calcium-sodium waters. The NO_3 and Cl anions increased greatly from zone B to C, being primarily recharged from fertilized and manured fields. The geochemistry stabilized in zone D (discharge-cropland) with convergence to a $Ca-Na-HCO_3$ ion type water. Here NO_3 and Cl anion concentrations decreased due to the resurfacing of zone B groundwater, while Na ion associated with weathering continued to increase. Thus, the groundwater chemistry substantiates the flow system hypothesized in Figure 4.

Wells and piezometers in the shallow-fracture layer exhibit some of the highest NO_3-N concentrations in WE-38, ranging up to 30 mg L^{-1}. Nitrate-N concentrations are definitely higher in the shallower than the deeper aquifer. However, the NO_3-N concentrations averaged over both layers are similar to those calculated by Pionke and Urban (18) using hypothesized recharge rates and the N excess computed by mass balance for cornland rotations. Also, groundwater about to discharge from the convergence zone (D) was at about the same NO_3-N concentration (7-8 mg L^{-1}) as a 5-year flow weighted mean concentration for the stream (6.5 mg L^{-1}). Thus, dilution of higher NO_3-N concentrations in discharging shallow groundwaters by discharging deeper groundwaters explains the NO_3-N concentrations observed in the resulting streamflow.

Implications for Subsurface Sampling. The multilayer geologic controls on subsurface flow discussed here have important sampling implications for many bedrock areas in the northeastern United States. Basically, a multilayer system requires a multilayer sampling scheme. Groundwater samples taken from the shallow-fracture layer represent short term, rapid, and often local recharge and/or lateral flow. Deeper wells, which case off the shallow-fracture zone, will sample longer term recharge, and local or regional groundwaters depending on well position in the watershed.

Away from the stream channels, the shallow-fracture layer is usually drained and can only be sampled during winter through spring. The deeper and near stream shallow aquifers can be sampled routinely, but usually the shallow aquifer sample is most likely to represent specific source areas and recharge events. In contrast, long-term trends in recharge quality are better sampled using the deeper aquifer.

Within each layer, the sampling should match the scale of the phenomenon controlling subsurface flow. For example, the length of piezometer opening needed to representatively sample NO_3-N or measure pressure head in fractured rock would depend on the fracture frequency.

Sampling NO_3-N Flux in Streamflow

Assuming a closed system, streamflow is the endpoint of all surface and subsurface flowlines within the watershed and is thereby an integrator of all flow and water-quality processes. Since streamflow in the humid-climate east is mostly from subsurface sources, it integrates infiltration and subsurface flow processes. This section examines the potential relationships between streamflow and the groundwater flow system, and their sampling implications.

Streamflow-Groundwater Relationships. To characterize watershed processes related to subsurface flow via stream sampling, we must know the areal and temporal recharge sources, the characteristics of the subsurface zone (aquifers) contributing streamflow, and the amount of subsurface flow contributing to streamflow. By using baseflow separation techniques, master baseflow curves, and simple mass balances, these sources and characteristics may be inferred.

Baseflow Separation. Baseflow separation techniques (e.g., 14, 19) are routinely used to separate stormflow into its subsurface and surface components. Applied to single storm hydrographs, some generalized characteristics of the contributing aquifer can be determined (20). Long-term records of streamflow can be used to estimate subsurface return flow, and to infer net recharge to groundwater. Layered geology, permeability changes with depth, ET, and varying size of the aquifer discharge zone can complicate but do not preclude the use of these techniques.

The baseflow component of streamflow for WE-38 was determined according to Walton (14). This showed groundwater discharge to dominate the streamflow regime, accounting for 76-86% of annual flow (Table I). High baseflows, following winter and spring storms decreased rapidly to very low flow rates until the next storm. This pattern following major storms indicates a transmissive aquifer with low sustainable capacity.

Table I. Yearly Water Balance for WE-38; Precipitation (P), Streamflow (Q), Baseflow (QB) and Evapotranspiration (ET)

Year	P	Q	QB	ET	QB/Q	QB/P
	---------- mm -----------				------ % -----	
1983	1328	693	540	635	77	41
1984	1140	605	491	535	81	43
1985	1036	338	287	698	84	28
1986[a]	1323	650	494	673	76	37
1987	953	432	373	521	86	39

[a]No streamflow record for Oct. 9 - Nov. 10, 1986.

Master Base-Stream Flow Recession Curve. A general descriptor of stream baseflow is the master recession curve (21). It assumes baseflow recession to be a property of the watershed as a whole, independent of storm characteristics, timing, duration, or antecedent conditions. It provides a continuous plot of the probable recession time for the watershed subsurface reservoir to be depleted. A master base-stream flow recession curve is constructed by fitting a linear regression model to storm recessions plotted

against stream discharge over time. The theoretical groundwater
storage depletion assumes a straight line decrease which is altered
by such factors as ET and multiple aquifer inputs. For the WE-38
watershed, the linear regression equation gave a good fit (r^2 =
0.92) over 14 day recessions plotted for 31 storms. However, the
regression slopes were greater for the higher compared to lower
baseflows suggesting a two-layered groundwater system.

The slope (α) of the recession curve can describe important
aquifer characteristics through (21):

$$\alpha = T(a^2 S)^{-1} = 0.933$$

where T (transmissivity) and S (storage coefficient) describe the
groundwater reservoir as a whole and a is the average distance from
stream to hydrologic divide. Assuming higher baseflows represent
drainage from the shallow fracture layer and the sustained lower
baseflows drain the regional aquifer, the ratio of slopes of these
two segments represents the ratio of transmissivities under the
same storage coefficient. Using the master base-stream flow recession curve developed for WE-38, the time for discharge to decline
by one log cycle was 4.8 and 20.6 d, leading to a T:S ratio of the
shallow to deep layer of approximately 4:1 which corresponds to
that reported by Gburek and Urban (Ground Water, in press). Thus,
this approach provides a means for subdividing the groundwater
system and sampling its discharge from the stream.

Mass Balances. Water and N balances at the watershed
scale provide a framework for examining the relationships between
land use, groundwater quality, and stream quality. The mass
balance approach is most usable where the watershed refills to
about the same hydrologic state each year, i.e., where Δ storage
approaches zero, causing the balance to become Q = P - ET as shown
for WE-38 in Table I. If this water balance exists and groundwater
can be sampled from streamflow, a N mass balance can be used to
estimate the long-term NO_3-N loss potential to groundwater. This
estimate could be verified from stream sampling.

The major flow pathways for NO_3-N in WE-38 are through the
subsurface flow system, with its sources being mainly fertilizer or
manure from intensively cropped lands. The NO_3-N loss in percolate
converts to NO_3-N loss in streamflow, suggesting that the rest of
the subsurface system acts mainly as a transmission and dilution
zone. Consequently, the primary control on NO_3-N losses to streamflow is the N quantity applied relative to that consumed in the
soil (1). However, NO_3-N concentrations in percolate will likely
be diluted upon entering the groundwater and stream (22). Streamflow sampling data, therefore, represents the integration of these
processes which will affect concentration, but not load unless
major NO_3-N sinks exist enroute.

The utility of mass balance approaches and sampling strategies
very much depends on the magnitude and position of N sinks in the
groundwater system (Pionke, H.B., Lowrance, R.R., Chapter 11,
Managing Nitrogen for Groundwater Quality and Quantity, in press).
Although 50% of the NO_3-N in the discharging riparian groundwater
of WE-38 could be denitrified under optimal conditions (23), the
total NO_3-N exported in streamflow was reduced by only 4% because

the conditions supporting denitrification are poorest when most NO_3-N is being exported and the riparian zone occupies a relatively small area within this watershed (24). Similarly, denitrification appeared to cause the much lower NO_3-N concentration observed in the deeper well waters within the shales of zone D (25), but these deeper shales also yield relatively little groundwater discharge. Thus, both sinks had no impact overall on NO_3-N exported from the watershed and could be ignored.

Areal and Time Controls on Streamflow Quality. A single stream-flow sample represents all contributing areas and times of travel upgradient from the sampling point. Thus, the flowlines contributing to that sampling point may represent different seasons, years, or even past land uses.

Flow rates during sampling are also critical for interpreting the NO_3-N output. Excluding storm periods, high flows generally represent drainage from all layers, i.e., the soil, shallow fracture layer, and the deeper regional aquifer. As flows decrease, streamflow originates deeper within the profile. If the system is strongly layered, drainage from individual layers may be identified by baseflow separation techniques. The NO_3-N concentrations observed in streamflow are a function of the concentration and volume contributed to each layer including the soil layer.

The main control on NO_3-N concentrations in baseflow relative to those of groundwater appears to be hydrologic rather than biological or chemical. By sampling baseflows from WE-38, Schnabel et al. (26) observed repeated cycles of NO_3-N concentrations, being highest following increases in flow and then decreasing during baseflow recession (Figure 6). A number of these cycles lasted 30-40 days during the growing season where baseflow receded to very low levels and the NO_3-N concentration approached a constant (3.5-4.5 mg L^{-1}). Because NO_3-N concentrations are greater in the shallow fracture layer (Gburek, W.J., Urban, J.B., Ground Water, in press), increased flow from this layer results in a more NO_3-contaminated baseflow. Subsequently, as groundwater levels drop, the NO_3-N concentration in baseflow decreases because the less contaminated regional aquifer dominates.

A complication on exploiting the linkage between baseflow and the soil zone for sampling purposes is that the timing of NO_3-N from soil to the aquifer depends on antecedent conditions. The correlation of flow and concentration patterns is generally weakest for winter through early spring when the soil profile and shallow aquifer is largely flushed of NO_3-N. After spring fertilization however, the relationship between flow and NO_3-N concentration is re-established. Finally, since recharge to the shallow aquifer occurs only after the soil water storage requirement is met, not all storms will cause the shallow aquifer to discharge.

Sampling Implications. Stream sampling represents the integrated results of all flow and N inputs and transformations occurring within the watershed. The inputs of all land uses and their variations can be best represented by a long-term continuing sampling program. However, if substantial watershed outflow does not resurface at the streamflow sampling point, then this underflow must be sampled as well. Also, to use a stream-based sampling approach,

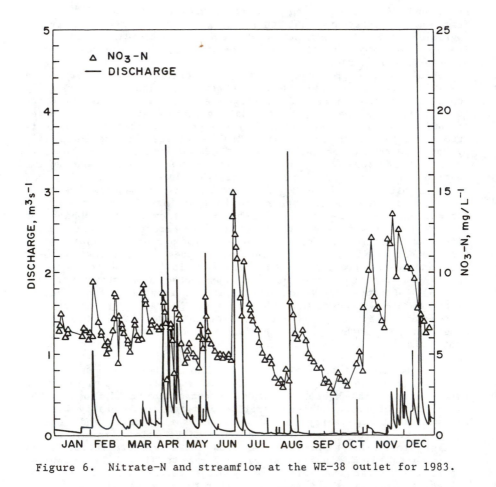

Figure 6. Nitrate-N and streamflow at the WE-38 outlet for 1983.

the water or N sinks within the watershed must be minor or the
sinks likely will need to be subsampled.

Interpretation of the NO_3-N patterns observed is made difficult by the variety of travel times represented, seasonal
differences, drainage from different layers depending on flow
rates, and the variety of flow lines contributing to a single
sample. However, simple techniques such as described can help
interpret areal and temporal sources of streamflow and thus aid in
developing of a flow-based sampling strategy.

Conclusions

The representability and interpretability of groundwater data
depends on proper sampling. Traditionally, the sampling issues
have been mechanical--well purging, sample storage and handling,
and analysis. Although sampling mechanics are very important, some
of the greatest errors on interpreting these data depend more on
the position and timing of sampling within groundwater systems.
Sampling strategies must first serve sampling objectives, but then
need to be developed in the context of the watershed and flow
system. The groundwater flow system not only includes the
aquifers, but groundwater recharge which begins as soil percolate,
and groundwater discharge that becomes streamflow. In humid
climates, where groundwater discharge to the stream occurs at small
spatial scales, a well thought out streamflow sampling strategy may
in effect sample groundwater.

In agricultural watersheds, soils and land use/management control the spatial variability of NO_3-N flux from soils, and thus,
recharge to groundwater. Soils exhibiting a large leaching index
because they are coarse-textured, thin or dominated by macropores,
and a high NO_3-N concentration because of cropping and fertilization practices create the most critical recharge areas. Their
extent and distribution dictate the sampling pattern needed.
Sampling percolate from the root zone as a surrogate for recharge
to groundwater assumes both are correlated which may not be true.
However, where travel times to groundwater are long, studies designed to be short term must sample percolate rather than groundwater.
To help guide sampling strategies, there are models, methods and
data bases that provide estimates of potential NO_3-N flux and
residence time in soil.

The aquifer system cannot be sampled properly unless the boundaries and basic structure controlling groundwater flow, as well as
residence times are known. In the WE-38 watershed, the side boundaries coincide with ridge tops, and the bottom boundary with the
closure of fractures at a depth of about 100 m. The flow system
can be viewed as two layered, in which the shallow layer (3-15 m)
transports most groundwater and the most NO_3-contaminated
groundwater. A sampling strategy that ignores the shallow layer
during the winter-spring period when flow rates and NO_3-N concentrations are highest would miss the bulk of the NO_3-N lost from farm
fields. Because this shallow layer rapidly transmits the most highly NO_3-contaminated groundwater to the stream, the deeper groundwater which is the source of most well water is protected or less
contaminated with NO_3-N. Thus, a sampling strategy based on
assuming a single homogeneous aquifer system or on sampling

production wells alone would provide misleading results on NO_3-N losses to groundwater and misleading interpretations regarding the impact of these NO_3-N losses on well water supplies. Where NO_3-N sources or sinks are important in the groundwater system, it may be necessary to isolate these critical volumes by sampling appropriately. In the WE-38 groundwaters, denitrification effects could be ignored, except when sampling the deeper shales.

Because groundwater from both the shallow and deeper layer discharges to the stream, stream sampling can provide much information on groundwater. This means sampling baseflows at landscape positions where most of the groundwater has resurfaced to become stream baseflow, and where groundwater systems behave conservatively, e.g., outflow NO_3-N not significantly altered by NO_3-N sinks and sources enroute. When these conditions are met, the effects of major features such as multilayer aquifer systems are determinable and to some degree separable using the stream baseflow analysis and sampling techniques described. Also, the NO_3-N concentrations and loads exported from the watershed at different baseflow rates provides measures of the NO_3-N content and contributions from the groundwater system overall and by layer or position which implies groundwater sampling strategies as well. Where the travel time between soil percolate and streamflow is rapid relative to rate of change in major land use or management, stream sampling can provide useful insights into the effects of management or land use on NO_3-N export from the watershed.

Literature Cited

1. Pionke, H.B., Urban, J.B., Ground Water 1985, 23, 68–80.
2. Madison, R.J., Brunett, J.O., In National Water Summary, 1984, U.S. Geological Survey Water Supply Paper 2275, 1984, 93–105.
3. Pionke, H.B., Glotfelty, D.E., Water Res. 1989, 23, 1031–1037.
4. Environmental Protection Agency, Chesapeake Bay Program Technical Studies: A Synthesis, Environmental Protection Agency, Washington, DC 20460, 1982, pp 634.
5. Trexler, J.P., Ph.D. Thesis, University of Michigan, Ann Arbor, MI, 1964.
6. Urban, J.B., Proc. Watershed Research in Eastern North America Workshop, Correll, D.L., Ed.; Smithsonian Institution, Edgewater, MD, 1977, 251–275.
7. Soil Survey Staff, Soil Taxonomy, Agric. Handbook No. 436, U.S. Government Printing Office, Washington, DC, 1975, pp 754.
8. Thornthwaite, C.W., Geographical Review 1948, 38, 55–94.
9. Knisel, W.G., CREAMS--A Field Scale Model for Chemicals, Run-Off, and Erosion from Agricultural Management Systems, USDA-ARS, Conservation Research Report 26, 1980, pp 640.
10. Williams, J.R., Jones, D.A., Dyke, P.T., Trans. ASAE 1984, 27, 129–144.
11. Schwab, G.O., Frevert, R.K., Edminster, T.W., Barnes, K.K., Soil and Water Conservation Engineering, John Wiley & Sons, Inc., New York, NY 1966, pp 683.
12. Rogowski, A.S., Wolf, J.K., Proc. Headwaters Hydrology Symposium, AWRA, Bethesda, MD, 1989, 665–674.
13. Davidson, J.M., Stone, L.R., Nielsen, D.R., LaRue, M.E., Water Resour. Res. 1969, 5, 1312–1321.

14. Walton, W.C., Groundwater Resource Evaluation, McGraw Hill Book Co., New York, NY, 1970, pp 664.
15. Urban, J.B., J. Soil Conser. Soc. Am. 1965, 20, 178-179.
16. Wyrick, G.G., Borchers, J.W., Hydrologic Effects of Stress Relief Fracturing in an Appalachian Valley, U.S. Geological Survey Water Supply Paper 2117, 1981, pp 50.
17. Urban, J.B., Gburek, W.J., In Ground-Water Contamination Field Methods, Collins, A.G. and Johnson, A.I., Eds.; American Society for Testing and Materials, Philadelphia, PA, ASTM SPC 963, 1988, 468-481.
18. Pionke, H.B., Urban, J.B., Proc. Eastern Regional Conference, National Water Well Assn., Worthington, OH, 1984, 377-393.
19. Barnes, B.S., Trans. Am. Geophys. Union 1939, 20, 721-725.
20. Singh, K.P., Water Resour. Res. 1968, 4, 985-999.
21. Rorabaugh, M.I., Simons, W.D., Exploration of Methods of Relating Ground Water to Surface Water, U.S. Geological Survey Open-File Report, 1966, pp 62.
22. Gburek, W.J., Pionke, H.B., In Water Resources in Pennsylvania: Availability, Quality and Management, Majumdar, S.K., Ed.; Typehouse of Easton, Phillipsburg, NJ, 1990, 354-371.
23. Schnabel, R.R., In Watershed Research Perspectives, Correll, D.L., Ed.; Smithsonian Institution Press, Washington, D.C., 1986, 263-282.
24. Gburek, W.J., Urban, J.B., Schnabel, R.R., Proc., Agricultural Impacts on Ground Water - A Conference, National Water Well Assn., Worthington, OH, 1986, 352-380.
25. Pionke, H.B., Urban, J.B., Ground Water Monit. Rev. 1987, 7, 79-88.
26. Schnabel, R.R., Urban, J.B., Gburek, W.J., Proc. Agricultural Impacts on Ground Water Quality, National Water Well Assn., Worthington, OH, 1990, 159-173.

RECEIVED September 28, 1990

Chapter 14

Water Quality Sampling Program at Low-Level Radioactive Groundwater Contamination Site

Wood River Junction, Rhode Island

Barbara J. Ryan[1] and Denis F. Healy[2]

[1]U.S. Geological Survey, Washington, DC 20240
[2]U.S. Geological Survey, Hartford, CT 06103

The U.S. Geological Survey conducted a three-year research study of ground-water contamination at a low-level radioactive waste site in southern Rhode Island. One goal of the study was to collect water samples that accurately represented water-quality conditions in the aquifer, while minimizing the variability due to sampling method, the potential for cross-contamination, and the time required for sample collection. The water-quality-sampling program consisted of establishing an observation well network, adopting and standardizing sampling procedures, and determining an optimum sampling frequency.

A network of 150 observation wells was used to determine direction of ground-water flow and spatial variations (horizontal and vertical) in ground-water quality. The sampling procedures involved the following three steps for each observation well: (1) insertion of a piece of 0.9 cm (inside diameter) polyvinylchloride suction tubing into the well and position its intake 0.6 to 0.9 m below the water level; (2) evacuation of approximately three times the volume of water in the well with either a centrifugal or peristaltic pump until steady-state conditions (stable specific conductance) were reached; and (3) attachment of a smaller variable-speed peristaltic pump to the tubing for sample collection and field measurements. From April 1981 through January 1984, a bimonthly sampling frequency for 30 to 75 observation wells was used to collect a total of 1,000 samples.

Spatial variations (horizontal and vertical) of gross-beta concentrations over very short distances suggest that samples collected correctly reflected water-quality conditions in the aquifer. The major drawback of the system was lack of control of pumping rates.

The collection of water samples representative of the water quality in an aquifer is an important part of ground-water-contamination studies. As part of a 3-year research study by the U.S. Geological Survey of ground-water contamination at a low-level radioactive-waste site in southern Rhode Island, ground-water sampling procedures were adopted and standardized to ensure that samples accurately represented chemical conditions in the aquifer while minimizing the variability due to sampling method, the potential for cross contamination, and the time required for sample collection. From April 1981 through January 1984, a bimonthly sampling frequency for 30 to 75 observation wells was used to collect a total of 1,000 samples. The purpose of this paper is to describe the methodologies and equipment used at this site for the collection of ground-water samples for radiological analyses. The methods presented here may have applicability to other similar ground-water contamination studies.

From 1964 through 1980, an enriched uranium cold-scrap recovery plant was operated near Wood River Junction, Rhode Island (Figure 1). The recovery process involved digestion of the scrap with hydrofluoric and nitric acids, and organic separation with tributyl phosphate and kerosene. Solid wastes from the process were shipped offsite for disposal. Liquid wastes were discharged to the Pawcatuck River through a drain pipe from 1964 through 1966, and to polyethylene- and polyvinyl-chloride (PVC)-lined evaporation ponds and trenches from 1966 through 1980. Overflow of the existing ponds due to precipitation runoff and high disposal flow rates led to periodic construction of additional ponds and trenches, which eventually encompassed approximately 2,300 m². Liquid contaminants from these ponds and trenches percolated to the water table and formed a plume of contaminated ground water.

Site Description

The study area, located within the lower Pawcatuck River basin, is approximately 3 km east of the confluence of the Pawcatuck and Wood Rivers. The aquifer is composed of unconsolidated sands and gravels of Pleistocene age. Sediments consist of predominantly medium to coarse sands and gravels to about 24 m below land surface and mostly fine sands and silts below a depth of 24 m. The water table slopes westward from the plant site at an average gradient of 14 m/km. Ground-water-flow velocities are estimated to range from 0.6 to 0.8 m/d (1). During low-flow periods (late summer and early autumn), the water table ranged from 0.6 m below land surface in a swamp at the western edge of the Pawcatuck River to approximately 8 m below land surface at the eastern side of the Pawcatuck River; water levels were higher during the rest of the year.
Well yields ranged from 1 L/min in wells screened in fine sands and silts to approximately 20 L/min in wells screened in coarse sands and gravels.

The plume of contaminated ground water extends a total of 700 m (Figure 2). From the source area, it extends northwestward approximately 460 m to the Pawcatuck River where it turns southwestward, extending approximately 240 m in a downstream direction through the swampy area west of the river. The plume is approximately 90 m wide and is confined to

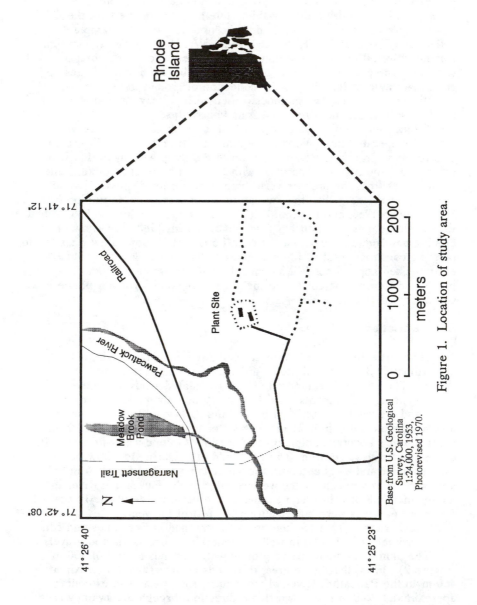

Figure 1. Location of study area.

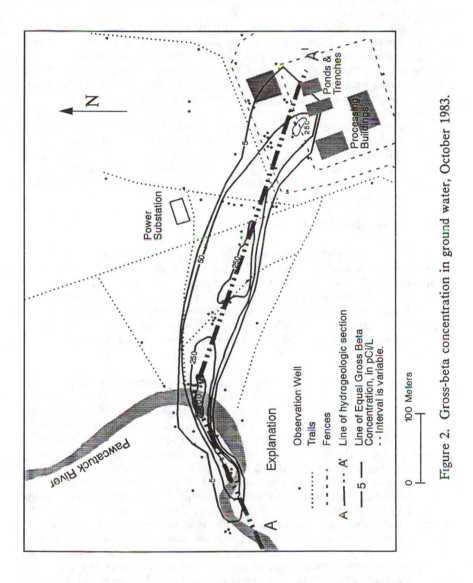

Figure 2. Gross-beta concentration in ground water, October 1983.

the upper 24 m of saturated thickness of sediments that consist of medium
to coarse sand and gravel (Figure 3). East of the Pawcatuck River the top
of the contamination plume is depressed below the water table, and its
depth increases away from the source area. The maximum depth of the
plume (24 m below land surface) is present 425 to 460 m from the source
area. Contaminants rise to land surface in the discharge area at the river
and adjacent swamp.

Chemical and radiochemical constituents in contaminated ground water
include nitrate, 5 to 1,200 mg/L; boron, 20 to 1,000 ug/L; potassium, 3 to
26 mg/L; strontium-90, 4 to 290 pCi/L (picocuries per liter); and
technetium-99, 75 to 1,350 pCi/L. Concentration of gross-beta emitters in
contaminated water range from 5 to 1,600 pCi/L, and specific conductance
ranges from 150 to 5,400 uS/cm (microsiemens per centimeter at 25
degrees Celsius).

Concentrations of chemical and radiochemical constituents in
uncontaminated water at the site are commonly below the following
detection levels: Nitrate, <0.1 mg/L; boron, <0.1 ug/L; potassium,
<0.1 mg/L; and strontium-90, 0.4 pCi/L. In uncontaminated water,
concentrations of gross-beta emiters are measured as low as 0.7 pCi/L, and
specific conductance is generally less than 100 uS/cm.

Water-Quality Sampling Program

To collect water samples that accurately represented water-quality
conditions in the aquifer while minimizing both cross contamination and the
time required for sample collection, three steps were taken: (1) an
observation-well network that would permit the collection of representative
samples was installed, (2) appropriate collection procedures for accurate
and efficient sampling were selected, and (3) an optimum sampling
frequency that would describe temporal variations in water quality was
chosen.

Observation-Well Network. An observation-well network consisting of 150
wells was installed at the site using hollow-stem auger, mud rotary, and
drive-and-wash drilling rigs. The hollow-stem auger rig was used to install
most of the wells in the network. Exceptions included wells deeper than
100 feet and wells that were larger than 3.8 cm in diameter, which were
drilled with a mud-rotary rig; and wells in the swamp which were drilled
with a drive-and-wash rig mounted on an all-terrain vehicle.

The observation-well network was used to determine the direction of
ground-water flow and the horizontal and vertical variations in water quality
in the aquifer. Water-level and water-quality data from approximately 20
existing wells and results of an electromagnetic survey (2) were used to
guide placement of observation wells. Observation wells used to determine
the direction of ground-water flow were installed approximately 150 m apart
in a grid-like pattern over much of the study area. These wells were
generally shallow (less than 9 m deep) and had one 3 m screened interval
located in the water-table fluctuation zone. The relatively long screen

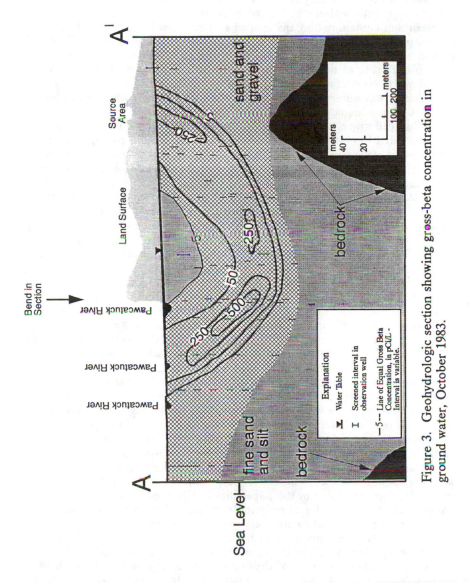

Figure 3. Geohydrologic section showing gross-beta concentration in ground water, October 1983.

ensured that most water-table fluctuations occurred within the screened
section of the well, thereby permitting a more representative measurement
of water-table altitude.

Observation wells used to sample the horizontal variation in water
quality were installed in lines approximately perpendicular to the long axis
of contamination. Spacing of each line of wells ranged from 30 to 150 m.
Observation wells within a given line were drilled to the anticipated depth
of maximum contamination and spaced so that areal boundaries of
contamination could be mapped. Generally, a minimum of five wells were
drilled along most lines - one along the axis of maximum contamination,
one on each side of the axis showing less contamination, and one on each
side showing no contamination. Additional wells were installed if, after
sampling, it was determined that the boundaries of the contaminant plume
had not been adequately defined.

Observation wells used to sample the vertical variation in water quality
were installed in clusters along the long axis of maximum contamination at
each line of wells previously described. Generally, a minimum of five wells
were drilled in each cluster - one well screened at the depth of maximum
contamination, one well screened above and one well screened below that
depth showing less contamination, and one well above and one below that
depth showing no contamination. Approximately 50 and 65 wells were
drilled to describe horizontal and vertical variations, respectively. Sampling
points are shown in Figures 2 and 3. These observation wells ranged from
1.5 to 70 m in depth and had screened intervals that generally ranged from
0.9 to 1.5 m.

Although diameters ranged from 0.9 to 13.3 cm, most wells were
constructed of 3.2 cm-diameter, rigid PVC pipe. Initially, one well was
installed in each borehole to eliminate cross contamination that might result
from vertical flow within the annulus (space between the well casing and
the borehole wall). Generally, five wells screened at various depth intervals
were installed in separate boreholes within a 3-m radius (Figure 4a).

As more wells were drilled, it became evident that collapse of the
formation around the casing had occurred, which reduced the likelihood of
flow within the annulus. It was determined, therefore, that water-quality
samples from multiple wells within a single borehole (Figure 4b) would
accurately represent vertical variations in the chemistry of the aquifer.
Smaller-diameter casings (0.9 to 2.5 cm) were used for this purpose.
Initially, a 0.3 to 0.6 m-long bentonite seal was installed in the annulus
above each well screen in each borehole. This procedure was eventually
abandoned, because the bentonite tended to bridge in the hollow-stem
auger, lodging the casing in the auger flight. Changes in chemical
concentrations of an order of magnitude within a 3-m vertical distance
convinced us that the bentonite seals were unnecessary.

Most wells were completed with 2.5 to 3.8 cm- diameter, rigid PVC
screens with either a 0.25 or 0.30 mm slot size. For a few wells, 0.9 cm-
diameter screens were made by drilling 0.6 cm holes in a 0.9 m length of
PVC pipe. A meshed polypropylene fabric was then sewn into a sleeve that
encased the perforated segment (A.D. Randall, U.S. Geological Survey, oral
commun., 1983).

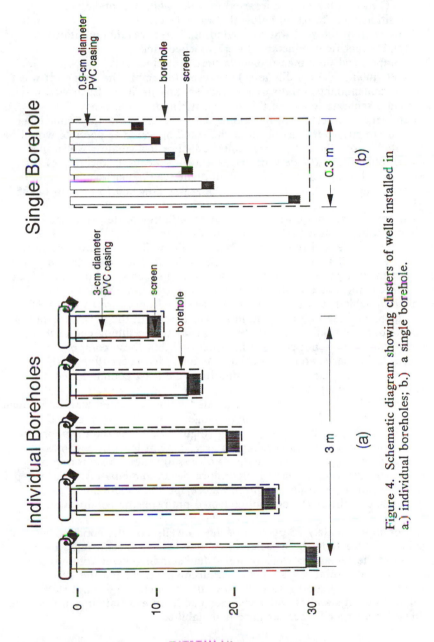

Figure 4. Schematic diagram showing clusters of wells installed in a.) individual boreholes; b.) a single borehole.

Sampling Procedures. The following procedures, which are illustrated in
Figure 5, were used to collect ground-water samples:
1. PVC suction tubing was inserted into the well; the intake was
    positioned 0.6 to 0.9 m below the water level;
2. Water from the well was evacuated until steady-state conditions
    (stable specific conductance) were reached; and
3. Samples and field measurements were collected.
    PVC tubing (0.9-cm diameter) was used to sample the water. It was
inert to contaminants, relatively inexpensive, and flexible. Each well was
assigned a separate length of this suction tubing to avoid cross
contamination. Insertion of the tubing 0.6 to 0.9 m below the water level
was done to ensure that as much of the standing water as possible would be
evacuated before the sample was collected. For low-yielding wells, the
suction tubing was extended further as drawdown caused lowering of the
water level. For approximately 15 small-diameter (0.9 cm) wells, adapters
were constructed so that the suction tubing was attached to the top of the
casing.
    A centrifugal pump, with a capacity to withdraw water at a rate of 20
L/min was used to evacuate water from wells east of the Pawcatuck River
where the vehicle had ready accessibility to the wells. Peristaltic pumps,
which could be hand carried to the wells, were used to evacuate water from
most of the wells west of the river where vehicle access was poor; a 1.6 cm
peristaltic pump was used to evacuate the larger volume wells, and a
smaller, variable-speed peristaltic pump was used on the remaining wells.
    Silicone tubing (4.5 to 0.6 m long) was used with the peristaltic pumps
because of its flexibility and durability. Polyethylene tubing connectors and
silicone tubing were not changed for each well, but these components were
flushed during parameter monitoring. As an added precaution, within any
given well cluster, sampling proceeded from the least contaminated well to
the most contaminated.
    The minimum volume of water needed to be evacuated was determined
early in the project from monitoring changes in specific conductance in
water from wells as they were pumped (Figure 6). These data support the
practice of evacuating the standing-water volume at least three times in
order to obtain representative samples of in-situ water. Monitoring of
specific conductance, pH, and temperature began immediately. Because pH
and temperature generally stabilized before specific conductance, samples
(filtered and unfiltered) were not collected until specific conductance
stabilized.
    For approximately 75 percent of the 75 wells, specific conductance had
stabilized by the time the evacuation volume had been removed. Twenty
percent of the remaining wells (4 wells) had standing-water volumes so
small that water removed for field measurements exceeded the evacuation
volume. In the remaining 16 wells, specific conductance did not stabilize
until six to eight standing-water volumes had been evacuated; samples were
collected once specific conductance had stabilized.

Sampling Frequency. Illustrations similar to Figures 2 and 3 generally were prepared after each sampling period. Comparisons of the results were used to choose an optimum sampling frequency to describe spatial variations in water quality with time. By tracking the movement of zones of contaminants and by monitoring water-quality changes within any given well, it was determined that an approximately bimonthly sampling schedule was suitable to describe changes in water quality for most wells. For several wells near the Pawcatuck River, a quarterly sampling scheme would have sufficiently described the spatial variations in water-quality conditions.

Evaluation Of Water-Quality Sampling Program

Spatial variations (both horizontal and vertical) in gross-beta concentration in ground water at the site occur over very short distances. This finding, and the tendency for the casings and drill stem to become sand locked during the drilling process - indicating complete collapse of the formation within the borehole - suggest that the water samples collected correctly reflected existing water-quality conditions in the aquifer.

Evaluation of the efficiency of the selected pumping apparatus and the ability to collect representative samples with the selected pumping apparatus was somewhat more difficult. The centrifugal pump, used to quickly remove the evacuation volume, coupled with the peristaltic pump, used to minimize cross contamination, seemed to be efficient. The 4.5 to 0.6 m length of silicon tubing used in the smaller variable-speed peristaltic pump and two polyethylene tubing connectors were the only parts of the sampling apparatus that came in contact with water from all observation wells.

The major drawback of the system was the lack of control of pumping rates. The centrifugal and the larger peristaltic pumps were not variable speed, and the pumping rates were determined by atmospheric pressure, depth to water, and well yield. Because 4 L/min was the rate obtained from the majority of the wells, the suction tubing was clamped to reduce flow when the pumping rate exceeded 4 L/min.

Summary

Establishment of an observation-well network, adoption and standardization of sampling procedures, and determination of an optimum sampling frequency comprised the water-quality sampling program for a 3-year research study conducted by the U.S. Geological Survey. Data from approximately 150 observation wells were used to study ground-water contamination at a low-level radioactive waste site in southern Rhode Island. Principal goals of the study were to (1) collect samples that accurately represented water-quality conditions in the aquifer, while minimizing the variability due to sampling method, and (2) minimize both the potential for cross-contamination and the time required for sample collection.

The wide range of gross beta emitters (0.7 to 1600 pCi/L) in ground water at the site made variability in water-quality conditions related to sampling method and cross contamination resulting from flow within the

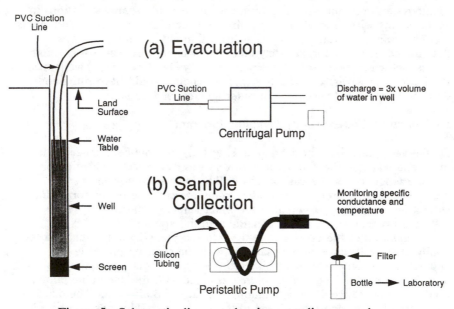

Figure 5. Schematic diagram showing sampling procedure.

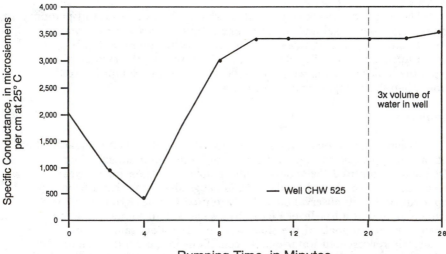

Figure 6. Typical changes in specific conductance that occur as water is evacuated from observation wells. *Continued on next page*

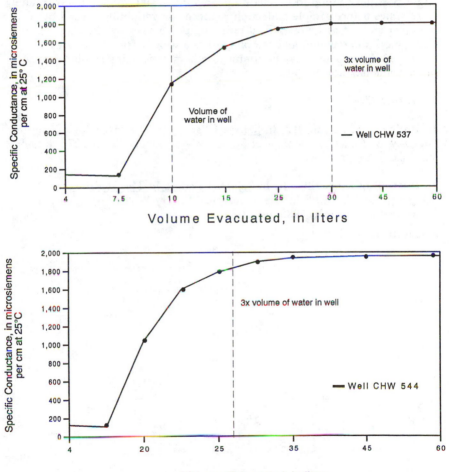

Figure 6 (continued). Typical changes in specific conductance that occur as water is evacuated from observation wells.

borehole or from sampling methodology a particular concern. Bimonthly samples were collected from 30 to 75 observation wells from April 1981 through January 1984 resulting in a total of 1,000 samples; therefore, minimizing sample collection time also was important.

Collapse of the formation around the well casing, use of individual pieces of suction tubing for each well, and use of a small variable-speed peristaltic pump for sample collection were steps taken to minimize the potential for cross contamination. Use of either a centrifugal pump or larger peristaltic pump to evacuate the standing-water volume approximately three times before sample collection reduced the variability caused by sampling method and streamlined sampling procedures by allowing simultaneous evacuation and sampling of two wells. The procedures presented in the paper may be useful to others conducting similar ground-water-contamination studies.

Literature Cited

1. Barlow, P.M.; Ryan, B.J. In Selected Papers In The Hydrologic Sciences; Subitzky, S., Ed.; U.S. Geological Survey Water-Supply Paper 2270, Reston, Virginia, 1985, pp 35-49.

2. Ryan, B.J.; Kipp, K.L., Jr. In Selected Papers In The Hydrologic Sciences; Subitzky, S., Ed.; U.S. Geological Survey Water-Supply Paper 2270, Reston, Virginia, 1985, pp 21-33.

RECEIVED January 22, 1991

Chapter 15

Economical Monitoring Procedure for Assessing Agrochemical Nonpoint Source Loading in Unconsolidated Aquifers

Roy F. Spalding[1,2], Mary E. Exner[3], and Mark E. Burbach[1]

[1]Water Center and [2]Department of Agronomy, Institute of Agriculture and Natural Resources, University of Nebraska, Lincoln, NE 68583–0844
[3]Conservation and Survey Division, Institute of Agriculture and Natural Resources, University of Nebraska, Lincoln, NE 68588–0517

Multilevel samplers (MLSs) consisting of piezometers and tube samplers are a logical approach for determining the direction of groundwater flow and chemistry in shallow (< 6 m) nonpoint source (NPS) groundwater investigations. These MLSs have evolved from fastening the tubing to conduit at specific depths while the conduit was lowered into the hollow stem auger train to the present method of installing pre-assembled MLSs in boreholes drilled by the reverse circulation rotary method without the use of drilling additives. This method allows the aquifer to be sectioned into discrete layers and provides an instantaneous snapshot of both flow and chemistry in three dimensions. The procedure has been used successfully at several sites in Nebraska. The method is cheap, fast, and accurate in areas where the depth to water is less than 6 m. While the same procedure can be used where depths to water exceed 6 m, the need for gas-driven samplers substantially increases the cost.

The problems with setting criteria for devising groundwater sampling strategies probably are best summarized by the statement that no two field sites or basins are identical. In spite of that fact, investigators still must follow logical protocols to attain their goals; otherwise, the comparative framework of their findings will be sacrificed. Guidelines for techniques, however, must be flexible if they are to be effective in a variety of scenarios.

While past sampling emphasis has been primarily on characterization of point source-contaminated sites (Superfund activities), the focus of the 1990s is rural America and its nonpoint source (NPS) agricultural problems. During the height of the Superfund boom in the 1980s many effective and accurate sampling techniques were developed. It is important that the knowledge gained during that decade serve as the basis for innovative approaches characterizing NPS problem areas. Monitoring NPS contamination, however, requires special sampling designs because the contamination results from very large input zones representing whole basins. Consequently, it lacks discrete centroids of high concentrations, obvious source areas, and high density loading effects. While extensive sampling installations are not critical in ambient monitoring networks used to delineate the boundaries of large areas of NPS contamination,

0097–6156/91/0465–0255$06.00/0
© 1991 American Chemical Society

they are an absolute necessity for defining the effects of prevention technology on groundwater contamination beneath field-size research areas.

The effects of improved agricultural management practices on groundwater quality no longer can be inferred solely from input data. Instead their impact on aquifer loading must be documented accurately and precisely by scientific measurements. The need for guidelines for sampling NPS-contaminated groundwater already has been recognized by investigators affiliated with several of the Management Site Evaluation Areas (MSEAs). Several of these research sites were selected to demonstrate that the use of prevention strategies (Best Management Practices) do affect agrochemical loading to the aquifer. In March 1990 the first MSEAs were selected in corn and soybean production areas of the five states represented in the north-central region of the United States where there is documented or suspected NPS agrochemical contamination. These initial MSEAs are federally funded through the United States Department of Agriculture (USDA), the Agricultural Research Service (ARS), and the Cooperative State Research Service (CSRS) and perhaps in the future by the United States Geological Survey (USGS). Other sites will be added in succeeding years. Many of these sites are in areas where the water table is relatively shallow; and in the final analysis, impacts of BMPs will be based on observations of the groundwater. Because data from these sites will be compared, it must be collected in a comparative manner; consequently, sampling guidelines are crucial.

Research Site NPS Sampling Methodology

Guidelines for all groundwater sampling programs should include pre-installation recommendations for siting monitoring equipment, drilling and logging boreholes, and constructing samplers, and post-installation recommendations for purging samplers and collecting samples. The recommendations presented here integrate methods previously reported for point source characterization with procedural modifications necessary to intensively monitor the fate and transformation of agrochemicals in NPS-contaminated groundwater systems.

The research goals and subsurface environment usually dictate the spatial distribution for sampling design. The direction of groundwater flow can be delineated by triangulation (1) with existing surveyed wells or with at least three piezometers. The piezometers may be driven or jetted sand points with minimum construction standards and used only to delineate the direction of lateral flow or they can be installed to meet higher performance standards and used as part of the site monitoring network. After determining the flow direction by triangulation, an array of multilevel samplers (MLSs) is installed. Instrument locations will vary from site to site and many can best be determined by phasing them in as information is gained from drilling, logging, sampling, and data interpretation. Experience has demonstrated that a step-by-step, phased approach of sampler installation is a wise allocation of time for making informed siting decisions. The final density of sampling sites is related to the lateral statistical variability of agrochemical measurements within discrete vertical layers. This variability may be due to complexities in unsaturated and saturated flow, upgradient agrochemical loading, prior on-site contamination, drilling access and safety, and economics.

The importance of the driller's experience with the drilling method, monitoring well construction, and sampler installation cannot be overstated. A contract driller who is receptive to delays caused by geologic sampling and logging and is attune to sampler installation protocol should be paramount.

There are a multitude of methods for installing sampling equipment including mud or air rotary, hammer drive, reverse circulation rotary, cable tool, jet drilling, and solid and hollow stem augering. Nested wells initially were used in NPS investigations (2-3). The materials and installation, however, are expensive and the wells sample a relatively large vertical interval. There is also the potential for cross-contamination because the wells generally are sampled with the same pump.

Driscoll (4) and the United States Environmental Protection Agency (USEPA) (5-6) list the advantages and disadvantages of methods used in well drilling and sampler installation but they do not mention multilevel sampler (MLS) installation which is quickly evolving as the preferred sampling method in most NPS investigations. Experience in the installation of MLSs in shallow unconsolidated aquifers, which are the most vulnerable to contamination, has shown that reverse circulation rotary drilling is a superior installation method. The benefits include (1) the absence of drilling additives (bentonite and organics) which tend to invade geologic formations and retard both flow and solute transport, (2) the ability to stop at will during drilling to collect geologic samples, (3) the ability to hold the borehole open during geophysical logging, (4) the ability to maintain large diameter boreholes (18 cm to 1 m) for elaborate system installation, and, most importantly, (5) the ability to seal between samplers where needed. Limitations of this method are (1) it relies on large quantities of water which must be free of all analytes of interest and (2) it is not appropriate for drilling in consolidated rock. In the latter situation air rotary or cable tool drilling can be substituted. Water should be analyzed for agrochemical residues and be free of analytes of interest prior to introduction into the borehole.

Although MLS installation with a hollow stem auger is relatively cheap, the method has several disadvantages (7). It is difficult to seal between layers and/or samplers; clays and silts smear against the borehole wall; and during retrieval of the auger flight, the auger tends to catch the MLS bundle, kinking the tubing and exhuming the samplers.

A variety of monitoring devices in a variety of materials are marketed for groundwater sampling. More conventional sampling techniques such as submersible pumps, auger screen samplers, packer pumps, and regular and Kemmerer bailers used in boreholes and existing wells tend to collect vertically composited samples rather than samples representative of thin discrete vertical intervals. The MLSs have very low pumping rates that do not significantly alter groundwater flow. The ability to economically obtain representative groundwater samples from thin vertical intervals falls almost exclusively to point samplers (MLSs). Since the introduction of MLSs (8-10), several modifications have been introduced. They include gas-driven MLSs for sampling moderate to deep groundwater (11-12) and modular dialysis MLSs (13). Several sophisticated multilevel systems for borehole investigations in bedrock and unconsolidated media are available commercially (14).

For most research site NPS agrochemical sampling in shallow groundwater, the less complicated the device, the better. A most appropriate MLS is a combination of tube samplers and piezometers that can be fabricated in the field in up to 30-m lengths. The tubes can discretely sample as many vertical intervals as are necessary to assess loading while the piezometers are used for manually monitoring the water-levels (Figure 1). The discrete samplers are composed of 9.52 mm (3/8-inch O.D.) high density polyethylene (HDPE), stainless steel (SS), or Teflon (PTFE) tubing with screened ports. Screens of SS are purchased locally, cut, and held in place with PTFE ferrules. The samplers are fastened at the appropriate depth to a piezometer that extends to the bottom of the borehole. The piezometer is constructed of 2.54-cm (1-inch) Schedule 40 PVC with a 61-cm (2-ft) slotted interval and capped at the bottom. Additional piezometers are fastened to this assemblage at the appropriate depth. The whole assemblage is put together at the site and lowered into the borehole as a continuous string.

Construction materials for the discrete samplers can be of HDPE, SS, PTFE, or some combination of the three. HDPE is much cheaper than SS or PTFE and is probably suitable for sampling chemicals with low sorptivities such as NO_3-N, atrazine (2-chloro-4-[ethylamino]-6-[isopropylamino]-s-triazine), and alachlor (2-chloro-2',6'-diethyl-N-[methoxymethyl]acetanilide). The USEPA (5) suggests that plastic

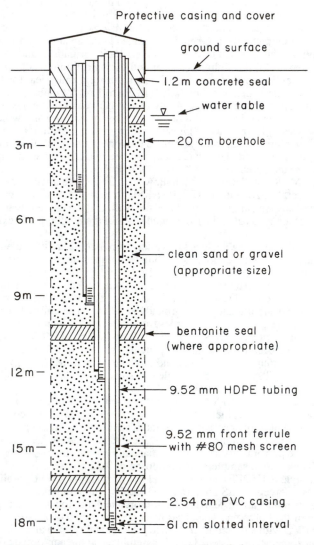

Figure 1. Multilevel sampler installation consisting of piezometers and tube samplers.

casings be checked for sorption by installing adjacent PTFE casings. The conclusiveness of this procedure in point source-contaminated areas, however, is being debated due to aquifer and solute transport variability. With MLSs, however, sorption can be checked by installing all three materials at the same sampling intervals within the same borehole and statistically evaluating the differences. Potential losses from sorption onto the tubing of the peristaltic pump also should be checked. Such experiments for the agrochemicals present in the groundwater beneath the Nebraska MSEA are planned for fall 1990.

The total time involved in drilling a borehole with a 21-m completion depth by the reverse circulation rotary method and installing a MLS with eight discrete samplers and four piezometers is approximately 2 h. This includes placing a filter pack of clean gravel or sand in the annular space around the eight discrete samplers and where necessary sealing between sampling intervals with at least a 30-cm bentonite lens. It is important to use precleaned filter packs that closely approximate aquifer hydraulic conductivities in order to prevent the introduction of contamination from chemical solution and desorption off the filter pack to the borehole and to minimize disruption of the natural flow system. The location and thickness of the bentonite seals are determined by the discrete sampling intervals, by the formation geology as interpreted from geologic and geophysical logs, and by potential vertical flow components. The seals are formed by chunk bentonite which is dropped slowly into the borehole until the desired thickness, as measured by displacement with a weighted line, is attained. Our experience has been that large chunk bentonite settles faster with less hydration than pellet bentonite.

Immediately after installation, the samplers are developed by pumping with peristaltic pumps in shallow (< 6 m) water-table areas or with gas-driven pumps in deeper applications. Development by pumping is an acceptable and well-documented technique (1). As many as 10 samplers can be pumped simultaneously with a multi-module peristaltic pump. A manifold attached to the gas-driven samplers also permits them to be pumped simultaneously. In sand and gravel aquifers sediment-free groundwater is produced quickly (< 20 min) when the reverse circulation rotary drilling technique is used. The same pumps are used for sampling. They are appropriate for sampling low volatile agrochemicals such as NO_3-N, atrazine, and alachlor; however, neither pump is recommended for sampling volatiles or gases.

Last year at a site near Grand Island, Nebraska, the total cost of a 20-m MLS with eight HDPE discrete samplers and four piezometers was less than $500 and included drilling, MLS materials, and installation. In areas where greater depths to groundwater (> 6 m) require gas-driven dedicated samplers, a similar MLS installation (eight gas-driven samplers and four piezometers) can be purchased and installed for about $4500.

At a sludge injection site in Nebraska where a strong NO_3-N concentration gradient exists (Spalding, R. F., University of Nebraska-Lincoln, 1989 contract report), the average differences in concentrations between four discrete samplers and the corresponding piezometers in eight MLSs were larger in the shallowest pair and decreased with depth. These average differences indicate that the tube samplers have a more discrete sampling capability than do the piezometers. This capability is important in discerning differences in concentration in the loading zone. The main advantages of MLSs, however, are lowered costs, reduced potential for cross-contamination, and shorter installation time.

MLSs have been used in two NPS investigations in Nebraska and the discrete samplers provided definitive results at both sites. A large MLS installation delineated a plume of nitrate contamination (Figure 2) whose source was sludge injected on an irrigated cornfield. MLSs were used near Oshkosh, Nebraska to document agrochemical contamination downgradient from irrigated cornfields (Exner, M. E., University of Nebraska-Lincoln, 1990 contract report).

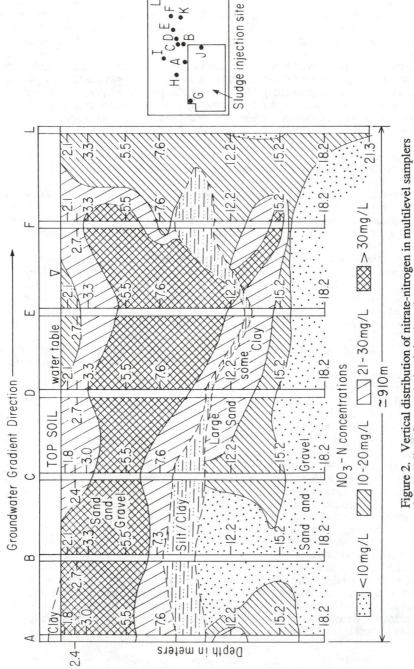

Figure 2.   Vertical distribution of nitrate-nitrogen in multilevel samplers downgradient from a sludge injection site

Conclusions

Prefabricated strings of tube samplers and piezometers in lengths up to 30 m can be easily assembled in the field and installed in logged boreholes drilled by the reverse circulation drilling procedure. These MLSs can be assembled and installed in a short period of time. The method provides a practical and economical approach to equip research sites with large numbers of discrete samplers for the assessment of NPS loading. The combination of tube samplers and piezometers provides a 3-dimensional snapshot of water quality and flow conditions. The distribution of nonvolatile agrochemicals can be characterized by this procedure. MLSs are essential for evaluating discrete differences in agrochemical loading with depth in shallow aquifers already stratified with agrochemicals and will be essential in the investigation at the Nebraska MSEA site.

Literature Cited

1. Heath, R. C. *U.S.G.S.Water-Supply Paper 2220.* **1983**, 84 pp.
2. Spalding, R. F.; Exner, M. E. *J. Environ. Qual.* **1980**, *9*, 466-479.
3. Fenn, D.; Cocozza, E.; Isbister, J.; Briads, O.; Yare, B.; Roux, P. *Procedures Manual for Ground Water Monitoring at Solid Waste Disposal Sites,* U.S. Environmental Protection Agency, Office of Solid Waste, 1977, 269 pp.
4. Driscoll, F. G. *Groundwater and Wells*; Johnson Division: St. Paul, MN, 1986, 1089 pp.
5. *RCRA Groundwater Monitoring Technical Enforcement Guidance Document*; U.S. Environmental Protection Agency: Washington, DC, 1986, 208 pp.
6. *Handbook of Suggested Practices for the Design and Installation of Ground-water Monitoring Wells;* Environmental Monitoring Systems Laboratory; U. S. Environmental Protection Agency: Washington, DC, 1989, 398 pp.
7. Hackett, G. *Ground Water Monitor. Rev.* **1988**, *8*, 60-68.
8. Hansen, E. A.; Harris, A. R. *Water Resour. Res.* **1974**, *10*, 375.
9. Pickens, J. F.; Cherry, J. A.; Grisak, G. E.; Merritt, W. F.; Risto, B. A. *Ground Water* **1978**, *16*, 322-27.
10. Cherry, J. A.; Anderson, E. G.; Johnson, P. E.; Gillham, R. W. *J. Hydrol.* **1983**, *63*, 31-49.
11. Nazar, A.; Prieur, J.; Threlfall, D. *Ground Water Monitor. Rev.* **1984**, *4*, 43-47.
12. Norman, W. R. *Ground Water Monitor. Rev.* **1986**, *6*, 56-60.
13. Ronen, D.; Magaritz, M.; Levy, I. *Ground Water Monitor. Rev.* **1987**, *7*, 69-74.
14. Ridgway, W. R.; Larssen, D. In *Ground Water and Vadose Zone Monitoring*; Nielsen, D. M.; Johnson, A. I., Eds.; American Society for Testing and Materials: Philadelphia, PA, 1990; pp. 213-237.

RECEIVED February 11, 1991

VADOSE ZONE SAMPLING

Chapter 16

Monitoring Agrochemical Transport into Shallow Unconfined Aquifers

K. W. Staver and R. B. Brinsfield

Wye Research and Education Center, Department of Agricultural Engineering, University of Maryland System, Queenstown, MD 21658

Recent documentation of agrochemical contamination of groundwater has suggested that agricultural practices need to be modified in order to reduce contaminant leaching from the root zone. Developing agricultural practices which maintain groundwater quality requires quantitative sampling approaches that allow determination of contaminant transport rates for specific practices. Increasingly widespread evidence of the transient and spatially variable nature of solute transport in the vadose zone suggests that sampling groundwater may provide the most reliable method for determining solute leaching rates, particularly where the water table is located close to the soil surface. Hydraulic gradients in the groundwater component of a vadose zone-unconfined aquifer flow system are generally lower and more stable than those in the unsaturated region, resulting in less transient flow conditions during recharge periods. As the thickness or water holding capacity of the vadose zone increases, the transport of solutes from the root zone to groundwater becomes less direct, requiring more solute data collection from the unsaturated region of the soil profile. Water and solute storage in the vadose zone immediately above the water table will alter leachate solute levels during recharge, to an extent determined by the water holding characteristics of the profile. Stratification of groundwater solute levels near the water table as a consequence of changes in root zone leaching rates requires discrete well screen placement based on water table fluctuation patterns if groundwater sampling is to be used to establish leaching rates for specific agricultural practices.

Recently, public concern regarding contamination of drinking water has focused attention on the impact of agricultural activities on groundwater quality. The presence of pesticides, as well as elevated nitrate concentrations in groundwater in widespread and varied agricultural settings has been documented (8,39). In addition to the potential for contamination of drinking water supplies, agrochemical transport into shallow aquifers can eventually result in surface water quality degradation as groundwater is discharged. Groundwater comprises a major flow component of perennial streams, especially in low relief regions such as the Atlantic Coastal Plain, and nitrate in groundwater discharge can be the dominant nitrogen component in streams draining primarily agricultural watersheds (4,14,18). In the major effort underway to reduce eutrophication of Chesapeake Bay, groundwater contributions of nitrogen have been identified as the major non-point source of nitrogen entering Bay waters in the Coastal Plain region of the watershed (27).

At present, nitrogen application to agricultural land is essentially unregulated and pesticide regulations deal primarily with application restrictions. However with increasing evidence of agrochemical transport into groundwater, the development of a regulatory system based directly on groundwater contaminant levels under agricultural land is becoming more likely (3,5). Regardless of whether groundwater quality standards are established in a voluntary or regulatory framework, before an effective strategy for improving water quality under agricultural land can be implemented, cause-effect relationships must be determined between various agricultural practices and rates of groundwater contamination. While the monitoring techniques necessary for developing these types of relationships for pollutants transported in surface water have been tested extensively since soil erosion was first recognized as an agricultural as well as environmental problem, much uncertainty remains regarding approaches for assessing rates of groundwater contamination in diffuse source settings. Simplistic solutions such as prohibiting the application of an agrochemical which has been detected in groundwater will certainly reduce future groundwater contamination from that compound, without requiring additional quantitative monitoring techniques. However this presence/absence approach has little utility for potential groundwater pollutants such as nitrate, which are essential for plant growth, and will become increasingly difficult to implement as advances in analytical techniques continue to lower detection limits for agrochemical residues.

Contaminant levels in groundwater are determined by the rate of contaminant transfer from unsaturated to saturated regions of the soil matrix. Determination of contaminant flux rates into groundwater in an agricultural setting where solute transport occurs across an aerially extensive and variable interface poses even more difficulty than assessment of the extent of groundwater contamination associated with point sources of pollution. In most crop production systems, fertilizers and pesticides are applied at or just below the soil surface. For most potential agrochemical groundwater pollutants, the ultimate extent of groundwater contamination will largely be determined by the rates of leaching from upper soil horizons. Nitrate appears to behave conservatively below the crop rooting zone (17,25), and pesticide microbial degradation rates as well as sorption sites generally decrease in the soil profile below the root zone (32). Thus, it is likely that persistent solutes which have been transported below the root zone will eventually reach groundwater. After contaminants have entered groundwater flow systems, opportunities for removal are

generally logistically or financially restricted. This suggests that
efforts to reduce groundwater contamination should focus on source
control, that is, development and implementation of agricultural
practices which reduce the rates at which pollutants move across the
lower boundary of the root zone. However, a prerequisite for
development of these practices is the ability to accurately quantify
the impact of specific agricultural practices on rates of pollutant
transport from upper soil horizons, through the vadose zone, and into
groundwater.

At present, the primary concern with regard to the leaching of
nitrate, as well as other potential groundwater pollutants used in
agricultural systems, is not the rate of leaching but the resulting
concentration of that pollutant in groundwater. Thus, even though the
leaching rates for most solutes ultimately determine their
concentrations in groundwater, existing and proposed standards are
based exclusively on concentration values rather than on leaching
rates. This is probably due in part to the greater ease with which
concentration data can be collected relative to the information needed
for determination of leaching rates, but also to the fact that it is
the concentration of a pollutant which will determine the health risk
associated with the consumption of a given volume of water or the
quantity of a pollutant in groundwater that is discharged into a
surface water body per unit volume of groundwater. As a result,
groundwater contamination studies generally rely heavily on groundwater
concentration data, with little consideration given to vadose zone
parameters. While this approach is useful for detecting the presence
of a groundwater contamination problem, and may give an indication of
average leaching rates where land use patterns stay the same through
many recharge cycles, in an agricultural setting where nitrogen and
pesticide application rates as well as tillage practices often change
every year, groundwater solute concentration data alone will provide
little information on rates of agrochemical leaching associated with
specific land use practices.

Sampling exclusively in the vadose zone also presents problems
when attempting to quantify contaminant leaching rates for specific
agricultural practices. Recent evidence from many differing
agricultural systems documenting rapid water and solute transmission
through large continuous soil pores complicates mass balance approaches
for determining solute flux rates in the vadose zone (7,15,29,30).
Non-uniform agrochemical application, microbial processes (degradation
of pesticides; uptake and release of nitrate), uptake of water and
solutes by plant roots, and many other factors combine to create rapid
temporal and spatial changes in solute concentrations in the root zone,
as well as uncertainty surrounding the causes for observed changes.
Thus, even for solutes which behave conservatively in the region of the
vadose zone below the root zone, the extreme variability of solute
levels in overlying soil horizons combined with the potential for high
rates of solute dispersion as water moves through the soil profile
limit the value of vadose zone sampling alone for determination of
solute transport rates to groundwater.

In regions where cropland overlies a thick layer of unsaturated
subsoil, sampling strategies designed to quantify leaching rates for
specific agricultural practices must be implemented in the vadose zone,
and results used to project future impacts on groundwater quality (25).
However in many agricultural regions the water table is located close

to the soil surface, thereby making the direct impact of agricultural activities on groundwater quality evident. These regions are also where groundwater is usually considered to be most vulnerable to contamination, although for water soluble contaminants which behave conservatively below the root zone, the thickness of the unsaturated zone delays, but does little to alter the total mass of solute that is eventually delivered to groundwater. Since geohydrologic systems with relatively short hydrologic flow paths from the root zone to the unconfined aquifer are most vulnerable to rapid contamination as a result of agricultural activities, these systems present ideal opportunities for establishing relationships between specific agricultural practices and rates of solute transport to groundwater. By combining standard saturated and unsaturated sampling techniques in these systems, a short-term mass balance approach can be applied to determine solute leaching rates on an annual basis. The most critical element in this, as well as in any other attempt to quantify solute transport, is to accurately describe the distribution and movement of water in the vadose zone-unconfined aquifer system. We will discuss the hydrologic considerations essential for development of sampling strategies for determining rates of agrochemical transport into shallow groundwater for specific agronomic practices.

Vadose Zone-Unconfined Aquifer Flow System

Before appropriate sampling strategies for quantifying rates of agrochemical transport to groundwater can be developed, the process of groundwater recharge must be considered. The hydrologic system within the soil profile can be divided into three functional components. The uppermost, the root zone, extends from the soil surface to the depth to which plant roots commonly penetrate and withdraw water and solutes from the soil matrix. In non-irrigated systems, evapotranspiration generally constitutes the single largest flow path of water from this component (38). Temperature and moisture levels fluctuate widely in this zone as do rates of microbial activity. The root zone can be viewed as the source zone for potential pollutants which are transported to groundwater. Between the root zone and the water table is the region of the unsaturated zone which is not directly affected by plant growth, and which conversely has little direct impact on crop production except in situations where drainage is severely restricted. The minor short-term role in crop production of this region of the soil profile, which recently has been referred to as the intermediate vadose zone (IVZ), probably explains the relative lack of information on physical and biological processes in the IVZ in comparison to the root zone. Recent concerns regarding groundwater pollution, and the obvious role of the IVZ as the conduit for contaminant transfer from the root zone to groundwater have dictated a need for a better understanding of solute transport processes in the IVZ (16,34). The third component of this simplified view of the soil hydrologic system is the saturated zone or unconfined aquifer, bounded above by the water table and below by a confining layer which restricts vertical flow. The relative dimensions of the three functional components of the soil hydrologic system have a direct bearing on the sampling approach required for quantifying rates of solute transport to groundwater.

Root Zone. The region of the soil profile from which plant roots remove water and solutes can generally be delineated, while the thickness of the IVZ and saturated zone will change as the water table fluctuates. The depth to which plant roots penetrate will depend on the crop and its stage of maturity, as well as soil structure, but generally will be limited to the top several meters of the soil profile, with significant water withdrawal usually limited to an even shallower depth. For example, in Figure 1, corn growth, as well as precipitation events, directly modified soil moisture levels at a depth of 0.3 m, while at 1.2 m there was little apparent effect. However despite the relatively shallow extent of direct influence of plant water uptake on soil moisture levels, the effect of plant water uptake on antecedent moisture conditions in upper soil horizons is critical in determining whether precipitation is stored in the root zone or results in water and solute flow through the root zone into the IVZ.

Intermediate Vadose Zone (IVZ). The hydraulic properties and vertical dimensions of the IVZ will determine its potential for solute storage, the time required for water and solute to move from the root zone into groundwater, and thus, the responsiveness of groundwater solute concentrations to changes in solute transport rates from the root zone. The water holding capacity of a soil horizon is primarily determined by soil texture, which is generally well-defined for surface soils in hydrologic transport studies, due to its role in determining the partitioning of precipitation between infiltration and surface runoff (37). Despite the uncertainties regarding the extent of solute dispersion as a consequence of non-uniform flow velocities as water moves through the IVZ (28,33), the water holding capacity of subsoil horizons will, never-the-less, give some indication of the time required for groundwater solute concentrations to reflect changes in solute transport rates from the root zone. Like surface soils, the water holding capacity of subsoil horizons can vary widely as a result of textural differences, thus affecting the total water content and turnover rate of the water stored in the IVZ. For example, Figure 2 demonstrates the wide range in water content that occurs in sub-soil horizons as a consequence of textural differences as well as position relative to the water table. The relatively low water content in the 60-120 cm region of both profiles corresponds to the presence of coarse grain sediments, which were most prevalent at the down-gradient site, resulting in a difference between the two sites of over 7 cm of water in a 60 cm interval of the soil profile. The effect of textural differences on soil water holding capacity is well documented (2,9), and when textural differences exist in soil profiles of several meters in thickness the differences in total water content above the water table can exceed annual recharge volumes by several fold. These differences in water volume will clearly affect the mass of solutes retained in the IVZ and the potential for dilution of leachate moving downward through the soil profile. When attempting to use groundwater solute data for determination of solute leaching rates for a particular land use practice, the potential for solute dilution in the IVZ must be considered.

The data presented in Figure 2 also demonstrate the increase in soil water content that occurs in the unsaturated matrix as the water table is approached. The high degree of water retention in the IVZ above the water table accounts for the large discrepancy between

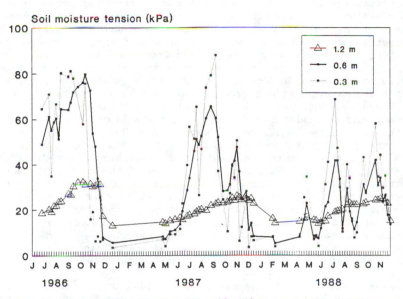

Figure 1.   Seasonal patterns of soil moisture tension under non-irrigated conventionally tilled continuous corn.

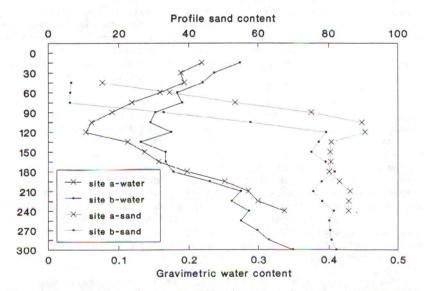

Figure 2.   Changes in sand (>63 micron) and water content with depth in the soil profile at adjacent sites (site a 100 m down-gradient of site b) in a Coastal Plain corn field (11/30/88).

specific yield and total pore space that typically exists for
unconfined aquifers. The relatively small percentage of air- versus
water-filled pore space immediately above the water table reduces the
impact of leachate solute concentrations on groundwater quality. The
largest volume of water added to groundwater during a recharge event
is water held in the unsaturated matrix from previous recharge events,
or from capillary rise, that is added to the groundwater pool by
inclusion as the water table rises. The water holding characteristics
of the IVZ matrix in the region of the water table, combined with the
discharge characteristics of the saturated component of the flow
system, will determine to what degree leachate is diluted by water held
in the unsaturated matrix as the water table rises.

The final point demonstrated in Figure 2 is the difference in
total water volume that can result from relatively minor differences
in the thickness of the unsaturated region of the soil profile. The
additional 60 cm of unsaturated profile at the up gradient site
increased the total water volume in the IVZ by over 25 cm, which is
approximately the average annual recharge volume for this location
(26). Spatial differences in groundwater quality responses to changes
in root zone solute leaching rates can be expected, even where vadose
zone sediments are laterally homogenous, if the thickness of the vadose
zone varies significantly across the monitoring site.

Unconfined Aquifer. The hydraulic properties and dimensions of the
saturated region of a flow system must be considered in a similar
manner as those in the IVZ when developing a sampling strategy for
quantifying the impact of a specific land use on groundwater quality.
However, from a monitoring standpoint, distinct differences exist
between water flow and distribution patterns in the saturated versus
unsaturated component of a flow system. In unconsolidated sediments,
total pore space is affected less by textural differences than is pore
size distribution. Thus, while a coarse sand horizon may hold several
times less water than a silt horizon at the same negative soil water
potential, differences in total water content of the two horizons will
be much less under saturated conditions. Water storage below the water
table will equal the total pore volume, which may be as high as 50
percent on a volume basis (disregarding temporarily entrapped air).
Unconfined aquifers in fractured rock formations will generally have
much lower and more variable volumetric water contents (31), and will
not be considered in this paper. Even though the water volume that can
be pumped from an unconfined formation will be much less than its total
water content, for evaluating solute concentrations in groundwater it
is the total volume rather than the yield or drainable porosity that
must be considered. Thus, even the top meter of the saturated zone may
contain a volume of water in excess of the annual recharge volume in
temperate non-irrigated agricultural systems, emphasizing the need for
judicious well screen placement when attempting to relate groundwater
solute concentrations to a specific set of conditions in the upper soil
profile.

The nature of hydraulic gradients differs dramatically between the
saturated and unsaturated components of a flow system. Since the rate
and direction of groundwater flow are directly dependent on hydraulic
gradients, an understanding of these gradients for a monitoring site
is essential when attempting to relate groundwater solute data to a
particular set of conditions in the root zone. The characteristics of

the profile matrix in the saturated zone will determine how the hydraulic gradient affects groundwater flow rates. Despite the relatively minor impact of textural differences on total water content below the water table in unconsolidated sediments, particle size distribution has a dramatic effect on the saturated flow characteristics of a water bearing formation which has been generally understood for some time (*11*). Saturated hydraulic conductivity typically increases with particle size (as a result of increasing average pore radius, even though total pore space may decrease), meaning that in coarse grained sediments, average cross sectional flow velocities will be higher than those in finer grained sediments for a given hydraulic gradient. Stated another way, the same volume of water can be moved through a coarse grained formation with a lower hydraulic gradient than would be required in a finer textured formation, analogous to the decreasing draw-down required in a well for a given pumping rate as the permeability of the formation increases. Thus, even though lateral velocities in unconfined aquifers increase proportionately at the same rate as a function of the hydraulic gradient, the absolute velocity will increase more rapidly in response to an increasing hydraulic gradient as hydraulic conductivity (K) increases (Figure 3). From a monitoring standpoint, flow velocities and discharge for unconfined aquifers will become more pulsed in response to recharge events as the conductivity of the matrix increases, requiring a more rigorous approach to sample scheduling. Accounting for lateral groundwater flow is critical when attempting to relate a specific land use to groundwater quality data, especially when land use treatments are applied on relatively small research areas. The problem of confounded groundwater solute data as a result of lateral flow under adjacent plots has been encountered in numerous studies (*1,12,24*).

Recharge Cycles. In most non-irrigated agricultural systems, temperature induced fluctuations in plant water uptake from the soil profile impart a seasonality to groundwater recharge patterns. Generally, evapotranspiration exceeds the average rate of precipitation during the middle and latter part of the growing season. During this period, crop water uptake creates soil moisture deficits in the primary root zone and a consequent hydraulic gradient favoring upward movement of water in the soil profile. However, the rapid decline in hydraulic conductivity that occurs when soil moisture levels drop well below field capacity will generally limit the volume of water that moves upward in the soil profile in response to high moisture tensions near the soil surface. Precipitation during this period serves to reduce soil moisture deficits in upper soil horizons but does not generally result in net water movement below the primary root zone, except where upper soil horizons have extremely low water holding capacities. Unusually intense precipitation during the growing season can produce macropore flow below the primary root zone (*23,30*) or surface runoff (*26*) despite the presence of soil moisture deficits in the upper soil profile. The relative importance of these processes will be determined by soil properties and precipitation patterns, but generally the volume of groundwater recharge will be greatly reduced during periods of crop water withdrawal from the soil profile. As the water table declines due to reduced recharge volumes, water held in the soil matrix immediately above the falling water table has by definition been

converted from groundwater to soil pore-water, even though its physical
location has changed very little. Conversely, as the water table rises
soil pore-water can enter groundwater by inclusion. Thus, the movement
of water and solutes into groundwater under conditions of a fluctuating
water table can not be treated as a discrete process, and both
saturated and unsaturated regions must be monitored to accurately
describe the process.

A situation requiring special consideration, is when the water
table rises, in the absence of recharge, as a consequence of an
increase in the hydraulic head of an adjacent hydraulically connected
water body, either surface or sub-surface. This can occur naturally,
where agricultural land is located adjacent to a surface water body
with a fluctuating free surface, or where localized differences in
slope and soil texture create concentrated runoff and recharge zones;
or artificially, where sub-irrigation is practiced. The impact of this
process will be most extensive in regions where the unconfined aquifer
has a relatively low horizontal hydraulic gradient and high hydraulic
conductivity. In these cases, water table rise results in upward
advective movement of groundwater and solutes, a very different
situation from where the water table rises as a result of leachate from
overlying soil horizons entering previously air-filled pore space. A
discussion of upward transport of groundwater solutes is beyond the
scope of this paper, but the critical point, as stressed in previous
sections, is that defining the hydrologic flow system is essential for
correct interpretation of solute data.

Although groundwater recharge is a relatively simple process in a
conceptual sense and can be monitored at one location with little
difficulty, quantifying the redistribution of water in the soil matrix
as the water table rises and falls is more problematic, but essential
for correct interpretation of solute concentration data from both the
saturated and unsaturated zones of the soil matrix. As the water table
declines during periods of high evaporative losses from upper soil
horizons, unless the water table is in the crop rooting zone, former
groundwater remaining in the lower vadose zone is little affected by
crop water withdrawal. As a result, profile moisture levels
immediately above the water table remain near equilibrium (pressure
head = -gravitational head), and the volume of air-filled pore space
remains well below that in upper soil horizons (assuming vertically
homogenous soil hydraulic properties, see Figure 2). The volume ratio
of water- to air-filled pore space in the IVZ is a critical parameter
affecting the short-term impact of leachate from the crop rooting zone
on groundwater quality. This ratio will affect the rate of dilution of
leachate leaving the root zone as it moves through the IVZ as well as
the ratio of "new" to "old" water added to groundwater as the water
table rises. For example, if the volumetric water content of the
vadose matrix immediately above the water table is 0.3 (0.3 m H_2O /m
soil profile) and the total pore volume is 0.4, this region of the
aquifer will have a specific yield of 0.1 (cm of H_2O released/cm drop
in the water table). In this system 2 cm of recharge will result in a
20 cm rise in the water table, and the addition of 8 cm of water to the
unconfined aquifer. In this case, of the 8 cm of water transferred
from the unsaturated to the saturated component of the flow system,
only 2 cm will have solute levels reflecting those in the overlying
soil horizons, while the solute level in 6 cm will reflect the
groundwater solute levels resulting from previous recharge cycles.

Even in profiles where solute levels in leachate leaving the root zone are little affected by matrix solute levels in the IVZ as a result of rapid movement through preferential flow paths, changes in groundwater solute levels will be buffered due to equilibration between leachate and matrix solute levels as vertical flow velocities decrease sharply at the water table. Knowledge of soil pore-water solute concentrations immediately above the water table prior to water table rise is essential when sampling the upper most layer of an unconfined aquifer for purposes of relating agrochemical transport rates to specific agronomic practices.

Groundwater Based Leachate Monitoring

Using shallow groundwater solute levels to determine solute transport rates from the root zone avoids the difficulty involved in trying to quantify the highly transient and variable flow processes in the vadose zone during recharge events. However, if groundwater solute data is to be used for quantifying solute leaching rates for a specific land use, it must be possible to associate a particular groundwater sample with a time and place in the root zone. Establishing this relationship becomes increasingly difficult as the water volume in the flow system between the root zone and point of groundwater sampling increases. Thus, the sampling approach should focus on reducing this volume to the extent possible. This means sampling in, as opposed to down-gradient of the treatment area, and as close as possible to the water table. In systems where the IVZ is relatively thin and coarse textured, a sampling approach based strictly in the saturated zone will be sensitive to changes in root zone solute transport rates (*12*). However, the sensitivity of groundwater sampling strategies to root zone leaching rates decreases rapidly as the depth of sampling below the water table increases (*1,13,22*). Figure 4 also demonstrates how rapidly solute concentrations can change with depth below the water table.

Stratification of solute concentrations near the water table becomes increasingly important if groundwater has a significant lateral component, especially where the treatment area is small. A well screened too far below the water table may be temporally insensitive to changes in solute transport rates from the root zone, if groundwater flow is primarily vertical, but may completely fail to sample leachate from the overlying root zone where groundwater flow is predominantly horizontal. In unconfined flow systems the location of the monitoring site relative to the water table divide will play a role in the ratio of vertical to horizontal flow as well as the velocity of water moving through the formation (*20,21*). Where practical, locating research sites as close as possible to the water table divide will reduce the confounding effects of up-gradient land uses on groundwater solute data, and allow more time for sampling of leachate after it has entered groundwater under a small treatment area.

The stratification of nitrate near the water table depicted in Figure 4 also demonstrates the need to consider annual fluctuations in the water table when trying to install a groundwater sampling system that is sensitive to changes in root zone leaching rates. Precipitation patterns, aquifer discharge rates, and the water holding characteristics of the IVZ in the boundary region between the saturated and unsaturated components of a flow system will determine the annual

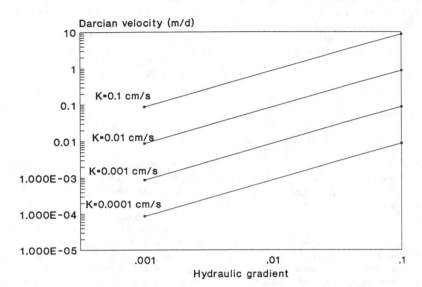

Figure 3. Darcian velocity as a function of the hydraulic gradient and saturated hydraulic conductivity (K). Note: Pore water velocity will be greater depending on effective porosity.

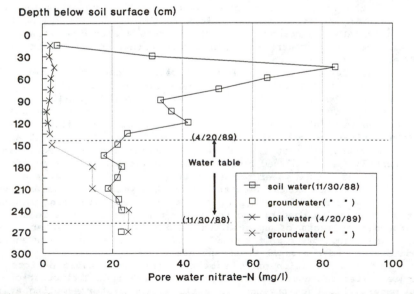

Figure 4. Changes in pore water nitrate concentration above and below the water table during a recharge cycle following the establishment of a rye cover crop (data from site a, Figure 2).

variation in water table elevation at a monitoring site. For sites where the water table fluctuates seasonally and vertical groundwater flow is limited, it may be necessary to screen wells in the transient saturated/unsaturated region of the profile to reduce the water volume, and resulting time lag, between root zone leaching events and collection of representative groundwater samples. Wells screened just under the annual water table minimum elevation may sample below several meters of saturated profile when the water table is elevated. These wells will be insensitive in the short-term to pesticide leaching events following early spring applications, when the water table is typically near its annual maximum in temperate non-irrigated agricultural systems. Using longer single well screens to sample across the annual range in water table elevation gives immediate representation to leachate entering groundwater as the water table rises. However, as screen length and the volume of groundwater represented by a single well increase, sensitivity to changes in solute leaching rates that create groundwater solute distribution patterns like those in Figure 4 is sacrificed. An alternative is to include wells in the monitoring network that are screened in the transient saturated/unsaturated region of the profile. Wells screened above the water table minimum will be dry for some part of the year, similarly to those installed for sampling perched groundwater (34). Another option is to install single casings modified to allow sampling at discrete depths (6,10,19), with sampling ports in the permanently saturated, as well as the transient saturated/unsaturated region of the profile.

As the thickness and water holding capacity of the IVZ increase, or as annual recharge volume decreases, sampling strategies for relating solute leaching rates to a specific land use must include a larger unsaturated zone component in order to properly interpret groundwater solute data. Sampling in the IVZ allows determination of changes in solute storage between the root zone and the water table while sampling the surface of the unconfined groundwater gives an indication of the quantity of solute moving through the IVZ. Sampling in the IVZ should not be undertaken during major periods of water percolation through the profile but rather during periods of relatively stable hydraulic gradients. This approach is especially applicable in non-irrigated systems where groundwater recharge is highly seasonal. The most crucial aspect of sampling the IVZ is for determination of water and solute storage, as well as the volume of unfilled pore space, immediately above the water table prior to recharge, since groundwater solute concentrations just below the water table during recharge will be highly influenced by these parameters. The importance of unsaturated zone sampling just above the water table dictates that the total depth of sampling should be based on the depth to the water table rather than on the depth below the soil surface. This will require sampling to differing depths in the unsaturated profile across a monitoring site and throughout the year to accommodate spatial and temporal changes in the depth to groundwater. Core sampling is the most straightforward single method for direct determination of the distribution of both water and solutes in the IVZ, and permits flexibility in sampling depth. However, many less labor-intensive and destructive alternative IVZ sampling methods are available (34-36), and several are discussed in detail elsewhere in this symposium.

Conclusions

Achieving improved groundwater quality under agricultural land will require implementation of practices which reduce rates of solute leaching from the root zone. However development of these practices will require methods to evaluate the rates of solute leaching associated with specific agricultural practices. The recent profusion of studies documenting rapid and highly variable rates of water and solute movement through the vadose zone suggests that groundwater solute data may give the most reliable indication of solute leaching rates, due to the lower and more stable hydraulic gradients that generally exist in the saturated versus the unsaturated component of a vadose zone–unconfined aquifer flow system. However, using groundwater solute data for determining rates of contaminant transport from the root zone as a result of specific agricultural practices requires different sampling methods than those used for monitoring solute transport in groundwater, as well as careful consideration of the hydraulic characteristics of the vadose zone–unconfined aquifer flow system. The distribution and volume of water in the IVZ will determine the sensitivity of groundwater solute concentrations to changes in rates of solute transport from the root zone. Thus, geohydrologic systems with a relatively thin layer of unsaturated subsoil are vulnerable to rapid groundwater contamination by agricultural activities, and therefore present ideal opportunities for using groundwater solute data to clarify the rates of contaminant transport from the root zone associated with specific agricultural practices. However, even an IVZ of less than 5 m in thickness can store a volume of water several times greater than the annual recharge volume in temperate non-irrigated agricultural systems, with potential water storage varying widely as a function of the pore size distribution in the IVZ matrix. In addition, even if solute levels in water leaving the root zone are little influenced by solute levels in the IVZ as a consequence of rapid movement through preferential flow paths, when this water slows dramatically as it reaches the lower boundary of the unsaturated zone, solute equilibration will occur between water held previously in the IVZ matrix immediately above the water table and leachate. The relative volume of water- to air-filled pore space in the unsaturated zone immediately above the water table will determine to what extent leachate solute levels are reflected in groundwater solute levels as the water table rises. Thus, even if stratified sampling methods are used to sample groundwater just below the water table as it rises during a recharge cycle, solute and moisture levels in the vadose zone above the water table prior to recharge must be taken into account in order to determine leachate volume and solute concentrations. Using stratified saturated sampling in the transient saturated/unsaturated region of the soil profile as the water table rises in conjunction with stratified sampling of the IVZ during relatively static flow conditions provides a method for quantifying solute leaching rates for a specific land use which circumvents the difficulty associated with attempts to directly measure highly variable and transient solute transport processes in the vadose zone.

Literature Cited

1. Bergstrom, L.; Brink, N. **Plant and Soil.** 1986, 93: 333-354.
2. Brady, N.C. **The Nature and Properties of Soils;** Macmillan Pub. Co.: New York, NY, 1984.
3. Bouwer, H. In: **Toxic Substances in Agricultural Water Supply and Drainage.** Summers, J.B.; Anderson, S.S., Eds.; U.S. Committee on Irrigation and Drainage. Denver, CO, 1989, pp 1-14.
4. Burwell, R.B.; Schuman, G.E.; Saxton, K.E.; Heinemann, H.G. **J. Environ. Qual.** 1976, 3: 325-329.
5. Creighton, J.L. In: **Toxic Substances in Agricultural Water Supply and Drainage.** Summers, J.B.; Anderson, S.S., Eds.; U.S. Committee on Irrigation and Drainage. Denver, CO, 1987, pp 1-21.
6. Galgowski, C.G.; Wright, W.R. **Soil Sci. Soc. Am. J.** 1980, 44: 1120-1121.
7. Germann, P.; Beven, K. **J. Soil Sci.** 1981, 32: 1-13.
8. Hallberg, G.R. **J. Soil and Water Cons.** 1986, 41: 357-364.
9. Hanks, R.J.; Ashcroft, G.L. **Applied Soil Physics;** Springer-Verlag: New York, NY, 1980.
10. Hansen, E.A.; Harris, A.R. **Water Resour. Res.** 1980, 16: 827-829.
11. Harr, M.E. **Groundwater and Seepage;** McGraw-Hill: New York, NY, 1962.
12. Hubbard, R.K.; Asmussen, L.; Allison, H. **J. Environ. Qual.** 1984, 13:156-161.
13. Hubbard, R.K.; Gascho, G.J.; Hook, J.E.; Knisel, W.G. **Trans. Amer. Soc. Agri. Eng.** 1986, 29: 1564-1571.
14. Hubbard, R.K.; Sheridan, J.M. **J. Environ. Qual.** 1983, 12: 291-295.
15. Kanwar, R.S.; Everts, C. **Amer. Soc. Agri. Eng.:** St. Joseph, MI, 1988 ASAE paper No. 88-2027.
16. Nielsen, D.R.; van Genuchten, M.Th.; Biggar, J.W. **Water Resour. Res.** 1986, 22: 89-108.
17. Parkin, T.B.; Meisinger, J.J. **J. Environ. Qual.** 1989, 18: 12-16.
18. Peterjohn, W.T.; Correll, D.L. **Ecology.** 1984, 65: 1466-1475.
19. Pickens, J. F.; Cherry, J.A.; Coupland, R.M.; Grisak, G.E.; Merritt, W.F.; Risto, B.A. **Ground Water Mon. Rev.** 1981, 1: 48-51.
20. Saines, M. **Ground Water Mon. Rev.** 1981, 1 (1): 56-61.
21. Schmid, G. In: **Finite Elements in Water Resources;** U. of Miss. Press: University, MS, 1980; Vol. 3, pp 2.45-2.48.
22. Sgambat, J.P.; Stedinger, J.R. **Ground Water Mon. Rev.** 1981, 1 (1): 62-69.
23. Smetten, K.R.J.; Trudgill, S.T.; Pickles, A.M. **J. Soil Sci.** 1983, 34: 499-509.
24. Spalding, R.F.; Exner, M.E. **J. Environ. Qual.** 1980, 9: 466-479.
25. Spalding, R.F.; Kitchen, L.A. **Ground Water Mon. Rev.** 1988, 8: 89-95.
26. Staver, K.W.; Brinsfield, R.; Magette, W. **Am. Soc. Agri. Eng.:** St. Joseph, MI, 1988, ASAE Paper No. 88-2040.
27. Staver, K.W.; Brinsfield, R.; Stevenson, J.C. In: **Toxic Substances in Agricultural Water Supply and Drainage.** Summers, J.B.; Anderson, S.S., Eds.; U.S. Committee on Irrigation and Drainage. Denver, CO, 1989, pp 163-179.

28. Steenhuis, T.S.; Hagerman, J.R.; Pickering, N.B. In: **Groundwater Issues and Solutions in the Potomac River Basin/Chesapeake Bay Region**. National Well Water Assoc.: Dublin, OH, 1989, pp 397–419.
29. Steenhuis, T.S.; Muck, R.E. **J. Environ. Qual.** 1988, 17: 376–384.
30. Thomas, G.W.; Phillips, R.E. **J. Environ. Qual.** 1979, 8: 149–152.
31. Urban, J.B. In: **Watershed Research in Eastern North America**; Correl D.L., Ed.; Smithsonian Institue, Washington, D.C., 1977, pp 468–481.
32. Wagenet, R.J. In: **Effects of Conservation Tillage on Groundwater Quality**. Logan, T.J.; Davidson, J.M.; Baker, J.L.; Overcash, M.R., Eds.; Lewis Publishers, Chelsea, MI, 1987, pp 189–204.
33. White, R.E.; Dyson, J.S.; Haigh, R.A.; Jury, W.A.; Sposito, G. **Water Resour. Res.** 1986, 22: 248–254.
34. Wilson, L.G. **Ground Water Mon. Rev.** 1981, 1 (3): 32–40.
35. Wilson, L.G. **Ground Water Mon. Rev.** 1982, 2 (1): 31–42.
36. Wilson, L.G. **Ground Water Mon. Rev.** 1983, 3 (1): 155–166.
37. Wischmeier, W.H. In: **Control of Water Pollution from Cropland.** Report No. ARS-H-5-2; A.R.S., U.S.D.A.: Washington, D.C., 1976, Vol. II, pp 31–57.
38. Woolhiser, D.A. In: **Control of Water Pollution from Cropland.** Report No. ARS-H-5-2; A.R.S., U.S.D.A.: Washington, D.C., 1976, Vol. II, pp 7–29.
39. Younos, T.M.; Weigmann, D.L. **J. Water Poll. Contr. Fed.** 1988, 60: 1199–1205.

RECEIVED September 17, 1990

Chapter 17

Experiences and Knowledge Gained from Vadose Zone Sampling

J. L. Starr[1], J. J. Meisinger[1], and T. B. Parkin[2]

[1]Environmental Chemistry Laboratory, Agricultural Research Service, U.S. Department of Agriculture, BARC–West, Beltsville, MD 20705–2350
[2]National Soil Tilth Laboratory, Agricultural Research Service, U.S. Department of Agriculture, 2150 Pammel Drive, Ames, IA 50011

Vadose zone sampling offers an opportunity for assessing the impact on groundwater quality of chemicals applied at the land surface. Many interacting factors control the fate of chemicals in the field cause major sampling problems even for experienced researchers. Underlying any sampling program is the absolute need to clearly define the study's objectives. The sampling procedure should then be developed with a clear conceptual view of the physical, chemical, and biological processes that affect the fate of the chemical(s) under investigation. Basic questions regarding the spatial, temporal, and statistical distributions of specific parameters must also be addressed in developing an efficient sampling plan. There is no "best sampling method" for all situations, rather, there are several techniques with attendant advantages and disadvantages. An efficient sampling plan considers: the underlying processes; spatial, temporal, and statistical distributions of important parameters; and limited resources to answer the study's objectives.

Increasing concern about the presence of nutrients, pesticides, and other chemicals in shallow and deep groundwater, along with the difficulty in quantifying and predicting their transformations and movement, has recently led to several symposia on the vadose zone (1-4). These symposia illustrate the cross-disciplinary nature of the problems that investigators face in attempting to characterize and quantify the fate of chemicals in the environment.

Vadose Zone Dynamics

The vadose zone represents the three-dimensional geological profile above the ground water table, extending to or through (depending on the chosen definition) the biologically active layer at the soil surface. Depending on the three-dimensional spatial distribution in hydraulic conductivities, portions of the vadose zone may become saturated for varying lengths of time following precipitation or irrigation events. The particular problems and opportunities associated with sampling the vadose zone are inherent in its nature.

The depth, hydraulic properties, and the nature of the chemical transformations in the vadose zone combine with the hydrologic recharge cycle to determine the extent to which chemicals applied to the land surface impact groundwater resources. Under conditions of one-dimensional downward flow, a deep vadose zone can potentially store a larger quantity of a particular solute than a shallow vadose zone, and therefore can supply solutes to groundwater for many years after the solute is no longer applied to the soil surface. For non-conservative chemicals this vadose zone condition may provide the time-space needed for significant solute attenuation to occur before it reaches the groundwater. In contrast, solutes in a shallow vadose zone may move quite rapidly to the groundwater, but the potential for residual effects is reduced. (Three-dimensional flow effects will be mentioned later in this chapter).

The concept of a representative elementary volume (REV) of a porous material can provide a useful framework for estimating the volume of individual samples needed to properly characterize vadose zone parameters (5-8). A REV is the sample volume, from a given domain, for which individual measurements of a given parameter (P) approach a statistical constant, independent of the sample volume. At very small sample volumes, the range of P values will fluctuate greatly due to extreme effects that can occur at the microscopic level. As the sample volume increases, the microscopic effects decrease and macroscopic effects of the domain increasingly dominate the variation of P. The REV is the sample volume at which the parameter variation is primarily controlled by macroscopic conditions. The REV may be quite different for different measured parameters, and may vary with time due to changing conditions that impact on that parameter.

Many factors (e.g., vadose zone properties and cultural treatments) affect the fate (transformations, adsorption-desorption, and movement) of chemicals in the vadose zone (9,10). A partial listing of dynamic and static vadose zone properties and model parameters is presented in Table 1 (compare Table 2-2 in Jury (9)). *Static* properties are associated with their location in the vadose zone. Measurements of the *static* properties near the land surface may change in time due to biological activity, temperature, tillage, soil traffic, etc. The dynamic vadose zone factors are largely associated with the water phase (and are affected by factors that move water in the vadose zone), and biological phenomena. Many of the factors are interdependent which can result in large spatial and temporal variation. This interdependency often gives rise to frequency distributions that are highly skewed and coefficients of variation in excess of 100% (11).

Table 1. Partial list of vadose zone properties and model parameters that can affect solute concentrations

DYNAMIC	STATIC*
Water content, θ	Texture
Evapotranspiration	Bulk density
Biological activity	Porosity
Solute concentration	Soil water characteristic
Solute velocity and	Adsorption parameters
dispersion coefficients	Cation exchange capacity
Structure	Saturated hydraulic
Hydraulic conductivity, K_0	conductivity, K_s

*or "Semi-Static" as most of these may be affected by biological activity, cultural practices, etc.

Sampling the Vadose Zone

Sampling the vadose zone is required in order to quantify and predict the rates of reaction, transformation, and movement of chemicals downward to groundwater, and laterally along inclined textural layers and in phreatic layers to drainage ditches, streams, etc. Proper development of a vadose zone sampling scheme requires knowledge of the principal factors controlling the fate of the chemicals in the vadose zone, their interdependencies, and the probable frequency distributions of the observations. An initial estimate of these items may be available in the literature, but a preliminary sampling survey will provide the best information for determining the best experimental methodology at a specific site.

A wide variety of techniques exists for sampling the liquid and solid phases of the vadose zone. Many of their advantages and disadvantages are presented elsewhere in this book as well as in recent literature reviews (1, 12-14). Several of these sampling methods are shown in Table 2. It is important to recognize that different vadose zone sampling methods may also measure different soil properties (e.g., solute or soil mass concentrations), and reflect different time increments (e.g., at a given time or averaged across time), and zones of influence (e.g., particle surfaces, micro to macro pores). Because different sampling methods sample different entities (Table 2), it is un-

derstandable that data produced by different research
studies will often produce seemingly conflicting interpretations. This is particularly true in the early stages
of a research thrust before a unified understanding of the
phenomena emerges.

**Table 2.   Characteristics of sampling methods for the
vadose zone, including saturated layers**

Method	Samples[a]	Time Step[b]	Pores[c]	Depth (m)
Suction Cups	C	P	b,c	0.1-3
Tile Lines	C, M	A,E	b,c	<1
Pan Lysimeter	C, M	E	b,c	1-2
Shallow Wells	C	A,E	b,c	1-5
Deep Wells	C	A	b,c	>5
Excavation:				
Augers, cores	C, M	P	a	0-1+
Dye tracers	FP	E	b,c	0-2+

[a] **C**: Pore Water Concentration; **M**: Mass; **FP**: Flow Path.
[b] **P**: Point in time; **A**: Average flow; **E**: By Event.
[c] Diameters (μm) **a**: <10; **b**: 10-1000; **c**: >1000.

Other chapters in this book have emphasized that underlying any sampling program is the absolute need to
clearly define the study's objectives, and to identify the
specific vadose zone information that is needed to achieve
the objectives. Even though there is a tendency to use
only one or two personally "tried and true" methods for
all investigations, the multiplicity of methods available
for sampling the vadose zone (12-14), suggests that there
is no "best" sampling method for all situations. A survey
of a watershed for the mass of chemical residing in the
vadose zone at a point in time necessitates a different
approach than that needed to monitor the changes in chemical
concentration in the mobile water phase as a function of
depth over some time interval. For example, a watershed
survey might be accomplished using soil core data while
monitoring solute chemical concentrations might be accomplished with suction cups or shallow wells. Several
approaches are usually needed to characterize and quantify
the processes controlling the fate and movement of a vadose
zone chemical to the groundwater (9).

Evaluating Sampling Techniques

In an attempt to determine the REV for several soil parameters from soil cores, e.g., denitrification (15) along

with several other chemical and physical properties (16), we conducted four vadose zone sampling experiments with six sample volumes (38, 58, 149, 216, 366, and 8770 cm^3) from the surface horizon. In the first experiment, 36 replicate samples were taken to a depth of 16 cm from within a 1.2 x 1.8 m area. Five soil cores were each taken inside each 8770 cm^3 sample (0.2 by 0.3 m template). In the first experiment, the large rectangular samplers were placed directly adjacent so that all the soil in the sampling area was removed from the sampling area. Hence the weighted mean of each measured parameter was equal to the population mean. Due to the disproportionate volume of the rectangular sample (size 6), the population mean was also approximately the same as the mean of size 6.

Figure 1 shows the chemical mass (relative to the population mean, dashed line) for the first five sample sizes for NO_3-N and ortho-P, with 90% confidence limits (solid lines). Land's method to compute the confidence intervals was used because it provides exact confidence intervals for lognormal distributions (17,18). The amount of skewness for each sample volume mean may be judged by the amount of displacement of the mean from the midpoint between Land's 90% confidence intervals. Although the relative variation for N was nearly twice that for P, the overlapping confidence intervals for the different sample and population means indicate that the REV for both parameters may be somewhat smaller than the smallest sample volume (38 cm^3).

Bootstrapping. Bootstrapping provides another way to estimate the REV for this data set because it provides a way to characterize the range of variances associated with the different sample sizes. Bootstrapping is a computer-intensive random resampling technique (19) that requires no assumptions regarding the underlying population distribution (19-22). For this method, data subsets are randomly drawn from the original data set with the number of observations (*n*) for each subset being the same as in the original data set. As an example of this method, 1000 NO_3-N variances, associated with the 1000 bootstrap means, are plotted for each sample volume (Figure 2). Based on overlapping 95% confidence intervals, the variances are not statistically different for the first five sample sizes. However, the 50% decrease in variance from the first to second sample size may indicate that the REV is better represented by sample volumes closer to 58 cm^3, rather than "somewhat < 38 cm^3" as suggested above.

Random subsampling. Random subsampling is another computer-intensive random resampling technique that can be used to determine the total mass of soil required to best estimate the population mean for variables with a lognormal distribution. Random subsampling differs from bootstrapping in that the number of observations randomly drawn

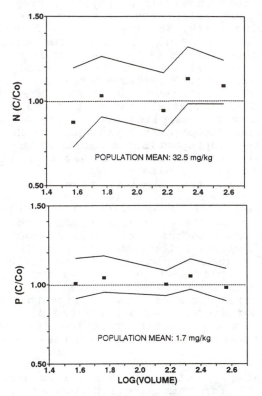

Figure 1.   Mean relative nitrate-N and ortho-P concentrations (solid squares) with 90% confidence window (solid line) vs $\log_{10}$[sample volume (cm^3)]. The dashed line represents the population mean.

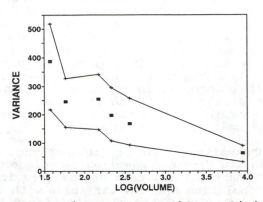

Figure 2. Mean variance and confidence limits based on 1000 bootstrap sample means for each sample size.

may be less than the number in the original data set. For this analysis, the data for sample sizes 1-5 were pooled into one data set (n = 180). Subsets of the data were randomly drawn from this pooled data set, with the number of samples drawn for successive subsets being incremented from 2 to 100. The mass of NO_3-N and soil mass were accumulated with each data subset. This procedure was repeated 10,000 times. Counts of the weighted sample means that came within 10, 20, and 50% of the measured population means are shown in Figure 3. Note that 2 kg of soil gave sample means that were within 10% of the mean only 45 and 58% of the time for NO_3-N and ortho-P, respectively. More than 10 kg of soil was required to approach the plateau for sample means within 10% of the true mean for both parameters.

Preferential Flow in the Vadose Zone

Vadose zone conditions may exist in which it is very difficult to determine the REV, e.g., conditions in which water and solutes can move rapidly through the vadose zone, by-passing most of the vadose zone matrix. This rapid flow along preferential flow paths in the upper part of the vadose zone is commonly caused by soil structure, decayed roots, worm holes, etc. (23-24). Fingering, another type of rapid flow along preferential paths, may occur at any vadose zone depth during saturated flow through a fine textured layer into a coarser textured layer below. At the textural interface the infiltrating water will break into fingers of saturated flow, and then moves at a speed of K_S through the coarser textured layer (23-26). Any type of preferential flow can result in a serious misunderstanding of adequate (or proper) sampling as demonstrated below. Evidence for preferential flow with the infiltrating solute by-passing most of the soil matrix under field conditions has been directly observed by infiltrating dye tracers followed by excavation by layers or vertical cuts (24,27), and inferred from leaching with bromide and atrazine (6-chloro-N-ethyl-N'-(1-methylethyl)-1,3,5-triazine-2,4-diamine) (28).

Some of the problems that can occur in terms of sampling and interpreting results from sites with preferential flow paths may be envisioned from Figure 4. This figure is a schematic cross-section of a 1.8-m diameter column at a depth of 1 m, following the ponding of a dye tracer solution on the soil surface (27). After infiltration, the column was excavated down through the fine sandy loam surface layer into the coarse sand layer. Dotted areas in this figure represent zones containing the dye tracer in the coarse sand layer. These dyed regions ranged from 5-20 cm diameter and occupied only 5% of the cross-sectional area of the column. Analysis of the shallow groundwater, 1 m below this cross-section, showed almost all the dye solution at position A and none at position C (Figure 4). Under

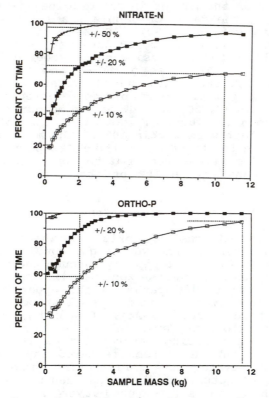

Figure 3. Percent of the time that sample means fall within 10, 20, and 50% of the true mean, as a function of soil mass.

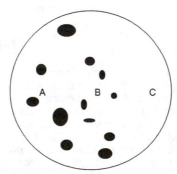

Figure 4. Schematic cross-section of the 1.8-m diameter column at the 1.0-m depth. The dotted areas represent dye tracer zones. Water samples were taken at positions A, B, and C in the shallow groundwater (2.1-m depth). (Adapted from ref. 27.).

such preferential flow conditions, it may not be possible to obtain a sample large enough to represent a REV. Hence, any sampling method for these conditions may lead to conflicting conclusions regarding the fate of a chemical in the vadose zone. At this site, with knowledge of the vadose zone factors controlling the movement of chemicals, it may be better to sample the shallow groundwater rather than to attempt to sample the vadose zone. The REV for groundwater samples at this site will be much larger than for sites where preferential flow does not occur. A good discussion related to the effects of different sample sizes under conditions of preferential flow is given by Nielsen et al. (5).

Statistical Analysis of vadose zone Data

At several points in this chapter, a high degree of variability was commonly exhibited in environmental data. High variability often results in a high degree of uncertainty with regard to the statistical estimation and inference processes, which then give rise to low discriminating power for hypothesis testing procedures. In recent years, many statistical techniques have been applied to soil parameters in an attempt to characterize their variability and which may be applicable to the vadose zone. Two techniques described earlier were Land's exact method for confidence intervals, and two computer-intensive random resampling techniques (bootstrapping and random subsampling). Other methods include geostatistics (29-31), spectral and autocorrelation analysis (32), state-space analysis (33), and the use of management models like PRZM and GLEAMS (34-35). Even as there is no "best" sampling method for vadose zone studies, there is no "best" statistical method for characterizing the variability that is inherent in environmental studies. Rather, extraction of the most accurate and useful information from vadose zone data often requires application of several methods of statistical analyses. Finally, with the aid of appropriate mathematical models one may begin to estimate the time, concentration, and mass of chemicals entering groundwater (2,8,10,30).

Need for cross-disciplinary research

The variety and interdependencies of the many factors that control the fate of chemicals in the vadose zone, the variety of sampling methods to measure the physicochemical properties of the vadose zone, and the characteristic spatial and temporal variability of most data, can lead to despair in accurately predicting the fate of chemicals in the vadose zone and groundwater. No one has all the knowledge and resources needed for vadose zone studies. Cross-disciplinary research efforts are required in order to gain the necessary understanding to provide the basis

for improved management of chemicals in the environment.
Many barriers often turn up to obstruct the implementation
of cross-disciplinary research.   Some of these barriers
may be overcome by encouraging team research at the start
of vadose zone projects, and by rewarding the contributions
of individuals on a team the same as done for solo scientific
contributions.   It is through such a process that the most
rapid progress can be made toward solving the complex
problems associated with groundwater quality.

LITERATURE CITED

1.   Germann, P.F., (ed.). J. of Contaminant Hydrology.
     1988, 3(No. 2-4), (Special Issue).
2.   Hern, S.C.; Melancon, S.M. (eds). Vadose Zone Mod-
     eling of Organic Pollutants; Lewis Publ., Inc.:
     Chelsea, MI. 1987.
3.   Nelson, D.W. (ed.). Chemical Mobility and Reactiv-
     ity in Soil Systems, SSSA Spec. Publ. 11. SSSA,
     Madison, WI. 1983.
4.   Proceedings of the NWWA Conference on Characteriza-
     tion and Monitoring of the Vadose (Unsaturated)
     Zone. Natl. Water Well Assoc., Dublin, Ohio. 1985
5.   Nielsen, D.R.; Wierenga, P.J.; Biggar, J.W. In
     Chemical Mobility and Reactivity in Soil Systems.
     SSSA Spec. Publ. 11. SSSA, Madison, WI. 1983; Chap-
     ter 5.
6.   Hubbert, M.K. In Trans. Am. Inst. Min. Met. Eng.
     1956. 207, 222-39.
7.   Bear, J. Dynamics of Fluids in Porous Media. Ameri-
     can Elsevier, N.Y. 1972.
8.   Bear, J; Verruijt, A. Modeling Groundwater Flow and
     Pollution. Reidel Pub. Co., Boston. 1987.
9.   Jury, W.A.; Valentine, R.L. In Vadose Zone Model-
     ing of Organic Pollutants; Hern, S.C. and Melancon,
     S.M., Eds.; Lewis Publ., Inc.: Chelsea, 1987; Chap-
     ter 2.
10. Nielsen, D.R.; Van Genuchten, M.Th.; Biggar, J.W.
     Water Resour. Res. 1986, 22, 89S-108S.
11. Parkin, T.B.; Meisinger, J.J.; Chester, S.T.;
     Starr, J.L.; Robinson, J.A. Soil Sci. Soc. Am. J.
     1988, 52, 323-329.
12. Everett, L.G.; Hoylman, E.W.; Wilson, L.G.; McMil-
     lion, L.G. Ground Water Monitoring Rev. 1984, 4,
     26-32.
13. Klute, A. (ed). Methods of soil analysis, Part 1.
     Agronomy No. 9. Am. Soc. Of Agron., Madison, Wi.
     1986.
14. Collins, A.G., and Johnson, A.I. (eds). Ground-
     Water Contamination: Field Methods. ASTM. Phila-
     delphia, PA. 1988; 491 pp.
15. Starr, J.L.; Parkin, T.B.; and Meisinger, J.J.
     Soil Sci. Soc. Am. J. 1987. 51, 1492-1501.

16. Parkin, T.B.; Starr, J.L.; and Kunishi, H.M. In Agron. Abstr. 1984. p. 172.
17. Land, C.E. Ann. of Math. Stats. 1971. 42, 1187-1205.
18. Parkin, T.B.; Chester, S.T.; Robinson, J.A. Soil Sci. Soc. Am. J. 1990, 54, 321-6.
19. Efron, B. Soc. Indus. Appl. Math. 1979. 21, 460-80.
20. Efron, B. In CBMS-NSF Regional Conference Series in Applied Mathematics, Soc. Indus. Appl. Math., No. 38, Philadelphia. 1982.
21. Diaconis, P.; Efron, B. Sci. Am. 1983. 248, 1116-30.
22. Tichelaar, B.W.; Ruff, L.J. EOS Transactions. Am. Geophy. Union. 1989. 71, 593.
23. Edwards, W.M; Shipitalo, M.J.; Dick, W.A.; Norton, L.D.; Owens, L.B. In Agrochemical Residue Sampling, Design and Techniques: Soil and Groundwater. Amer. Chem. Soc., Boston, MA, April 22-27, 1990. Agri-chemicals Div., No. 53.
24. Steenhuis, T.S. J. Irrig. Drain. Eng. 1990, 116, 50-66.
25. Hill, D.E.; Parlange, J.-Y. Soil Sci. Soc. Am. J. 1972, 36, 697-702.
26. Glass, R.J.; Steenhuis, T.S.; Parlange, J.-Y. Soil Sci. 1989, 148, 60-70.
27. Starr, J.L.; DeRoo, H.C.; Frink, C.R.; Parlange, J.-Y. Soil Sci. Soc. Am. J. 1978, 42, 376-91.
28. Starr, J.L.; Glotfelty, D.E. J. Environ. Qual. 1990, 19:552-558.
29. Burgess, T.M.; Webster, R. J. Soil Sci. 1984, 35, 127-40.
30. Rao, P.S.C., Wagenet, R.J. Weed Sci. 1985, 33, 18-24.
31. Wagenet, R.J.; Rao, P.S.C. Weed Sci. 1985, 33, 25-32.
32. Kachanoski, R.G.; Rolston, D.E.; De Jong, E. Soil Sci. Soc. Am. J. 1985, 49, 804-12.
33. Morkoc, F.; Bigger, J.W.; Nielsen, D.R.; Rolston, D.E. Soil Sci. Soc. Am. J. 1985, 49, 798-803.
34. Carsel, R.F.; Nixon, W.B.; Gallantine, L.G. Envi-ron. Toxicol. and Chem. 1986, 5, 345-363.
35. Smith, M.C.; Campbell, K.L.; Bottcher, A.B.; Tho-mas, D.L. Amer. Soc. Agr. Eng./Can. Soc. Agr. Eng., 1989, paper no. 89-2072.

RECEIVED September 28, 1990

Chapter 18

Tension Lysimeters for Collecting Soil Percolate

J. Scott Angle, Marla S. McIntosh, and Robert L. Hill

Department of Agronomy, University of Maryland, College Park, MD 20742

Tension lysimeters are widely used to sample soil percolate. A vacuum is applied to the interior of a porous ceramic cup and soil percolate is pulled into the cup and held until collection. Many questions, however, exist as to the proper use of lysimeters. Foremost among the questions is the source of water which is pulled into the lysimeter. Lysimeters generally collect larger volumes of percolate during peak flow events when soil water is being retained at lower suctions, and thus may not accurately estimate the magnitude of solute losses. Problems also exist in the use of lysimeters to measure specific pollutants. Many pesticides are volatile, especially under reduced pressure, and concentrations are likely to be underestimated using tension lysimeters. Nutrient analysis of percolate collected with lysimeters is often skewed due to adsorption or desorption of inorganic ions. An additional problem exists with the analysis of resulting data. Since sampling times are not randomized, usual assumptions for analyses, such as independence of error, may not be valid. Measurements are often lognormally distributed and thus require transformation.

The presence of nutrients and pesticides in soil water provides evidence of potential leaching and contamination of groundwater. Numerous methods have been used in the past to collect soil water; however, each of these methods have been limited by cost, technical or theoretical limitations. The use of suction or tension lysimeters to collect and monitor soil water from the unsaturated or vadose zone has been practiced on a routine basis since the early 1960's.

0097–6156/91/0465–0290$06.00/0
© 1991 American Chemical Society

The first use of a tension lysimeter was reported in 1904 *(1)*. The authors used a small porous ceramic cup to which a vacuum could be applied. The cup, called an "artificial root", was used to study soil water availability and the composition of the water. A variety of designs have since been investigated including porous disks or plates *(1-4)*, fritted glass filters and cups *(5-7)*, and ceramic cups *(8-13)*. In 1961, G.H. Wagner of the University of Missouri and the SoilMoisture Equipment Company (Santa Barbara, CA) assembled and tested the first commercially-available tension lysimeter. The design, which has changed only slightly during the last 30 years, has become the single most popular method of collecting soil water and will be the topic of subsequent discussion *(8)*. In in a field comparison of soil solution samplers, the commercial lysimeter was the "best" cup-type sampler for soil solution in terms of minimum alteration of soil solution, low failure rates, and adequate sample volumes *(7)*.

Lysimeter Description

The commercial tension lysimeter consists of a 4.8 cm (OD) polyvinyl chloride chamber tube attached to a ceramic cup. The top of the cylinder is sealed and polyethylene inlet and outlet tubing allow pressure regulation and sample removal. One tube extends to the bottom of the cup for sample collection while the other tube extends just below the lysimeter top and is used to regulate the pressure. It is important to ensure that the long tube extends to the bottom of the cup *(14)*. If the tube is not properly placed at the cup bottom, residual sample may remain in the cup after sample removal, resulting in a "dead space" of as much as 80 mL. This residual sample would dilute and contaminate subsequent samples.

The ceramic cups are typically available in two different porosities which have different flow properties. The bubbling or air-entry pressure of a ceramic cup is the air pressure necessary to force air through a porous cup which has been saturated with water. A low-flow ceramic cup (standard 0.2 MPa cup) has a bubbling or air-entry pressure of 0.2 MPa and has a maximum pore diameter of approximately 1.4 microns. A high-flow ceramic cup (standard 0.1 MPa cup) has an air-entry pressure of 0.1 MPa and a maximum pore diameter of approximately 2.8 microns. While the maximum vacuum that may be applied is 0.1 MPa, the size of the maximum pores within each cup will control the sample flow rate through the cup. Flow rates into the cup are most closely correlated to the applied vacuum followed by pore size of the ceramic cup *(14)*.

Prior to installation, lysimeter cups should be washed with dilute acid to remove contaminates. Several mg L^{-1} Ca, Mg, and Na could be removed by cleaning with a dilute acid *(15)*. Nitrogen and P may also be in the cup prior to washing *(16)*. Pulling 250 to 500 mL 0.01 N HCl solution through the ceramic wall of the cup followed by deionized water is usually sufficient to remove contaminates.

Lysimeter Installation

To install the lysimeter, a bucket auger (7.5 cm OD) or hollow-stem auger is used to core a hole to the desired depth. Lysimeters should be soaked in water before installation to saturate the ceramic cup. Lysimeters may be installed to any depth, although the minimum depth requires that the top of the lysimeter be located below the soil surface to prevent channeling along the chamber. Samplers must also be of adequate depth to prevent mechanical damage from wheel or livestock traffic. For deep installations that exceed the potential lift, it may be necessary to install lysimeters with a check valve removal system so that samples may be removed using positive pressure without pressurizing the porous cup *(9)*.

It is essential that good contact between the soil and the ceramic cup wall be established. A sieved soil slurry (consisting of soil collected from the bottom of the hole) is poured back into the hole. The ceramic cup is then seated into the soil slurry, which ensures good contact between the soil and cup. Silica flour (200 mesh) can be poured into the hole in place of soil to seat the cup to prevent plugging of pores in the ceramic cup by fine soil particles *(14)*. Use of a silica sand is also essential in highly structured or cracked soils. In these soil types, channeling may rapidly transport the percolate to the cup area where it is preferentially absorbed. If a channeling situation exists, then the percolate collected is not typical of the average percolate at the desired depth *(17)*. The silica flour also helps to maintain contact during periods of freeze-thaw. During the winter, freeze-thaw can potentially break contact between the cup and soil. Silica sand allows for slight flexion of the basal material. After seating the cup in a soil slurry or sand, screened soil is then backfilled into the hole with gentle tamping to prevent channeling. To maintain the profile continuum, soil should be backfilled in the same order in which it was removed. If channeling is a potential problem due to excess gravel in the soil, it may also be desirable to backfill several centimeters with a bentonite clay-soil mixture, keeping in mind that the clay mixture may interfere with some subsequent chemical solution determinations. Access tubes should extend above the soil surface and should be protected to eliminate contamination from the surrounding soil. Covering with a plastic bag is usually suitable for this purpose.

Lysimeter Operation

To collect soil water samples, a vacuum of 0.01 to 0.08 MPa is applied inside the sampler via a single access tube. Soil water will be pulled into the lysimeter when the soil water suction by which water is retained within the soil is exceeded by the suction internally applied to the porous cup. Lysimeters cannot be used to sample water retained by soils at suctions greater than 0.1 MPa since the maximum amount of pressure within the ceramic cup is limited to -0.1 MPa. The time required to collect a sufficiently large sample may range from 2 to 72 h, depending upon i) the soil moisture content, ii) the suction at which water is retained in the soil, iii) the hydraulic conductivity of the soil, iv) the flow properties of the ceramic cup, and v) the vacuum applied. The vacuum applied should be the lowest possible value that will allow adequate sample collection within a reasonable time period. A vacuum may be applied continuously or intermittently, using either a hand- or motor-driven pump. Intermittent vacuum application using a hand-pump is satisfactory for sample collection, although several applications may be necessary to collect sufficient sample volume. Motor-driven pumps, which greatly increase sampling costs, are usually required when it is desired to maintain constant vacuum conditions.

Soil-water flow to the cup will be radial from around the cup, although the flow amount from any one direction will depend upon how tightly water is being retained by the soil in that respective direction. If soil conditions are uniform in all directions, then flow to the cup will be relatively uniform. If the lysimeter is positioned where soil conditions are uniform with the exception of increasing soil water content in a given direction, then water movement to the cup will be greater from the region of increasing water content.

After an appropriate sample volume has entered into the cup, the soil percolate is removed using either suction or positive pressure techniques previously mentioned. Samples should be held in sterile containers and transported on ice immediately back to the laboratory for analysis. At this point, samples should be immediately analyzed or stored frozen if the contaminants of interest are amenable to storage.

Advantages and Limitations

Nutrients. Attempts have been made to examine nearly all agronomically important nutrients in percolate collected by tension lysimeters. Unfortunately, limitations in the collection method have made interpretation of results difficult. Cationic contaminant analysis in percolate is limited by the contribution of the ceramic cup to cation concentrations in the percolate. Significant quantities of the cations Ca, Na and K were leached from the ceramic cup and into the percolate (7,18). This problem is especially important when the cation concentration in the percolate is low. To correct this potential error in the measurement, the authors suggested that the ceramic cups be washed in 0.1 N HCl as previously discussed. This procedure was found to reduce Na and K to acceptable levels; however, it had no significant effect on contamination by Ca. The concentration of K was higher in the percolate, although concentrations of Ca and Na were not affected (19).

The presence of P in percolate is both environmentally and agronomically important. Numerous attempts have been made to analyze soil percolate collected by tension lysimeters, though most attempts have led to false low P concentrations. Phosphorus adsorbs very tightly to the ceramic cup as it passes through the pores. Up to 110 mg P could be adsorbed by a single ceramic cup (20). Adsorbed P could not be desorbed when leached with deionized water. Forty-three percent of the P from a test solution was adsorbed when compared to Teflon cups (21). Therefore, it is recommended that tension lysimeters, as described in the current paper, should not be used when P is the nutrient of interest.

While downward NH_4-N leaching is not generally an important consideration, the use of ceramic cups to collect percolate is not desirable if NH_4-N is to be examined. Ammonium-N as with any other cation, can potentially adsorb to the ceramic cup, thus producing false low results. Significant quantities of NH_4-N are removed from the percolate by the cups (19,20). Coeffiecient of variation NH_4-N in percolate was reported as 160% (22).

The most common nutrient analyzed in percolate collected with tension lysimeters is NO_3-N. Nitrate-N, as an anion, does not interact with the ceramic cup and thus is not subject to chemical reductions. Nitrate-N concentrations within a full lysimeter were representative of amounts present in soil (8,23). Hundreds of references are available where tension lysimeters have been used to study NO_3-N concentration in percolate.

However, if the percolate was allowed to remain in the lysimeter for a long period of time (i.e. more than two days), then the sample variance increased with time (20), because of N transformation occurring within the lysimeter. It has been shown that in finer textured soils, up to three days are required to collect an adequate sample volume. Substantial microbial changes of the N could occur during this time. Samples should be collected several hours after vacuum application to eliminate this problem (20). This period of time would not be sufficient for microbial immobilization of NO_3-N. If the soil texture prevents rapid collection of samples, then alternative collection methods should be considered.

Very little information is available on the use of tension lysimeters for monitoring heavy metals in percolate. Most metals were reduced by 5 to 10% during passage through ceramic cups (24). Allowing the percolate to remain in the cup may further reduce concentrations by precipation, ion pairing reactions, and chelation.

Pesticides. Although tension lysimeters have been used in the past to estimate pesticide leaching losses, several important problems prevent quantitative use of this data. The most significant limitation is that many pesticides have relatively high vapor pressures. Application of a vacuum to the system increases the vapor pressure to the point where much of the pesticide may volatilize within the lysimeter (25).

An additional concern for pesticide monitoring with tension lysimeters is whether the pesticides are adsorbed to the polyvinyl chloride walls of the lysimeter. Adsorption is generally very strong and it is difficult to desorb pesticides from the surface. Atrazine [6-chloro-N-ethyl-N[1]-(1-methylethyl)-1,3,5,-triazine-2,4-diamine], for example, is known to be adsorbed to a variety of repeating polymers. Hence, tension lysimeters made of PVC are not appropriate for use to sample this compound. Chlorinated hydrocarbons are adsorbed very strongly to the interior of tension lysimeters *(24)*. Concentrations of DDD [1,1-dichloro-2,2,bis(p-chlorophenyl)ethane], DDE [1,1-dichloro-2,2,bis(p-chlorophenyl)ethene], and DDT [1,1,1-trichloro-2,2-bis(p-chlorophenyl)ethane] were reduced 90, 70, and 94%, respectively. Therefore, while tension lysimeters have been used to monitor pesticide losses through the soil, care should be taken in the interpretation of research results when lysimeters are the primary means of sample collection for pesticide loss evaluation.

Interpretation of Lysimeter Data

Tension lysimeters are commonly used to determine the magnitude of chemical leaching losses. If the water movement rate through a soil layer is known and the concentration of the chemical contaminate in the soil water is also known, it is theoretically possible to estimate the loss of the chemical through that soil depth. Unfortunately transient conditions exist in soil profiles, not only for soil water conditions, but also for soil solution concentrations. The tension lysimeter by its nature adds to these transient conditions by creating a hydraulic gradient towards the ceramic cup and acting as a sink for the soil solution. In a steady-state water regime simulation, a 2-cm ceramic cup acting as a point sink would cause deflection of uniform water flow within 10-cm of the cup *(26)*. Van der Ploeg and Beese *(27)* concluded that there "was no useful relation between the extracted amount of soil water and freely percolating soil water . . ." The sample chemical composition does not reflect the depth from which it was taken, but is a composite of the soil solution which contributed to the sample. This chemical composition can vary depending on the size of the cup, amount of vacuum applied, volume of sample collected, and moisture content of the surrounding soil. These interacting phenomena help explain previous research findings which have shown a relationship between solution concentration of chemical constituents and the extraction time length *(28)*.

To mathematically model movement of agrochemicals through a soil profile, the amount of chemical at a given soil depth must be treated either as volume-averaged resident concentrations or as flux-averaged effluent concentrations because these two types of concentrations require different mathematical treatments *(29)*. Problems arise when tension lysimeters are used to determine solute concentration profiles because these chemical concentrations are neither volume-averaged resident concentrations nor flux-averaged effluent concentrations *(30,31)*.

Statistical Analysis of Lysimeter Data

Data collected with lysimeters can be used to 1) establish baseline values for particular sites or situations, 2) determine whether the chemical concentrations in the percolate exceed some standard value or 3) test for significant differences in concentrations of chemicals among treatments. Unfortunately, many monitoring programs lack an appropriate experimental design and statistical analysis *(32)*.

One basic problem with many monitoring studies is the lack of any true replication. What constitutes a replication depends on the use for the data. Data collected as baseline data or for comparing sample values to standards, require that percolate samples be collected and analyzed separately for replicated lysimeters at chosen time intervals. Additionally, to test for differences among treatments, not only lysimeters but also treatments must be replicated. For example, to compare the chemical concentration of soil percolate in no-till versus conventional-till fields, plots should be randomized and replicated, and one or more lysimeters would be sampled per plot. If the treatments were not replicated and a conventionally-tilled field was compared to a no-till field, the effects of the tillage method would be confounded with other differences between the fields, and there would be no valid estimate of experimental error. If it is not practical to replicate treatments, analysis of baseline data for both control and treated plots would aid in the interpretation of treatment effects.

Another shortcoming in many lysimeter studies is the lack of adequate lysimeter numbers. Chemical concentrations in soil percolate tend to be quite variable when the concentrations are high. When the data are highly variable, many lysimeters are needed per plot to provide the precision desired to estimate the mean and to find significant differences among means. The variability problem arises because concentrations of nutrients in water samples often follow a lognormal distribution which is skewed and more variable than a normal distribution. A log transformation of the data should result in the data following a normal distribution *(33)*. The minimum lysimeter number needed per plot and the minimum number of plots per treatment (for the log transformed data) can be calculated for a chosen level of precision *(34)*. However, the number of lysimeters needed will be greater than the calculated minimum number because samples often cannot be collected from all lysimeters because of mechanical lysimeter failure or variations in the uniformity of the soil moisture regime. Largely due to spatial variability in the field, the calculated minimum number of lysimeters required may be too large to be feasible. In order to reduce the number of lysimeters required, statistical methods and experimental designs have been developed that address the problem of spatial variability and improve experimental efficiency. These include using incomplete block designs *(35)*, nearest neighbor analysis *(36)* or trend analysis *(37)*.

The greatest potential use of lysimeters is the repeated collection of soil percolate at a given location and depth over time. Lysimeter data collected over time should be analyzed in a repeated measures analysis to determine whether there are changes in chemical concentrations over time. A univariate split plot analysis *(38)* can be used unless assumptions about independence of error are violated. If the assumptions are not satisfied because measurements that are from adjacent times are more highly correlated than those of more distant times, a multivariate repeated measure analysis may be more appropriate. Unfortunately, the multivariate analysis is lower in power than the univariate analysis.

Conclusions

Tension lysimeters are widely used to sample soil percolate since they are relatively inexpensive and easy to install. The soil solution may be removed from unsaturated soils and potential pollution sites may be sampled repeatedly and non-destructively over time near the potential pollution source. Unfortunately, many times the solute concentration profiles determined using tension lysimeters have been interpreted as actual representations of chemicals being lost in soil percolate. Attempts to relate contaminant concentrations in percolate to loading rate losses have led to erroneous conclusions. Concentrations in percolate collected with tension lysimeters may share no consistent relationship to actual concentrations in percolate moving through the soil. Agrochemical concentrations in percolates collected with tension lysimeters should therefore be considered as qualitative data which are useful to rank, or order, treatment effects with respect to a control or other treatments of interest.

An additional, yet basic, error commonly made in data interpretation is that soil percolate and groundwater are one and the same. For example, several studies have noted that when the NO_3-N concentration in soil percolate exceeds the U.S. Public Health Service limit of 10 mg/L, then the percolate should be considered contaminated. These NO_3-N concentrations could well be due to the freely moving fraction of the soil water which may be sampled during peak flow periods using tension lysimeters without considering the redistribution of soil water which will occur in the profile. Consideration is also rarely given to the numerous chemical and microbiological transformations that occur as percolate moves through the soil profile such that near-surface concentrations in percolate have little relationship to groundwater concentrations. Several studies have shown that the NO_3-N concentration in percolate rapidly decreased as the water flows down through the soil profile *(39-41)*. Thus, statements that the percolate NO_3-N concentration exceeds accepted limits are not appropriate since definitions of percolate and groundwater have little relationship to each other.

Although tension lysimeters have some limitations which must be considered prior to use and in the interpretation of data, they offer a viable alternative for the sampling of soil solution. By following recommended installation procedures, using uniform techniques in the collection of samples, and recognizing limitations in data interpretation, they can be useful tools in the analysis of contaminant leaching through the soil.

Literature Cited

1   Briggs, L.J.; McCalla, A.G.. *Science*. 1904, *20*, 566-569.
2   Tanner, C.B.; Bourget, S.J.; Holmes, W.H. *Soil Sci. Soc. Am. Proc.* 1954, *18*, 222-223.
3   Cole, D.W. *Soil Sci.* 1958, *85*, 293-296.
4   Cole, D.W.; Gessel, S.P.; Held, E.E. *Soil Sci. Soc. Am. Proc.* 1961, *25*, 321-325.
5   Chow, T. *Soil Sci. Soc. Am. J.* 1977, *41*, 19-22.
6   Long, L.F. *Soil Sci. Soc. Am. J.* 1978, *42*, 834-835.
7   Silkworth, D.R.; Grigal, D.F. *Soil Sci. Soc. Am. J. 45*, 440-441.
8   Wagner, G.H. *Soil Sci.* 1962, *94*, 379-386.
9   Wood, W. *Water Resour. Res.* 1973, *9*, 486-488.
10   Bell, R. N.Z.J. *Exp. Agric.* 1974, *1*, 173-175.
11   David, M.; Struchtemeyer, R. *Maine Life Sci. Agric. Exp. Stn.* 1980, *773*, 1-16.
12   Johnson, T.M.; Cartwright, K. *U.S. Geo. Sur. Div. Circular 154.*
13   Everett, L.G.; Wilson, L.G.; Hoylman, E.W. *Vadose Zone Monitoring for Hazardous Waste Sites*. Noyes Data Corp.: Park Ridge, NJ, 1984.
14   Morrison, R.D.; Lowery, B. *Soil Sci.* 1990, *149*, 308-316.
15   Wolff, R.G. *Am. J. Sci.* 1967, *265*, 106-117.
16   Linden, D.R. *U.S.D.A. Agric. Res. Ser. Tech. Bull.* 1562.
17   Schaffer, K.A.; Fritton, D.D.; Baker, D.E. 1979. *J. Environ. Qual.* 179, *8*, 241-246.
18   Grover, B.L.; Lamborn, R.E. *Soil Sci. Soc. Am. Proc.* 1970, *34*, 706-708.
19   Haines, B.L.; Waide, J.B.; Todd, R.L. *Soil Sci. Soc. Am. J.* 1982, *46*, 658-661.
20   Hansen, E.A.; Harris, A.R. *Soil Sci. Soc. Am. Proc.* 1975, *39*, 528-536.
21   Zimmermann, C.F.; Price, M.T.; Montgomery, J.R. *Estuar. Coastal Mar. Sci.* 1978, *7*, 93-97.
22   Starr, M.R. *Soil Sci.* 1985, *140*, 453-461.
23   Joslin, J.D.; Mays, P.A.; Wolfe, M.H.; Kelly, J.M.; Garber, R.W.; Brewer, P.F. *J. Environ. Qual.* 1987, *16*, 152-160.
24   Morrison, R.D.; Tsai, T.C. Cal-Sci Res. Inc. 1981, Huntington Beach, CA.
25   U.S. Environmental Protection Agency. *Test methods for analysis of solid waste. Physical/chemical methods.* SW-846 2nd ed. Office of Solid Waste and Emergency Response, Wash., DC, 1982.
26   Warrick, W.; Amoozegar-Ford, A. *Water Resour. Res.* 1977, *13*, 213-217.

27   Van der Ploeg, R.R.; Besse, F. *Soil Sci. Soc. Am. J.* 1977, *41*, 466-470.

28   Severson, R.C.; Grigal, D.F. *Water Resour. Bull.* 1976, *12*, 1161-1170.

29   Van Genuchten, M.T.; Parker, J.C. *Soil Sci. Soc. Am. J.* 1984, *48*, 703-708.

30   Shulford, J.W.; Fritton, D.D.; Baker, D.E. *J. Environ. Qual.* 1977, *6*, 255-259.

31   Parker, J.C. *Soil Sci. Soc. Am. J.* 1984, *48*, 719-724.

32   Millard, S.P. 1987. *The Am. Statistician.* 1987, 251-253.

33   Parkin, T.B.; Meisinger, J.J.; Chester, S.T.; Starr, J.L. and Robinson, J.S. *Soil Sci. Am. J.* 1988, *52*, 323-329.

34   Gomez, K.A.; Gomez, A.A. *Statistical procedures for agricultural research.* John Wiley and Sons, New York, NY, 1984.

35   van Es, H.M.; van Es, C.L.; Cassel, D.K. *Soil Sci. Soc. Am. J.* 1989, *53*, 1178-1183.

36   Wilkinson, G.N.; Eckert Sr., T.; Hancock, T.W.; Mayo, O. *J.R. Stat. Soc.* 1983, *45*, 151-211.

37   Tamura, R.N.; Nelson, L.A.; and Naderman, G.C. *Agron. J.* 1988, *80*, 712-718.

38   Steel, R.G.; Torrie, J.H. *Principles and procedures of statistics.* McGraw-Hill, New York, NY, 1980.

39   Angle, J.S.; Gross, C.M.; McIntosh, M.S. *Agric. Ecosyst. Environ.* 1989, *25*, 279-286.

40   Linden, D.R.; Clapp, C.E.; Larsen, U.E. *J. Environ. Qual.* 1984, *13*, 256-264.

41   McLaughlin, R.A.; Pope, P.E.; Hansen, E.A. *Trans. Am. Soc. Agric. Eng.* 1980, *23*, 643-648.

RECEIVED October 12, 1990

Chapter 19

Compendium of In Situ Pore-Liquid Samplers for Vadose Zone

David W. Dorrance[1], L. G. Wilson[2], L. G. Everett[3], and S. J. Cullen[4]

[1]ENSR, 3000 Richmond Avenue, Houston, TX 77098
[2]Department of Hydrology and Water Resources, University of Arizona, Tucson, AZ 85721
[3]Metcalf and Eddy, 816 State Street, Santa Barbara, CA 93102–0551
[4]Institute for Crustal Studies, University of California, Santa Barbara, CA 93106

In recent years, there has been increasing emphasis on monitoring contaminant transport in the vadose zone. Vadose zone monitoring relies on a variety of in situ samplers to collect pore-liquids under saturated and/or unsaturated conditions. This compendium describes these samplers together with their advantages and disadvantages.

The vadose zone is the hydrogeological region extending from the land surface to the principle water table. Other commonly used terms for this region are the "unsaturated zone" and the "zone of aeration". These alternative terms do not take into account the existence of saturated flow above the principle water table. Saturated or near-saturated flow occurs primarily under the influence of gravity (referred to as free drainage). Under some conditions, pore-liquids may collect on perching layers, and locally saturated conditions (perched ground water) may develop. Unsaturated pore-liquid flow through the vadose zone under unsaturated conditions is controlled primarily by negative pore-liquid pressure gradients (negative pore-liquid pressures are referred to as pore-liquid tensions or matric potential).

Chemical species released at or near the land surface will be transported to some degree through the vadose zone. Extraction and chemical analyses of vadose-zone liquids is regarded as an early warning approach to potential ground-water pollution from such releases. This information can be used to mitigate potential problems prior to ground water degradation (1–2). Vadose-zone liquids can be extracted from cores in the laboratory or, alternatively, pore-liquids may be sampled directly from "undisturbed" soils by installing in situ pore-liquid samplers.

0097–6156/91/0465–0300$09.00/0

The most obvious difference between these two techniques is that vadose zone sampling is a destructive process which prevents repetitive sampling from the same location. More importantly, the two techniques do not sample the same types of liquid (3-4). In situ samplers are only capable of sampling pore-liquids held at tensions of up to about 60 kPa (5). Soil sampling with subsequent pore-liquid extraction provides liquids which may be held at tensions of up to several bars, depending on the extraction technique. Extraction under several bars of pressure may strip off cations preferentially sorbed in electrical double layers, sorbed organics, and even components of the soil. These species may not be present in the same concentrations (absolute or relative) in samples provided by in situ pore-liquid samplers.

This chapter reviews various in situ samplers and includes relevant literature citations. Some of the described samplers are not commercially available at this time. However, they may have been available in the past, and may be encountered at sites with established vadose zone monitoring programs. Some of the samplers can be fabricated. There are numerous qualifiers, hints, and warnings which should accompany the description of each sampler. We depend on the reader to review cited references to obtain complete expositions of the covered samplers. The applications and limitations of many of the samplers presented here were described previously (5-9).

In Situ Pore-Liquid Sampler Categories

In situ samplers extract liquids from saturated and unsaturated zones. Most samplers designed to sample from unsaturated soils also sample from saturated soils. This is useful in areas where the water table fluctuates, resulting in alternating saturated and unsaturated conditions. In contrast, samplers designed for sampling from saturated zones cannot be used in unsaturated conditions. This is because the negative pore-liquid pressures in unsaturated zones prevent liquid from moving into air-filled cavities at atmospheric pressures (Richard's Outflow Principle). Also, the openings in saturated samplers are too large to prevent air from entering the samplers when suctions are applied. Using this distinction, the types of pore-liquid samplers have been categorized as follows:

- o suction samplers (unsaturated/saturated sampling)
- o experimental suction samplers (unsaturated/saturated sampling)
- o experimental absorption samplers (unsaturated/saturated sampling)
- o free drainage samplers (saturated sampling)
- o perched ground water samplers (saturated sampling).

The term "pore-liquid" could be applicable to any
liquid residing in soil ranging from aqueous pore-liquids
to oil. However, all of the samplers described in this
paper were designed to sample aqueous pore-liquids only.
The abilities of these samplers to collect other pore-
liquids may be quite different than those described.

Vadose Zone Monitoring Program Design

The choice of appropriate sampling devices for a particular
location is dependant on various criteria (Table I). Well-
structured soils have two distinct flow regions including
macropores (e.g. interpedal openings, cracks, burrows, and
root traces) and micropores (e.g. intrapedal openings
between soil grains). Under saturated conditions, liquids
move more rapidly through macropores than through
micropores. Because of this, contaminants transported by
free drainage may bypass the finer pores. Consequently,
pore-liquids in macropores may have different chemistries
than those in micropores (10). This is enhanced by the fact
that oxygen contents of macropores can change in a matter
of hours during an infiltration event, whereas micropores
may remain suboxic regardless of flow conditions (11). In
addition, micropores are less susceptible to leaching than
macropores (2, 12-14). Because of these differences, sample
chemistry can vary widely from location to location and
from time to time depending on the amount of liquid drawn
from these two flow systems. Therefore, it is prudent to
consider using both unsaturated and free drainage samplers
in a sampling program, depending on site characteristics.

Table I. Criteria for Selecting Pore-Liquid Samplers
----
1. Required Sampling Depths
2. Required sampling Volumes
3. Soil characteristics
4. Chemistry and biology of the liquids to be sampled
5. Moisture flow regimes
6. Required durability of the samplers
7. Required reliability of the samplers
8. Climate
9. Installation requirements of the samplers
10. Operational requirements of the samplers
11. Commercial availability
12. Costs

Specific guidelines for designing vadose zone
monitoring programs have been discussed (1-2, 7-9, 15-20).

Suction Samplers (Unsaturated/Saturated Sampling)

Table II presents suction samplers and some of their
operational constraints. In general, a suction sampler

Table II: **Suction Sampler Summary**

Sampler Type	Porous Section Material	Max. Pore Size (um)	Air Entry Value (kPa)	(HB)* (HL)*	Operational Suction Range (kPa)	Max. Operational Depth (m)
			Commercially-Available Suction Samplers			
Vacuum Lysimeters	Ceramic	1.2-3.0	>100	HL	<60-80	<7.5
	PTFE	15-30	5-10	HB	<5-10	<7.5
	Stainless Steel	7	20	HL	<20	<7.5
Pressure-Vacuum Lysimeters	Ceramic	1.2-3.0	>100	HL	<60-80	<15
	PTFE	15-30	5-10	HB	<5-10	<15
	Stainless Steel	7	20	HL	<20	<15
High-Pressure-Vacuum Lysimeters	Ceramic	1.2-3.0	>100	HL	<60-80	<90
	PTFE	15-30	5-10	HB	<5-10	<90
	Stainless Steel	7	20	HL	<20	<90
Filter Tip Samplers	Ceramic	1.2-3.0	>100	HL	NA (a)	Unlimited
			Experimental Suction Samplers			
Cellulose-Acetate Hollow-Fiber Samplers	Cellulose-Acetate	<2.8	>100	HL	<60-80	<7.5
	Non-Cellulosic Polymer	<2.8	>100	HB	<60-80	<7.5
Membrane Filter Samplers	Cellulose Acetate	<2.8	>100	HL	<60-80	<7.5
	PTFE	15-30	5-10	HB	NA	<7.5
Vacuum Plate Samplers	Alundum	7	20	HL	<20	<7.5
	Ceramic	1.2-3.0	>100	HL	<60-80	<7.5
	Fritted Glass	4-5.5	50	HL	50	<7.5
	Stainless	7	20	HL	<20	<7.5

*(HB) - Hydrophobic;   *(HL) - Hydrophilic
(a) NA = Not Available

consists of a hollow, porous section attached to a sample vessel or a body tube (Figure 1). Samples are obtained by applying a vacuum within the sampler and collecting pore-liquid in the body tube. Samples are retrieved by a variety of methods.

The principles of suction sampler operation are as follows. Unsaturated portions of the vadose zone consist of interconnecting soil particles, interconnecting air spaces, and interconnecting liquid films. Liquid films in the soil provide hydraulic contact between the saturated porous section of the sampler and the soil (Figure 1). When a vacuum greater than the pore-liquid tension is applied within the sampler, a pressure-potential gradient is created toward the sampler. If meniscuses of the liquid in the porous segment are able to withstand the applied suction, liquid moves into the sampler. The ability of the meniscuses to withstand a suction decreases with increasing pore size and also with increasing hydrophobicity of the porous segment. This relationship is defined by the capillary rise equation (19,21). If the maximum pore sizes are too large, and/or they are hydrophobic, the meniscuses are not able to withstand the applied suction. As a result, they break down, hydraulic contact is lost, and only air enters the sampler.

The ability of a sampler to withstand applied suctions is gaged by its bubbling pressure (19,22). The bubbling pressure is measured by saturating the porous segment, immersing it in water, and pressurizing the inside of the porous segment with air. The pressure at which air starts bubbling through the porous segment into the surrounding water is the bubbling pressure. The magnitude of the bubbling pressure is equal to the magnitude of the maximum suction which can be applied to the sampler before air entry occurs (see air entry values in Table II). Because the bubbling pressure is a direct measure of how a sampler will perform, it is more useful than measurement of pore size distributions.

As pore-liquid tensions increase (low pore-liquid contents), pressure gradients toward the sampler decrease. Also, the soil hydraulic conductivity decreases exponentially. These result in increasingly lower flow rates into the sampler. At pore-liquid tensions above about 60 kPa (for coarse grained soils) to 80 kPa (for fine grained soils), the flow rates are effectively zero and samples cannot be collected (5). Samplers which have air entry values exceeding the 60-80 kPa range are preferred (Table II).

New samplers may be contaminated with water-soluble cations during manufacturing (23). In order to reduce chemical interferences from these and other substances on the porous sections, a variety of pre-installation procedures have been developed, including acid flushing (24-29). It is recommended to discard the first one or two sample volumes when sampling dilute solutions with newly

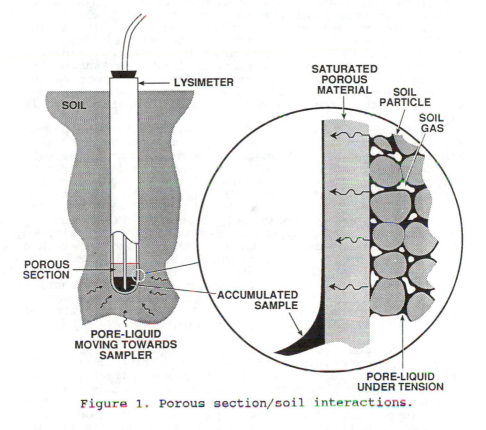

Figure 1. Porous section/soil interactions.

acid-flushed, installed samplers (27). This allows cation
exchange between the porous segment and the pore-liquid to
equilibrate following acid flushing. Stainless steel
samplers used in virus studies are chlorinated and rinsed
with a 10% solution of sodium thiosulfate to neutralize
free chlorine (30). Pressure testing, another pre-
installation procedure is recommended (5,31). Additional
installation and sampling procedures for suction samplers
are described elsewhere (2,19,22,31-38).

Vacuum Lysimiters. Lysimeters are defined in two contexts,
either as soil monoliths, used to characterize physical or
chemical changes within a bounded mass of soil, or as
devices for collecting percolating water for analyses. The
units described in this chapter belong in the last
category. Vacuum lysimeters generally consist of a porous
cup mounted on the end of a tube, similar to a tensiometer
(Figure 2a). A stopper is inserted into the upper end of
the body tube and fastened in the same manner as the porous
cup or, in the case of rubber stoppers, inserted tightly
(19).
    A variety of materials have been used for the porous
segment including nylon mesh (39), fritted glass (40),
sintered glass (41), Alundum^R, stainless steel (42,43),
polytetrafluorethylene (PTFE) (31) and ceramics (22). The
sampler body tube has been made with PVC, ABS, acrylic,
stainless steel (44) and PTFE (31). The stopper is
typically made of rubber (19), neoprene, or PTFE. The
outlet lines are commonly polycarbonate, PTFE, rubber,
polyethylene, polypropylene, Tygon^R, nylon, stainless
steel, and historically, copper. Fittings and valves are
available in brass, stainless steel, PVC, and PTFE.
    Vacuum lysimeters transfer samples directly to the
surface via a suction line. Because the maximum suction
lift of water is about 7.5 m, these samplers cannot be
operated below this depth. In practice, suction lifts of
even 7.5 m may be difficult to attain.

Pressure-Vacuum Lysimeters. These samplers, depicted in
Figure 2b, were developed for sampling pollutants moving in
the vadose zone beyond the reach of vacuum lysimeters (45).
Again, the porous segment is usually a porous cup at the
bottom of a body tube. Two lines are forced through a two-
hole stopper sealed into the upper end of the body tube.
The discharge line extends to the base of the sampler and
the pressure-vacuum line terminates a short distance below
the stopper. At the surface, the discharge line connects to
a sample bottle and the pressure-vacuum line connects to a
pressure-vacuum pump. The sampler and its components are
commonly made out of the same materials used for vacuum
lysimeters. Pressure-vacuum lysimeters first collect pore-
liquid in the body tube by application of vacuum through
the pressure-vacuum line. The sample is then retrieved by
pressurizing the sampler through the same line; this

pushes the sample up to the surface through the discharge line (Figure 2b).

Because samples are retrieved under pressure, these samplers can be used below 7.5 m. However, when positive pressure is applied for sample retrieval, some of the sample may be forced back out of the cup. At depths of over about 15 m, the volume of sample lost in this manner may be significant. In addition, pressures required to bring the sample to the surface from depths greater than 15 m may be high enough to damage the cup or to reduce its hydraulic contact with the soil (46-47). Rapid pressurization causes similar problems. Morrison and Tsai (48) developed a tube lysimeter with the porous section located midway up the body tube instead of at the bottom (Figure 2c). This design mitigates the problem of sample being forced back through the cup. However, it does not prevent problems with porous segment damage due to over pressurization or rapid pressurization. The sleeve lysimeter (which is not presently available commercially) was a modification of this design for use with a monitoring well (2). Another modification is the casing lysimeter which consists of several tube lysimeters threaded into one unit (Figure 2d). This arrangment allows precise spacing between units (30).

Figure 2e shows a design which allows incoming samples to flow into a chamber not in contact with the basal, porous ceramic cup (49). The ceramic cup is wedged into the body tube without adhesives or threading. The sampler was used to sample the vadose zone, the capillary fringe and the fluctuating water table in a recharge area. A sampler with the porous cup mounted on the top of a chamber (Figure 2f) is reported (50). These designs also allow pressurization for sample retrieval without significant liquid loss. However, because the porous cups are exposed to pressure, possible damage due to over pressurization or rapid pressurization remain a problem.

High Pressure-Vacuum Lysimeters. High pressure-vacuum lysimeters operate in the same manner as pressure-vacuum lysimeters. However, they include one-way check valves and a transfer vessel or chamber between the sampler and the surface (Figure 2g). These accessories prevent sample loss through the porous section during pressurization, and possible cup damage due to over pressurization. The samplers are generally manufactured using the same materials as vacuum lysimeters (22,30).

Filter Tip Samplers. Filter tip samplers consist of two components: a permanently installed filter tip, and a mechanically-retrievable glass sample vial (Figure 2h). The filter tip includes a pointed end to help with installation, a porous section, a nozzle, and a septum. The tip is threaded onto riser pipes which terminate at the surface. The sample vial includes a second septum. When in use, the vial is seated in an adaptor which includes a

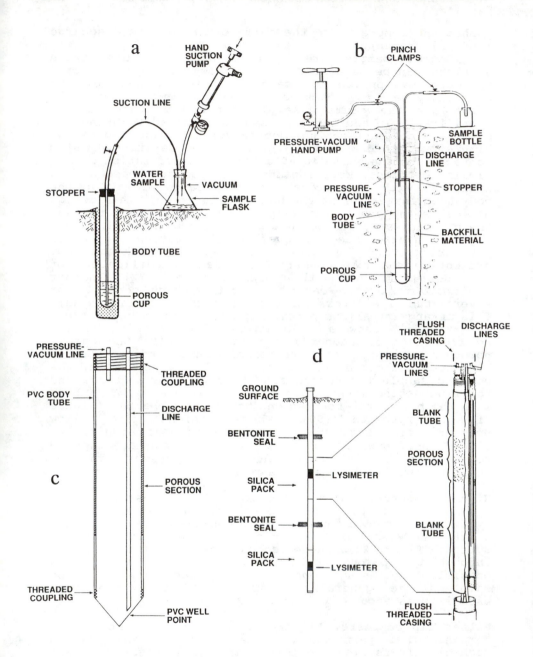

Figure 2. (a) Vacuum lysimeter and (b) pressure vacuum lysimeter (both adapted from ref. 22); (c) tube pressure vacuum lysimeter and (d) casing lysimeter (adapted from refs. 48 and 30, respectively).

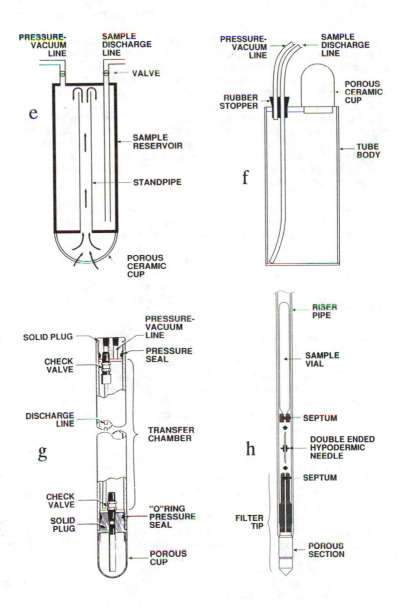

Figure 2 (continued). (e) Modified pressure-vacuum lysimeter, (f) Knighton–Streblow type vacuum lysimeter, (g) high-pressure-vacuum lysimeter (adapted from refs. 49, 50, and 22, respectively); (h) filter tip sampler (adapted from ref. 51).

disposable hypodermic needle to penetrate both of the
septa, allowing sample to flow from the porous segment into
the vial.

The body of the filter tip is constructed from a
variety of materials, including thermoplastic, stainless
steel, or brass. The attached porous section is available
in high density polyethylene, porous ceramic, or sintered
stainless steel. The septum is made of natural rubber,
nitrile rubber, or fluororubber (51,52).

A sample is collected from a filter tip sampler by
lowering an evacuated sample vial down the access tube to
the porous tip. The vial is coupled with the porous tip via
the hypodermic needle and sample flows through the porous
section into the vial. Once full, the vial is mechanically
retrieved (Figure 2h).

**Experimental Suction Samplers (Unsaturated/Saturated
Sampling).** Experimental samplers, described in the
literature, are usually limited to research applications
because of their fragility. For the most part, these
samplers are not commercially available. However, most of
these samplers may be easily fabricated. Experimental
suction samplers operate on the same principles as vacuum
lysimeters, and are also limited to depths of less than 7.5
m (Table II).

Cellulose-Acetate, Hollow Fiber Samplers. These samplers
consist of a bundle of cellulose-acetate hollow fibers
(Figure 3a). The bundle of flexible fibers is pinched shut
at one end and attached to a suction line at the other end.
The suction line leads to the surface and attaches to a
sample bottle and source of suction in the same manner as
a vacuum lysimeter. Similar fibers have been made from a
noncellulosic polymer solution (53) .

Membrane Filter Samplers. Figure 3b shows that membrane
filter samplers consists of a membrane filter of
polycarbonate, cellulose acetate, cellulose nitrate or PTFE
mounted in a "swinnex" type filter holder (2,19, 54-56).
The filter rests on a glass fiber prefilter. The prefilter
rests on a glass fiber "wick" which in turn sits on a glass
fiber collector. The collector is in hydraulic contact with
the soil, extending the sampling area of the small diameter
filter (Figure 3b). A suction line leads from the filter
holder to the surface. At the surface, the suction line is
attached to a sample bottle and suction source in a manner
similar to vacuum lysimeters.

Barrel Lysimeter. There are two limitations with suction
samplers. First, they may not sample from macropores
(unless the macropores are directly intercepted). Second,
their results cannot be used in quantitative mass balance
studies. Figure 3c shows an installation which overcomes
these limitations (57). A barrel-sized casing (e.g., 57 cm

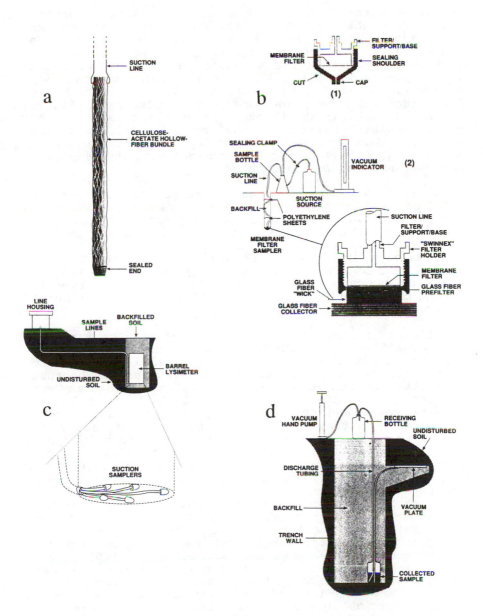

Figure 3. (a) Cellulose-acetate hollow-fiber sampler. (b) Membrane filter sampler, (c) barrel lysimeter, and (d) vacuum plate sampler installation (adapted from refs. 55, 58, and 62, respectively).

outside diameter by 85.7 cm high) is placed in a support
device and gently pushed into the soil with a backhoe. As
the casing is pushed, soil is excavated around it to ease
it into place. The process results in an encased monolith
of undisturbed soil. The monolith is then rotated and
lifted, pressure-vacuum lysimeters are placed in its base,
and the bottom is sealed. Subsequently the assembly is
placed back into the ground at the monitoring site (Figure
3c). All fluid draining through the monolith is collected
by the samplers. Inasmuch as the boundaries of the system
are sealed, the flux of liquid through the system requires
maintaining a vertical hydraulic gradient by applying
continual suction to the samplers.

Vacuum Plate Samplers. A vacuum plate sampler consists of
a flat porous disk fitted with a nonporous backing attached
to a suction line which leads to the surface (Figure 3d).
Plates are available in diameters ranging from 4.3 to 25.4
cm and custom designs are easily arranged ([2],[22]). Plates
are available in Alundum[R], porous stainless steel ([43]),
ceramic (1.2 to 3.0 um max. pore size) or fritted glass (4
to 5.5 um max pore size)([22],[36],[59]-[64]). The non permeable
backing can be a fiberglass resin, glass, plastic or butyl
rubber.

Operational Constraints of Suction Samplers. The inherent
heterogeneities of unsaturated pore-liquid movement and
chemistry limit the degree to which samples from these
devices can be considered representative. This is because
the small cross sectional areas of suction samplers may not
adequately integrate for spatial variabilities in liquid
movement rates and chemistries ([27],[65]-[67]). It has been
suggested ([67]) that results of chemical analyses from
suction sampler samples are good for qualitative but not
quantitative comparisons, unless the variabilities of the
parameters involved are established. Additionally, results
from suction sampling cannot be used for quantitative mass
balance studies ([3]).

   Chemical interactions between porous segments and the
liquids which pass through them affect the validity of
pore-liquid samples collected with suction samplers ([68]).
Potential interactions can include sorption, desorption,
cation exchange, precipitation, and screening ([69]). These
interactions can also occur with all other parts of the
samplers which liquids contact. However, the much higher
surface area within the pores of porous segments makes them
the most critical element chemically. Table III presents
the results of a literature review for porous section/pore-
liquid interactions. An attempt has been made to document
the pertinent features of the listed studies. However, the
reader should refer to the original papers to determine if
experimental techniques are applicable to the situation of
interest. The absence of entries for a constituent relative
to a material does not infer absence of interactions.

Table IIIa Porous Material Interactions[a]

	Absorbs Species	Desorbs Species	Screens Species	No Signif. Inter-action	No Inter-action
Al[b]		[c]C(2)[d]		C(16)	
Alkal-inity				SF(11)	
Ca		C(1,2, 18), CAF(18) A(14)		C(3,6, 10, 11,25) PTFE(3) A(3) FG(18, 22) CAF(10)	
C		FG(22)			
CO_3		C(2)			
HCO_3		C(2)			
Cd	C(11)			C(3) PTFE(3) A(3)	
Cl				C(11, 25) SF(11)	PTFE (13)
Cr	C(19)	C(3) PTFE (3) A(3)			
Cu	C(11)	C(3) PTFE (3)		A(3)	
Fe	C(11)	PTFE (3) A(3)		C(3,25)	PTFE (13)
H				SF(11)	
K	C(5,6, 15)	C(18)[e] A(14)		C(1,25) CAF(18) FG(18, 22)	

Continued on next page

Table IIIa Porous Material Interactions[a] (Continued)

	Absorbs Species	Desorbs Species	Screens Species	No Signif. Inter-action	No Inter-action
Mg	C(6)	C(2,3, 11,18) A(3,14) CAF(18)		C(10, 25) PTFE (3) CAF(10) FG(18, 22)	PTFE (13)
Mn	C(11)	A(3)		C(3) PTFE (3) A(14)	PTFE (13)
Na	C(6)	C(12, 18) A(14) CAF(18) FG(18, 22)		C(1,11, 25)	PTFE (13)
NH_4	C(4,12)			PTFE (4)	
N		FG(22)			
NO_2				C(4,5) PTFE (4)	
NO_3			CAF(10)	C(4,8) PTFE (4)	
NO_3-N			C(10) CAF(10)		
$(NO_2+ NO_3)-N$				C(5)	
P	C(1,5, 8,15,18)			CAF(18) FG(18)	
PO_4	C(4,5, 7)			PTFE (4) CAF(10)	
PO_4-P				C(10) CAF(10)	

Table IIIa Porous Material Interactions[a] (Continued)

	Absorbs Species	Desorbs Species	Screens Species	No Signif. Inter-action	No Inter-action
Pb					PTFE (13)
SiO_2		C(2)			
Si				C(4) PTFE (4)	
SO_4				C(11)	
Sr		C(11)			
Zn		C(11)			PTFE (13)
High Molec. Wt. Cmpds.			C(17, 21) CAF(10)		
4-nitro-phenol	PTFE (23)				
Chlorin-ated Hydro-carbons	PTFE (23,24)				
Diethyl Phthal-ate				PTFE (23)	
Naphth-alene	PTFE (23)				
Acen-aphth-ene	PTFE (23)				

Notes on Table IIIa:
a: Comparisons of materials based on this table should be made cautiously. Differing experimental techniques should be considered as a source of differing conclusions. Undocumented factors often include material age and sampling history.
b: Valence states are often not reported in studies.
c: Abbreviations:
    1. C = porous ceramic
    2. PTFE = porous PTFE
    3. A = porous Alundum
    4. CAF = cellulose acetate fibers
    5. FG = fritted glass or glass fibers
    6. SF = silica flour
d: Numbers in parenthesis refer to references in Table IIIb
e. Example: Reference 18 in Table IIIb (i.e., citation number 113 in text) found that there is no significant interaction of cellulose acetate fibers with potassium in solution. The porous section was washed prior to testing and results were found to be a function of several factors.

Table IIIb   References and Notes on Experimental Techniques (a)

Reference Number in Table	Citation Number in Text	Porous Section was Washed	Results are a Function of Several Factors	Dilute Solutions Were Tested	Experiments Were Performed on Non-porous Materials
1	108	X			
2	24	X			
3	106	X			
4	109	X			
5	110		X		
6	27	X	X		
7	28	X			
8	69	X			
9	33		X		
10	53		X		
11	29	X			
12	111				
13	2	X			
14	23	X	X	X	
15	112	X			
16	68		X		
17	3				
18	113	X	X		
19	11		X		
20	114				
21	34				
22	56	X	X		
23	115		X	X	X
24	116		X	X	X
25	105	X	X		

(a) Absence of information on experimental technique means
    that the techniques were not specified in the citation

Experimental Absorption Samplers (Unsaturated/Saturated Sampling)

Absorbent samplers depend on the ability of a material to absorb pore-liquids (2). Samples are collected by placing the sampler in contact with soil. Liquid is allowed to absorb into the sampler material over time. The sampler is then removed, and liquid is extracted for analyses. The simplicity of these samplers have made them attractive to some investigators.

Physically, absorbent methods are limited to soils approaching saturation. Sampling requires removing the device and bringing it to the surface. Because of this requirement, repeat sampling at the same location is difficult. Although the sampler may be placed back at its original location, identical hydraulic contact with the soil cannot be guaranteed.

Sponge Samplers. This sampler includes a cellulose-nylon sponge seated in a galvanized iron trough (70). Samples are collected by pressing the dry sponge against a soil surface with a series of lever hinges. The sponge is left in place until a sufficient volume of pore-liquid has been collected for analyses. Theoretically, there is no maximum sampling depth for sponge samplers. However, because access trenches are required for operation, installations are restricted to shallow depths dictated by excavation equipment and safety considerations.

Ceramic Rod Samplers. These samplers consist of solid, tapered ceramic rods. Prior to installation, the rods are boiled in distilled water, dried, and weighed. The rods are simply installed by driving them into the soil. After a period of time, the rods are withdrawn, weighed, and again boiled in distilled water. The water is then analyzed (71).

Problems With Experimental Absorption Samplers. As with other samplers, there are problems with chemical absorption, desorption, precipitation, cation exchange and screening of various pore-liquid components as a function of the sampler materials (70-71). A discussion of the limitations when sampling for NO_3-N with ceramic rod samplers has been discussed (71).

Free Drainage Samplers (Saturated Sampling)

A free drainage sampler consists of some sort of collection chamber which is placed in the soil. Pore-liquid in excess of field capacity is free to drain through soil (usually through macropores) under the influence of gravity. Hence, these samplers collect liquid from those portions of the vadose zone which are intermittently saturated because of rainfall, flooding, or irrigation. This gravity drainage creates a slightly positive pressure at the soil-sampler

interface causing fluid to drip into the sampler. Some free
drainage samplers apply a small suction in order to break
the initial surface tension at the soil-sampler interface.
Samples are retrieved either by accessing the samplers at
depth or by drawing samples to the surface through a
suction line.

Suction samplers can also be used to sample free
drainage flow. However, the small area of those samplers
compared to the spacing of macropores limits their
usefulness for this application. In addition, suction must
be applied to suction samplers to collect samples, even
under saturated conditions. Free drainage samplers are
passive collectors which automatically collect the
percolating liquids.

Free drainage samplers are classified differently by
various authors, depending on the installation methods.
Many free drainage samplers are installed in the side walls
of trenches and are referred to as trench lysimeters.
However, free drainage samplers are also installed in the
walls of vertical caissons. The principle behind each of
the samplers is essentially the same.   However, the
materials and construction differ. Free drainage samplers
include the following:

> o Pan Lysimeter
> o Glass Block Lysimeters
> o Caisson Lysimeters
> o Wicking Soil Pore-Liquid Samplers
> o Trough Lysimeters
> o Vacuum Trough Lysimeters
> o Sand Filled Funnel samplers.

Pan Lysimeters. A pan lysimeter generally consists of a
galvanized, metal pan of varying dimensions (Figure 4a).
A copper tube is soldered to a raised edge of the pan.
Plastic or Tygon tubing connects the copper tube to a
collection vessel. Any liquid that accumulates on the pan
drains through the tubing into the vessel (19,45).

Glass Block Lysimeters. Figure 4b shows a free-drainage
sampler made from a hollow glass brick (72). These glass
bricks, which are produced as ornamental masonry, have
dimensions of 30 by 30 by 10 cm and have a capacity of 5.5
L. To build a sampler, nine holes, 0.47 cm in diameter,
are drilled along the perimeter of one of the square
surfaces of a brick. Nylon tubing is inserted into one of
the holes to allow for sample removal. The collecting
surface is fitted with a fiberglass sheet to improve
contact with the soil. Pore-liquid collection is enhanced
by a raised lip along the edge of the surface.

Level blocks are critical for retrieving the bulk of
the sample. However, the inside glass surface is uneven and
has low spots ("dead spots") where residual sample collects
between sampling cycles. This leads to cross-contamination
of samples.

Caisson Lysimeters. A caisson lysimeter consists of collector pipes, radiating from a vertical chamber (2). Figure 4c shows a unit consisting of nearly horizontal, half-screened PVC casing (73). Another design consists of the following components: (1) a stainless steel tube extending diagonally upward through the caisson wall into the native soil, (2) a screened plate assembly within the tube to retain the soil, (3) a purging system used to redevelop the sampler when it becomes clogged, (4) an airtight cap that prevents exchange between the air in the caisson and the soil air (74).

Wicking Soil Pore-Liquid Samplers. Figure 4d shows a wicking sampler which combines the attributes of free drainage samplers and pressure-vacuum lysimeters (58). The sampler collects both free drainage liquid and liquid held at tensions to about 4 kPa. A hanging "Hurculon" fibrous column acts as a wick to exert a tension on the soil pores in contact with a geotextile fiber which serves as a plate covering a 30.5 by 30.5 by 1.3 cm pan. The terminus of the fibrous column is sealed into the cap of a tubular chamber. This chamber also contains an inlet pressure-vacuum line and a sample collection tube. Materials for the sample collection tube depend on the constituents being sampled. Glass and PTFE were recommended materials when sampling for organics (58).

Trough Lysimeters. Trough lysimeters, also known as Ebermayer lysimeters, rely on a trough or pail to collect pore-liquid. A fiberglass screen is suspended inside the trough to maintain a firm contact with the edges of the sampler and the soil. The screen is lined with glass wool and covered with soil until the soil is even with the top of the trough (75).

Figure 4e illustrates a trough lysimeter in which two parallel metal rods are inside the trough, in contact with the bottom side of the screen, and bent toward the collection tube (2). Liquid that enters the trough migrates along these rods towards the collection tube in response to capillary forces. A modification of this design consists of a metal trough with a length of perforated PVC pipe mounted inside. The trough is filled with graded gravel so that coarse material is immediately adjacent to the PVC pipe and fine sand is at the edges and the top of the trough. The pipe is capped at one end while the other end is connected to a sample container via a drainage tube (2).

Vacuum Trough Lysimeters. The vacuum trough lysimeter consists of a metal trough equipped with two independent strings of ceramic pipe, each 13 mm in diameter (76). The primary purpose of this design is to sample free drainage.

However, the device also apparently allows extraction of samples under applied suctions of up to 50 kPa. The ceramic pipes act as a vacuum system, and samples are extracted through a suction line.

Sand Filled Funnel Samplers. Figure 4f shows a sand-filled funnel for collecting freely draining liquid (77). The funnel is filled with clean sand and inserted into the sidewall of a trench. The funnel is connected through tubing to a collection bottle. Application of suction to a separate collection tube pulls the sample to land surface.

Perched Ground Water Sampling (Saturated Sampling)

Perched water occurs where varying permeability layers in the vadose zone retard downward movement of liquid. Over time, liquid collects above lower permeability layers and moisture content may increase to saturation (9,78). Once soil becomes saturated, wells and other devices normally installed below the water table can be used to collect samples.

Sampling perched liquid is attractive because the perching layer collects liquid over a large area. Such integrated samples are more representative of areal conditions than suction samples (78). This also allows the sampler to potentially detect contaminants which may not be moving downward immediately adjacent to the sampler. In addition, larger sample volumes can be collected than those which can be obtained by suction samplers. The incorporation of perched ground water sampling into monitoring programs has been proposed (7,9).

Perched water systems can be difficult to find and delineate. Surface and borehole geophysical methods (e.g neutron logging) and video logging of existing wells are often used. Also, perched systems tend to be ephemeral. Therefore, suction samplers are sometimes required as backups. As with all samplers, potential chemical interactions between sampler materials and the constituents of interest should be considered. Because these samplers are usually installed for other purposes, incompatibility of materials with monitoring objectives is often a problem (9, 79-80).

Following are some of the methods for sampling perched ground water:

        o Point Samplers
        o Wells
        o Cascading water samplers
        o Drainage samplers

Point Samplers. Point samplers are open ended pipes or tubes, such as piezometers or wells with short screened intervals, installed for the purpose of collecting samples from a discrete location in saturated material (Figure

5a). Samples are collected by bringing liquid which flows freely into the device to the surface by one of a variety of methods. Figure 5a presents various point sampler configurations which have been used (2,81-83).

Wells. A monitoring well is similar to a point sampler except the screened interval is longer. Therefore, samples are averaged over the screened length (84). Samples are collected by bringing liquid which flows freely into the well to the surface by one of a variety of methods. Figures 5b, 5c, and 5d present examples of well designs which may be used under different conditions (2,9,19,86-96).

Cascading Water Samplers. Cascading water occurs when a well is screened throughout a perched layer and the underlying water table (Figure 5e) or when water leaks through casing joints at the perched layer. Because the water table is lower than the perched layer, water flows into the well in the portion open to the perched layer, and cascades downward to the water table. This situation is common in some areas where the practice has been to install water wells with large screened intervals (44). Samples are collected by capturing liquid flowing into the well from the perched layer before it cascades down to the water table (78). Alternatively, water samples pumped from a well that has been shut down for a period of time represent ground water that has been influenced by cascading water (Figure 5e).

Drainage Samplers. Shallow perched systems may spread contamination, cause problems with structures, or interfere with agriculture. Drainage systems are installed to alleviate these problems. These systems cause gravity flow of perched ground water to a ditch or sump from which it is pumped out. This outflow can be sampled. Typical drainage systems include tile lines, half perforated pipes, synthetic sheeting, or even layers of gravel and sand. Depending on the design of the system, it may be possible to sample outflows which drain different areas such as agricultural areas (96-104) and sanitary landfills (104).

Concluding Remarks

A review of the cited literature reveals that it is difficult, and perhaps impossible, to obtain pore-liquid samples which are not altered by the sampling process. Investigators should choose sampling devices and methods which provide the least altered samples. However, cost considerations will dictate a point at which increased sample representativeness is not practical. At this point, it is the investigator's responsibility to document the types of alteration caused by the sampling process. The research necessary to quantify these alterations has increased in recent years as vadose zone monitoring concepts have matured.

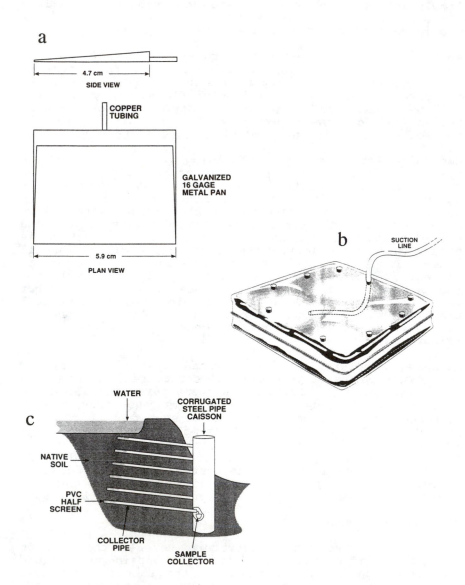

Figure 4. (a) Example of pan lysimeter and (b) glass block lysimeter (both adapted from ref. 80), and (c) example of caisson lysimeter (adapted from ref. 73).

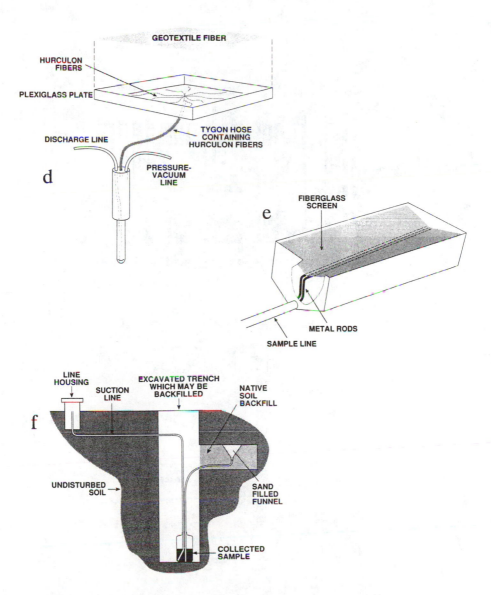

Figure 4 (continued). (d) Wicking type soil-pore liquid sampler, (e) trough lysimeter, and (f) sand filled funnel sampler installation (adapted from refs. 58, 75, and 80, respectively).

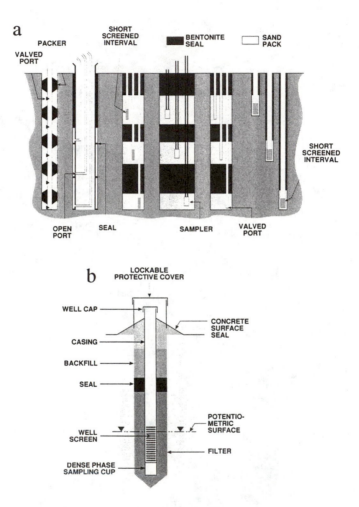

Figure 5. (a) Examples of point sampling systems (adapted from ref. 83). (b) A monitoring well with the uppermost ground-water level intersecting the slotted well screen (adapted from ref. 94).

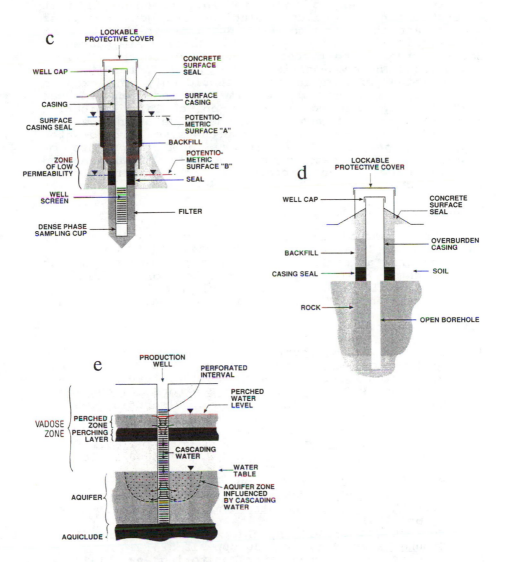

Figure 5 (continued). (c) A monitoring well installed to sample from the lower of two ground-water zones and (d) an open-hole ground-water monitoring well in rock (both adapted from ref. 94). (e) Conceptualized cross section of a well showing cascading water from perched zone (adapted from ref. 78).

Liquid alterations caused by the sampling process are important. However, a newly forming consensus is that alterations of non-dilute solutions are generally less significant than the inherent, spatial variabilities of pore-liquid chemistries. Such variabilities are caused by natural vadose zone processes such as preferential flow, and physical heterogeneities (105-107). As a result, even an unaltered pore-liquid sample should only be viewed as temporally representing the sampling location and not spatially representing any other point. This limitation does not detract, however, from the value of vadose zone monitoring systems in a comprehensive ground-water monitoring program. After all, a primary purpose of installing such systems is for **detecting** pollutants moving from a source and not necessarily for determining their exact concentrations.

Acknowledgments

The work reported in this chapter was funded by the Institute for Crustal Studies (No. 0044-11HW), University of California at Santa Barbara, through a grant from the Environmental Monitoring Systems Laboratory, Office of Research and Development, U.S. Environmental protection Agency, Las Vegas, NV, Project Officer L. A. Eccles.

Literature Cited

1.   Wilson, L.G., Monitoring in the Vadose Zone: A Review of Technical Elements and Methods, U.S. Environmental Protection Agency, EPA-600/7-80-134, 1980.
2.   Morrison, R.D., Ground Water Monitoring Technology, Timco MFG., Inc., Prairie DU Sac, Wisconsin, 1983.
3.   Law Engineering Testing Company, Lysimeter Evaluation Study, American Petroleum Institute, May 1982, 103 pp.
4.   Brown, K.W., Efficiency of Soil Core and Soil Pore-Liquid Sampling Systems, U.S. EPA/600/52-86/083, Virginia, February, 1987.
5.   Everett, L.G.; McMillion, L.G., Ground Water Monitoring Rev., 1985, vol 5, pp 51-60.
6.   Everett, L.G., Ground Water Monitoring Rev., 1981, vol 1, pp 44-51.
7.   Everett, L.G.; Wilson, L.G.; and McMillion, L.G., Ground Water, 1982, vol 20, pp 312-324.
8.   Wilson, L.G., Ground Water Monitoring Rev., 1982, 2, 31-42.
9.   Everett, L.G.; Hoylman, E.W.; Wilson, L.G.; McMillion, L.G., Ground Water Monitoring Rev., 1984, vol 4, pp 26-32.
10. Thomas, G.W., and Phillips, R.E., J. Environ. Qual., 1979, vol 8, pp 149-152.
11. Anderson, L.D., Ground Water, 1986, vol 24, pp 761-769.

12. Severson, R.C.; Grigal, D.F., Water Resources Bull., 1976, vol 12, pp 1161-1169.
13. Shuford, J.W.; Fritton, D.D.; and Baker, D.E., J Environ Qual, 1977, vol 6, pp 736-739.
14. Tyler, D.D.; Thomas, G. W., J Environ Qual, 1977, vol 6, pp 63-66.
15 Wilson, L.G., Thirteenth Biennial Conference on Ground Water, September 1981, p 38.
16. Wilson, L.G., Ground Water Monitoring Rev, 1983, vol 3, pp 155-165.
17. Robbins, G.A.; Gemmell, M.M., Ground Water Monitoring Rev, 1985, vol 5, pp 75-80.
18. Merry, W.M.; Palmer, C.M., Proceedings of the NWWA Conference on Characterization and Monitoring of the Vadose Zone, Nat Water Well Assoc, 1985, p 107.
19. Permit Guidance Manual on Unsaturated Zone Monitoring for Hazardous Waste Land Treatment Units, U.S. Environmental Monitoring Systems Laboratory, Office of Solid Waste and Emergency Response, EPA/530-SW-86-040, 1986a, 111 pp.
20. Ball, J.; Coley, D.M., Proceedings of the Sixth National Symposium and Exposition on Aquifer Restoration and Groundwater Monitoring, NWWA/EPA, 1986, p 52.
21. Hillel, D., Fundamentals of Soil Physics, Academic Press, Inc., Orlando, Florida, 1980, p 413.
22. Soilmoisture Equipment Corporation, Sales Division, Catalog of Products, Santa Barbara, California, 1988.
23. Neary, A.J.; Tomassini, F., Can J Soil Sci, 1985, vol 65, pp 169-177.
24. Wolff, R.G., Am J of Sci, 1967, vol 265, pp 106-117.
25. Wood, W. W., Water Resources Res, 1973, 9, pp 486-488.
26. Wilson, L.G., in Ground Water and Vadose Zone Monitoring; Nielsen, D.M.; Johnson, A.I., Eds; Am Soc Testing Materials, Philadelphia, Pa, 1990, pp 7-24.
27. Debyle, N. V.; Hennes, R. W.; Hart, G.E., Soil Sci, 1988, vol 146, pp 30-36.
28. Bottcher, A.B.; Miller, L.W.; Campbell, K.L., Soil Sci, 1984, vol 137, pp 239-244.
29. Peters, C.A.; Healy, R.W., Ground Water Monitoring Rev, 1988, vol 8, pp 96-101.
30. Timco Manufacturing, Inc., Timco Lysimeters, Sales Division, Prairie DU Sac, Wisconsin, 1988.
31. Linden, D.R., U.S. Dep. Agric. Technical Bull., 1977, 1562.
32. Rhoades, J.D.; Oster, J.D., In Methods of Soil Analysis, Agronomy; Klute, A. Ed., Am Soc Agron, Soil Sci Soc Am, Madison, Wisconsin, 1986, Vol.9, pp 985-1006.
33. Methods of Soil Analyses, Klute, A. Ed., Agronomy, American Society of Agronomy, Soil Sci Soc Am, Madison, Wisconsin, 1986, Vol. 9.

34. Brose, R.J.; Shatz, R.W.; Regan, T.M., _Proceedings of_
     the Sixth National Symposium and Exposition on Aquifer
     Restoration and Ground Water Monitoring, Nat Water
     Well Assoc, 1986, pp 88-95.
35. Cole, D.; Gessell, S.; Held, E., _Soil Sci Soc Am_
     Proceed, 1968, vol 25, pp 321-325.
36. Wengel, R.W.; Griffin, G.F., _Soil Sci Soc Am J_, 1971,
     vol 35, pp 661-664.
37. Brown, K.W., Thomas, J.C., and Aurelius, M.W., _Soil_
     Sci Soc Am J, 1985, vol 49, pp 1067-1069.
38. Chow, T.L., _Soil Sci Soc Am J_, 1977, vol 41, pp 19-22.
39. Quin, B.F.; Forsythe, L.J., _New Zealand J Sci_, 1976,
     vol 19, pp 145-148.
40. Long, F.L., _Soil Sci Soc Am J_, 1978, vol 42, pp
     834-835.
41. Starr, M.R., _Soil Sci_, 1985, vol 140, pp 453-461.
42. Mott Metallurgical Corporation, Sales Division,
     Catalog of Products, Farmington, Conn., 1988.
43. Smith, C.N.; Carsel, R.F., 1986, _Soil Sci Soc Amer J_,
     vol 50, pp 263-265.
44. Smith, S.A.; Small, G.S.; Phillips, T.S.; Clester, M.,
     Water Quality in the Salt River Project, A Preliminary
     Report, Salt River Project Water Resource Operations,
     Ground Water Planning Division, Phoenix, Arizona,
     1982.
45. Parizek, R.R.; Lane, B.E., _J Hydrol_, 1970, vol 11, pp
     1-21.
46. Trainor, D.P., M.S. Thesis/Independent Report, The
     University of Wisconsin, Madison, Wisconsin, 1983.
47. Young, M., _Proceedings of Monitoring Hazardous Waste_
     Sites, Geotechnical Engineering Division, Am Soc Civil
     Engin, Detroit, Mich., 1985.
48. Morrison, R.D.; Tsai, T.C., _Modified Vacuum-Pressure_
     Lysimeter for Vadose Zone Sampling, Calscience
     Research Inc., Huntington Beach, California, 1981.
49. Nightingale, H.I.; Harrison, D.; Salo, J.E., _Ground_
     Water Monitoring Rev, 1985, vol 5, pp 43-50.
50. Knighton, M.D.; Streblow, D.E., _Soil Sci Soc Am J_,
     1981, vol 45, pp 158-159.
51. BAT Envitech, Inc., Sales Division, _Catalog of_
     Products, BAT Envitech Inc., Long Beach California,
     1988.
52. Haldorsen, S.; Petsonk, A.M.; Tortensson, B.A.,
     Proceedings of the NWWA Conference on Characterization
     and Monitoring of the Vadose Zone, Nat Water Well
     Assoc, 1985, p 158.
53. Levin, M.J.; Jackson, D.R., _Soil Sci Soc Am Jo_, 1977,
     vol 41, 535-536.
55. Stevenson, C.D., _Environ Sci Technol_, 1978, vol 12, pp
     329-331.
56. Wagemann, R.; Graham, B., _Water Res_, 1974, vol 8, pp
     407-412.
57. Sales Division, _Catalog of Products_, Cole-Parmer
     Instrument Company, 1988.

58. Hornby, W.J.; Zabick, J.D.; Crawley, W., Ground Water Monitoring Rev, 1986, vol 6, pp 61-66.
59. Sales Division, Catalog of Products, Corning Glass Works, New York, 1988.
60. Duke, H.; Kruse, E.; Hutchinson, G., USDA Agricultural Research Service, ARS, 1970, 41-165.
61. Tanner, C.B.; Bourget, S.J.; Holmes, W.E., Soil Sci Soc Am Proc, 1954, vol 18, pp 222-223.
62. Cole, D.W., Soil Sci, 1958, vol 85, pp 293-296.
63. Nielson, D.; Phillips, R., Soil Sci Soc Am Proc, 1958, vol 22, pp 574-575.
64. Chow, T.L., Soil Sci Soc Am J, 1977, vol 41, 19-22.
65. Amoozegar-Fard, A.D.; Nielsen, D.R.; Warrick, A.W., Soil Sci Soc of Am J, 1982, vol 46, 3-9.
66. Haines, B.L.; Waide, J.B.; Todd, R.L., Soil Sci Soc Am J, 1982, vol 46, pp 658-660.
67. Biggar,J.W.; Nielsen, D.R., Water Resources Res, 1976, vol 12, pp 78-84.
68. Litaor, M.I., Water Resources Res, 1988, vol 24, pp 727-733.
69. Hansen, E.A.; Harris, A.R., Soil Sci Soc Am Proc, 1975, 39, 528-536.
70. Tadros, V.T.; McGarity, J.W., Plant and Soil, 1976, vol 44, pp 655-667.
71. Shimshi, D., Soil Sci, 1966, vol 101, pp 98-103.
72. Barbee, G.C.; Brown, K.W., Soil Sci, 1986, vol 141, pp 149-154.
73. Schmidt, C.; Clements, E., Reuse of Municipal Wastewater For Groundwater Recharge, U.S. Environmental Protection Agency, 68-03-2140, 1978, Ohio, p 110.
74. Schneider, B.J.; Oliva, J.; Ku, H.F.H.; Oaksford, E.T., Proceedings of the Characterization and Monitoring of the Vadose (Unsaturated) Zone, National Water Well Assoc, Las Vegas, Nevada, 1984, p 383.
75. Jordan, Carl F., Soil Sci, 1968, vol 105, pp 81-86.
76. Montgomery, B.R.; Prunty, L.; Bauder, J.W., Soil Sci Soc Am J, 1987, vol 51, pp 271-276.
77. Brown, K.W., Hazardous Waste Land Treatment, U.S. Environmental Protection Agency, Office of Research and Development, SW-874, Cincinnati, Ohio, 1980.
78. Wilson, L.G.; Schmidt, K.D., Establishment of Water Quality Monitoring Programs, Proceedings of a Symposium, Am Water Resources Assoc, 1978, p 134.
79. Dunlap, W.J., Some Concepts Pertaining to Investigative Methodology for Subsurface Process Research, U.S. Environmental Protection Agency, 1977, p 167.
80. U.S. Environmental Protection Agency, RCRA Ground-Water Monitoring Technical Enforcement Guidance Document, Office of Waste Programs Enforcement, Office of Solid Waste and Emergency Response, OSWER-9950.1, 1986b, p 208.

81. Reeve, R.C.; Doering, E.J., Soil Sci, 1965, vol 99, pp 339-344.
82. Pickens, J.F.; Cherry, J.A.; Coupland, R.M.; Grisak, G.E., Merritt, W.F.; and Risto, G.A., Ground Water Monitoring Rev, 1986, vol 6, pp 322-327.
83. Patton, F.D.; Smith, H.B., Ground-water Contamination Field Methods, Am Soc of Testing Materials, STP963, Philadelphia, PA, 1988.
84. Pickens, J.F.; Grisak, G.E., Ground Water, 1979, vol 17, pp 393-397.
85. Campbell, M.; Lehr, J., Water Well Technol, McGraw-Hill Book Co., New York, New York, 1973, p 681.
86. Scalf, M.R.; McNabb, J.F.; Dunlap, W.J.; Cosby, R.L.; Fryberger, J., Manual of Ground Water Sampling Procedures, Nat Water Well Assoc, Ohio, 1981, p 93.
87. Minning, R.C., Proceedings of the Second National Symposium on Aquifer Restoration and Ground Water Monitoring, Nat Water Well Assoc, Columbus, Ohio, 1982, p 194.
88. Richter, H.R.; Collentine, M.G., Proceedings of the Third National Symposium on Aquifer Restoration and Ground Water Monitoring, Nat Water Well Assoc, Columbus, Ohio, 1983, p 223.
89. Gass, T.E., Water Well J, 1984, vol 38, pp 30-31.
90. Driscoll, F.G., Groundwater and Wells, Johnson Division, St. Paul, Minnesota, 1986, p 1089.
91. Keely, J.F.; Boateng K., Ground Water, 1987, vol 25, pp 3-4.
92. Riggs, C.O., Proceedings of the Workshop on Resource Conservation Recovery Act Ground Water Monitoring Enforcement: Use of the Technical Enforcement Guidance Document and Compliance Order Guide, Am Soc of Testing Materials, 1987.
93. Hackett, G., Ground Water Monitoring Rev, 1987, vol 7, pp 51-62.
94. Riggs, C.O.; Hatheway, A.W, Proceedings of the ASTM Conference on Ground Water Technol, Am Soc of Testing Materials, 1986.
95. Hackett, G., Ground Water Monitoring Rev, 1987, vol 7, pp 51-62.
96. Taylor, T.W.; Serafini, M.C., Ground Water Monitoring Rev, 1988, vol 8, pp 145-152.
97. Drainage for Agriculture, Schilfgaarde, J.V., Ed.,Agronomy Series, American Society of Agronomy, Madison, Wisconsin, 1974, Number 17.
98. Donnan, W.W.; Schwab, G.O. In Drainage for Agriculture, Schilfgaarde, J.V., Ed., Agronomy Series, American Society of Agronomy, Madison, Wisconsin, 1974, Number 17, pp 93-114.
99. Gilliam, J.W.; Daniels, R.B.; Lutz, J.F., J Environ Qual, 1974, vol 2, pp 147-151.
100. Gambrell, R.P.; Gilliam, J.W.; Weed, S.B., J Environ Qual,1975, vol 4, pp 311-316.

101. Eccles, L.A.; Gruenberg, P.A., Proceedings: Establishment of Water Quality Monitoring Programs, Am Water Resources Assoc, 1978, p 319.
102. Gilliam, J.W.; Skaggs, R.W.; Weed, S.B., J Environ Qual, 1979, vol 8, pp 137-142.
103. Jacobs, T.C.; Gilliam, J.W., J Environ Qual, 1985, vol 14, 472-478.
104. Wilson, L.G.; Small, G.G., Hydraulic Engineering and the Environment, Proceedings 21st Annual Hydraulics Specialty Conference, Am Soc Civil Engin, 1973, 427.
105. Johnson, T.M; Cartwright, K., Monitoring of Leachate Migration in the Unsaturated Zone in the Vicinity of Sanitary Landfills, Illinois State Geological Survey Circular 514, Urbana, Illinois, 1980.
106. Creasey, C.L.; Dreiss, S.J., Proceedings of the NWWA Conference on Characterization and Monitoring of the Vadose Zone, National Water Well Assoc, pp. 173-181.
107. Peters, C.A.; Healy, R.W., Ground Water Monitoring Rev, 1988, vol 8, pp 96-101.
108. Grover, B.L.; Lamborn, R.E., Soil Sci Soc Am Proc, 1970, vol 34, pp 706-708.
109. Zimmermann, C.F.; Price, M.T.; Montgomery, J.R., Estuarine and Coastal Marine Sci, 1978, vol 7, pp 93-97.
110. Napgal, N.K., Canad J Soil Sci, 1982, vol 62, pp 685-694.
111. Wagner, G.H., Soil Sci, 1962, vol 94, pp 379-386.
112. Faber, W.R.; Nelson, P.V, Communications in Soil Sci Plant Analysis, 1984, vol 15, pp 1029-1040.
113. Silkworth, D.R.; Grigal, D.F., Soil Sci Soc Am J, 1981, vol 45, pp 440-442.
114. Barbarick, K.A.; Sabey, B.R.; Klute, A., Soil Sci Soc Am J, 1979, vol 43, pp 1053-1055.
115. Jones, J.N.; Miller, G.D., Ground-Water Contamination Field Methods, Am Soc of Testing Materials, STP 963, Philadelphia, PA, 1988, p 185.
116. Barcelona, M.J.; Helfrich, J.A.; Garske, E.E., Ground-Water Contamination Field Methods, Am Soc of Testing Materials, STP 963, Philadelphia, PA, 1988, 221.

RECEIVED December 10, 1990

SOIL SAMPLING TECHNIQUES

Chapter 20

Aseptic Sampling of Unconsolidated Heaving Soils in Saturated Zones

L. E. Leach and R. R. Ross

Robert S. Kerr Environmental Research Laboratory, U.S. Environmental Protection Agency, P.O. Box 1198, Ada, OK 74820

Collecting undisturbed subsurface soil samples in noncohesive, heaving sandy environments below the water table has been extremely difficult using conventional soil sampling equipment. Several modifications of the conventional hollow-stem auger coring procedures were adapted, which allowed collection of depth-discreet soil samples in very fluid, heaving sands. These methods were used where accurate subsurface characterization of the contamination of RCRA and CERCLA sites was essential. Cohesionless cores were consistently retrieved, aseptically extruded from the core barrel inside an anaerobic environmental chamber, and preserved in the field. The physical, chemical, and biological integrity of discreet soil intervals was maintained for laboratory analysis. Statistical analysis of repeated collection of soil samples from the same depth intervals in nearby boreholes was documented.

An accurate characterization of subsurface materials and ground water is essential for successful and efficient design of monitoring or remediation of hazardous waste sites. Shallow water table aquifers historically have the highest incidence of contamination, yet they are often the most difficult to physically characterize or in which to construct monitoring wells. The unstable nature of unconsolidated, noncohesive sediments, particularly in the saturated zone, continues to present a challenge to the engineer where conventional well construction and sediment sampling is attempted.

There have been a number of recent articles written describing hollow-stem auger procedures for monitoring well construction and coring in unconsolidated water table aquifers (1-6). Equipment developed by essentially all the major hollow-stem drill manufacturers performs well even below the water table where unconsolidated materials contain sufficient clay to be cohesive and maintain stability (Central Mine and Equipment Company, cat. prod. lit., St. Louis, MO, 1987; Mobile Drilling Company, cat. prod. lit., Indianapolis, IN, 1983).

However, the use of conventional hollow-stem auger equipment in heaving sediments continues to plague drillers during well construction and depth discrete sampling. When the inner string of tools are raised inside the hollow auger, hydrostatic pressure forces cohesionless sand into the annulus of the hollow auger. Once this occurs, conventional sampling methods such as split spoon, barrel or shelby tube coring can no longer be managed since the sediments are too fluid to be retained in the sampler during retrieval.

New hollow-stem auger drilling and sampling techniques which resolve these difficulties have recently been described (7). The conventional lead hollow auger is equipped with a special clam-shell designed cap which seals the auger annulus, preventing annular blockage before sampling can be accomplished. Special modification of the thin wall barrel sampler, routinely used in hollow-stem auger sampling, was required to hold slurried samples in the sampler during retrieval. In order to retain the collected core inside the sampler, modifications of a special internal sampler tube vacuum piston developed by the University of Waterloo, Ontario, Canada, was used. This piston is held stationary as the sampler is forced through the auger clam-shell doors and into the sediment, thus creating enough negative pressure inside the sampler to retain fluid samples during retrieval (8).

Additional field equipment was also developed which allowed sample retrieval and preservation as found in nature. Retrieved samples were transferred from the piston sampler into sealed sterile containers by inserting one end of the sampler through an iris diaphragm into a plexiglass anaerobic nitrogen filled flow-through glove box designed after similar laboratory equipment.

Quality control studies were conducted to determine the credibility of the innovative sediment sampling technique. The reliability was tested by collecting samples from closely spaced boreholes at discrete depths and statistically comparing the physical and chemical data of 10 cm cores at three identical depths from each of the boreholes. The physical data (grain size distribution) indicated a very high correlation for each depth.

Conventional Hollow-Stem Auger Drilling

Conventional monitoring well construction and soil sampling in recent years has most often been performed with hollow-stem auger drill rigs. These tools perform well in both unsaturated and saturated materials containing sufficient clay to maintain the cohesive nature of the subsurface strata during drilling and sample collection. Hollow-stem auger drilling and coring have several distinct advantages over other methods. Drilling is performed without lubricants on the drill string and without circulating fluids which could impact the subsurface geochemistry. In addition, the hollow-stem auger serves as a temporary casing to maintain the borehole in unstable formations during sampling and placement of well screens and casing.

The major disadvantages of this drilling technique are that the equipment is limited to unconsolidated or minimally consolidated sediments. In addition, most auger equipment is limited to depths of about 50 meters. Some vendors have recently manufactured a limited number of rigs capable of drilling to depths of 100 meters.

Soil Sampling.   During conventional sampling the borehole can be
advanced to a specific depth by using a lead pilot bit assembly on the
center rods which rotate at the same speed as the outer hollow auger
drill string as shown in Figure 1. This inner pilot assembly serves
as part of the bit and also as a plug for the hollow auger annulus.
Once the desired depth has been reached for either soil sampling or
monitoring well installation, the lead pilot assembly can be retrieved
providing access through the auger annulus for either sampling or well
screen placement.
     The pilot bit assembly is often replaced with a non-rotating,
capped sample tube when only a shallow depth must be drilled before
deeper sample profiling is performed as shown in Figure 2.   When the
plugged sample tube is in use, the spindle assembly, which rotates and
drives the augers, is equipped with an in-line bearing allowing the
augers to rotate while the inner string is held stationary during
vertical advancement. Equipping the spindle assembly in this manner
allows quick replacement of the capped sample tube with a conventional
thin walled barrel sampler, split-spoon or shelby tube sampler, and
soil coring can proceed with minimum equipment assembly time.
     Following drilling with either procedure, the borehole is readily
accessible for sampling, whichever inner tool is in use. Sampling may
be accomplished by hydraulically pressing or percussion driving the
sample tube, shelby tube or split-spoon into the soil beyond the lead
auger.   Most driving is done by reciprocally dropping a 62 kilogram
weight onto the hammer drive head (anvil) attached to the center rods
using a cat head or hydraulically operated trip hammer.
     If continuous sampling is desired, a standard thin walled barrel
sampler, split spoon or shelby tube sampler attached to the lead end
of the center rods can be coupled to the in-line bearing spindle
assembly.   This assembly allows samples to be collected without
rotating the sampler tube, thereby minimizing sample disturbance while
simultaneously advancing the augers (Central Mine and Equipment
Company, cat. prod. lit., St. Louis, MO, 1987).
     Sequential samples from desired intervals can be collected simply
by uncoupling the augers and retrieving the sampler while the auger
flighting remains in the borehole. During advancement of the sampler,
the cutting shoe pares the sample as it is pushed into the tube.
Fluids and air trapped inside the tube above the sample are vented out
through a ball valve inside the sampler drive cap as shown in Figure
3. The next sample interval may be immediately collected by inserting
another sample tube and drilling to the next desired depth. This is
repeated until samples from all desired depths have been collected.

Monitoring Well Installation.   One of the greatest advantages of
hollow-stem auger drilling is in ground water monitoring well
installation.   When wells are being installed in unconsolidated
cohesive or semicohesive material and the borehole walls are relatively
stable, the auger string should be removed and the well constructed in
an open borehole. The borehole should be prepared by first bailing
until the cuttings are removed along with most of the turbid water.
A 60 to 100 cm sand pack should be placed in the bottom of the borehole
to stabilize the remaining slurried material before installing the
screen. Normal open hole well construction procedures should be
performed as quickly as possible before the wet borehole walls begin
to collapse.

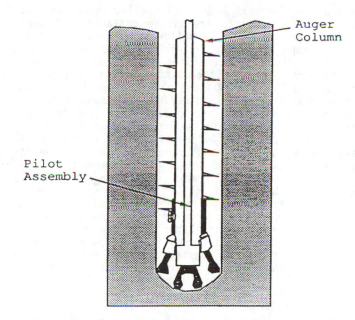

Figure 1. Auger Column Containing Pilot Assembly.

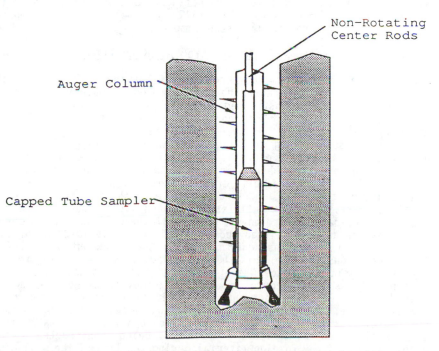

Figure 2. Capped Tube Sampler Plugging Auger Annulus.

When a monitoring well is being installed in unstable noncohesive sediments, but the sediments are not so unstable that they begin to heave up the auger annulus, the hollow auger should be left in place to serve as a temporary casing. A well can then be constructed inside the auger using the same procedures as in open hole construction. However, the auger must be intermittently retrieved in 30 to 60 cm increments as the filter pack and sealant materials are placed inside the auger, around the screen and easing. Frequent extraction of the auger prevents bridging of the gravel pack and borehole sealant materials between the screen or casing and inner wall of the auger.

The above procedures for hollow-stem auger soil sampling and well installation work extremely well in unconsolidated sediments in both the unsaturated and saturated zones when sediments are sufficiently cohesive to remain relatively stable. However, attempts to construct wells in totally cohesionless aquifer material below the water table are often unsuccessful. During hollow-stem auger drilling in cohesionless sediments below the water table, hydrostatic pressure can force sand up inside the hollow auger when the inner string of tools are raised. Once this occurs, conventional sampling methods or proper screen placement during well construction can no longer be performed due to annular blockage of the auger. The sediment materials forced into a core barrel are too fluid to be retained during retrieval, and the sample integrity is destroyed by soil movement.

Since more than 80 percent of ground water contamination incidents occur in shallow water table aquifers, it is imperative that sampling technology be developed to accurately characterize and remediate such sites. Continued frustration with heaving sands prompted innovative modifications of hollow-stem auger drilling as well as the development of a special wireline vacuum piston sampler.

Technical Modifications for Hollow-Stem Auger Drilling

In an attempt to overcome the problems of heaving sand blocking the auger annulus when the center head is removed, a number of drillers in recent years have constructed a special wood or metal disc shaped knock-out plate to cover the hollow auger annulus. This plate is carefully fitted inside the auger bit and held in place by continuous vertical pressure during drilling. No internal string of tools are used when the knock-out plate is used. This device allows borehole construction to desired depths in heaving soils, preventing hydraulic movement of soils until a well screen or soil sampler can be correctly placed inside the auger. When a monitoring well is installed using this procedure, the knock-out plate is pushed out of the lead auger by applying vertical pressure on the plate with the well screen and casing string as the auger is lifted about 30 cm. Normally 10-20 cm of fine sand is poured inside the auger on top of the knock-out plate before the plate is pushed out. This thin sand pack serves as a plug to hold heaving material in place while a gravel pack is placed around the screen. The same procedure is used for sampling except no sand is placed on top of the knock-out plate before it is displaced using vertical force. There are two distinct disadvantages of using a knock-out plate. First, it is undesirable to leave any foreign material in a borehole when a monitoring well is being constructed, since decomposition of any foreign material during the life of the well could have considerable impact on ground water quality. Second, when the disc is dislodged from the lead auger, it will often not move laterally into the borehole wall and continue to block the entrance of the sampler, thus preventing sample collection.

Retrievable Annular Cap.   To prevent knock-out plate contamination, a special retrievable steel clam-shell cap was designed to cover the annulus of the lead auger as illustrated in Figure 4.   This device, as with the knock-out plate, prevents hydraulic movement of soils until sampling or screen placement can be completed.   The hinged clam-shell doors are closed at the surface and carefully held in place as the auger is forced into the soil during initial drilling of the borehole. Constant vertical pressure is necessary to keep the doors closed until the desired sampling depth is reached.   During drilling with this device, none of the inner tools, such as the center head or sample tube, are attached to the center rods.   When the borehole is completed, the augers are decoupled from the spindle assembly just above surface and left open until a special sampler can be inserted.

Monitoring well installation using this clam-shell capped bit is similar to methods described earlier utilizing the knock-out plug.   A sand pack about 10 cm thick is placed inside the augers on top of the clam-shell doors before the well screen is inserted.   The clam-shell doors are then pushed open by holding vertical force on the casing and screen assembly as the auger string is retracted about 30 cm.   In order to prevent damage to the open clam-shell doors, the auger string must not be rotated and must be carefully pinned with an auger fork on each incremental lift without allowing it to move back down the borehole. Once the doors have been opened, well construction or core sampling may continue inside the augers as described in conventional methods.

It is not possible to close the clam-shell doors on the lead auger and continue drilling as presently designed, nor is it desirable since contaminated soils generally move inside the doors and annulus of the lead auger once the sampler is retrieved.   Therefore, if deeper samples are desired, the entire flight of augers must be carefully removed from the borehole without rotation.   The annulus of the augers, exterior flighting, and the clam-shell doors must be thoroughly high pressure steam cleaned to ensure the integrity of sequential samples.   The borehole can then be backfilled with clean sand or uncontaminated cuttings and then redrilled to the next desired sampling depth.   In many situations, researchers prefer to move the rig a few feet and drill a new hole to the next sample depth with the clam-shell auger bit.   Admittedly, the process is slow, but the tools must be clean and the annulus sealed if high integrity samples are to be consistently obtained.

Special Wireline Piston Sampler Design.   In order to overcome the chronic difficulties of sampling cohesionless sediments, development of innovative sampling equipment was mandatory.   A special wireline piston sampler originally designed and tested by the Institute of Water Research, University of Waterloo, Ontario, Canada, has been reasonably successful in collecting heaving aquifer sediments (8).   After numerous field tests of the Waterloo sampler, several modifications were made in the basic design to improve sample recovery and provide the capability to collect samples aseptically in the field.

The aluminum canister sleeve used inside the sample barrel in the Waterloo sampler was discarded since special field techniques of sample

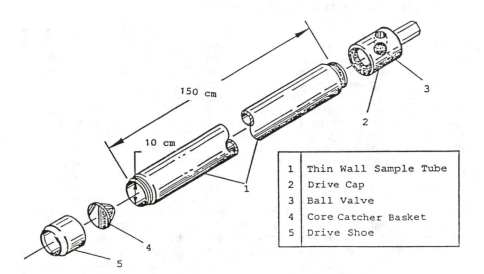

1	Thin Wall Sample Tube
2	Drive Cap
3	Ball Valve
4	Core Catcher Basket
5	Drive Shoe

Figure 3.   Central Mine and Equipment Standard Thin Wall Sample Tube.

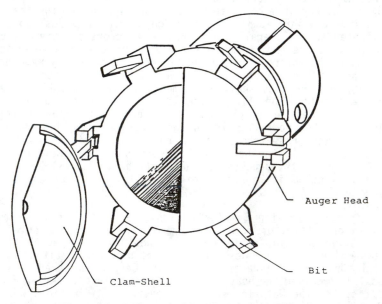

Figure 4. Clam-Shell Fitted Auger Head.

collection and preservation were desired.   A piston similar to Waterloo's was built to fit tightly inside the conventional thin wall 10 cm diameter by 150 cm long barrel sampler tube as shown in Figure 5.

The Waterloo sampler does not have a ball valve in the sampler drive cap to relieve internal pressure between the top of the piston and the sampler drive cap when the piston is moved up the interior of the sample tube.   Internal compression of fluids or gasses between the top of the piston and inner face of the drive cap creates short period shock and recoil when reciprocally hammering the sampler.   This compressive shock retards sampling and tends to dewater the sample. Retaining the pressure relief ball valve in the conventional drive cap in the modified piston sampler design reduces the compressive effects.

Frequent sampling of a wide variety of sediments and ground water aquifer material containing organic pollutants mandates that all components of the sampling equipment in contact with the sample material be inert.   Thus, teflon and stainless steel plates were added to the bottom of the piston to prevent organic contamination of sediment samples from the neoprene seals shown in Figure 5.   Additional allen-head screws were also added to the design providing more uniform compression of the neoprene seals.

Operation of the special internal piston is controlled by a wireline which passes through the sampler drive cap to the surface. The wireline is coupled to the piston with a swivel which allows the sampler to be assembled and disassembled without twisting or fouling the wireline.   The wireline is fixed rigid at the surface to hold the piston when the sampler is pushed or driven downward, thus creating a suction which holds the sample inside the core barrel during retrieval to the surface.

Initially, a hardened steel cutting shoe without a core catcher basket was tested with the piston positioned flush with the cutting edge of the shoe.   However, when tested in very fluid heaving sands, the piston would not consistently create sufficient suction to hold the cored sample in the sample barrel when raised above the heaving material.   Assembly with the original manufacturers core catcher basket and cutting shoe and initially positioning the piston on top of the core catcher basket resolved this problem.   As a result, an excess of 95 percent core recovery in saturated unconsolidated heaving sands is routine with the modifications as described.

Actual operation is accomplished by lowering the piston sampler inside the auger with the center rods while maintaining slack in the wireline attached to the piston inside the sampler.   The sampler is slowly lowered until it contacts the inner face of the clam-shell doors.   The center rods are then decoupled and attached to the drill spindle on the rig to prevent upward movement of the sampler as the clam-shell doors are opened as previously described.   This procedure allows the piston sampler to instantly contact the soil interface as the clam-shell doors are opened before aquifer heaving can occur.   The aquifer can then be sampled by hydraulic percussion hammering as previously described.

Once the clam-shell doors have been opened and the piston sampler has made contact with the fluid soil sample, slack in the piston wireline is taken up.   The wireline is held taut by maintaining tension with a wireline reel or fixing the wireline firmly to the rig.   The wireline is then marked at some reference point, usually at the top of

the open auger, so that during sampling the piston's fixed position
can be assured. The desired sample interval is marked on the center
rods, usually using the top of the auger as the lower reference. It
is generally advisable to collect no more than 90 - 100 cm of sample
since longer samples of wet sand are extremely difficult to
hydraulically extrude from the core barrel. When a wet sandy sample
is pressed from the core barrel, it begins to dewater and adhere to the
inner walls of the core barrel creating tremendous wall friction. If
the piston moves as the sampler is driven downward, less sample will
be collected than indicated by the depth of sampler penetration.

Once the sampler is filled by percussion driving, it is retrieved
using a technique very similar to that described for conventional
sampling. The sampler is slowly removed from the soil with a wireline
attached to the center rods. The sampler should not be retrieved with
the piston wireline. If the piston moves during retrieval, unwanted
sample may be sucked into the core barrel, or if the sampler is in the
borehole water or air column, piston movement could aerate or introduce
water into the sample. The slack in the piston wireline is retrieved
with minimum tension to prevent fouling in the borehole as the sampler
is raised. The cutting shoe on the sampler is immediately wrapped with
plastic as the tool is lifted from inside the auger, thus minimizing
aeration of the exposed sample. Next, the sampler drive cap is removed
and the piston is pulled from the sampler while maintained in the
vertical retrieved position. Maintaining the sampler in a vertical
position keeps the fluidized sample intact until a 10 cm long, tightly
fitted stainless steel plug can be quickly inserted and pressed down
onto the sample, tightly trapping it inside the core barrel.

The temporarily preserved sample can then be extruded from the
core barrel and collected for analysis. A special hydraulic powered
core extruder is routinely used as part of the thin wall barrel
sampling equipment. The extruder can be portable or fixed to the frame
of the rig. The core barrel can be screwed into the sample extruder,
and the stainless steel plug can be used as a foot to extrude the fluid
sample. However, no plug is required for extruding unsaturated cores.
During routine geotechnical sampling there is little concern for
protecting cores from exposure to the atmosphere. In these
circumstances, the cutting shoe is removed from the core barrel and is
replaced with a specially designed stainless steel paring device shown
in Figure 6 which peels away the outer 2.5 cm of core as it is
extruded. Geotechnical samples can be described, collected, and
preserved in the field as they are pared.

Aseptic Collection of Sediment Samples

Precise characterization of the distribution of organically
contaminated sediments and soil biota requires a special aseptic and
oxygen free environment for capturing samples as they are extruded
from a sampler (9). If samples are extruded in the natural atmosphere,
unstable organics instantly volatilize, many inorganics can be
oxidized, and soil biota can be inactivated or killed, thus destroying
in situ integrity. This problem can be overcome by using a specially
designed anaerobic glove box in which to extrude samples from the core
barrel. The sealed cutting shoe end of the core barrel can be inserted
through a self-closing iris diaphragm on one end of a specially
constructed portable 1.0 cm thick plexiglass glove box with dimensions

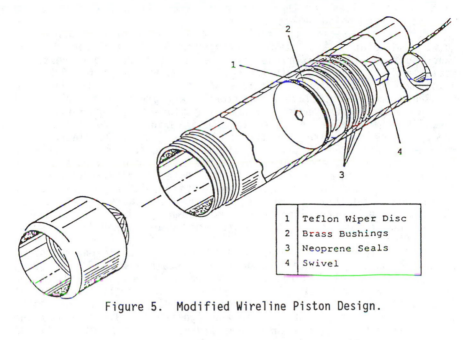

1	Teflon Wiper Disc
2	Brass Bushings
3	Neoprene Seals
4	Swivel

Figure 5. Modified Wireline Piston Design.

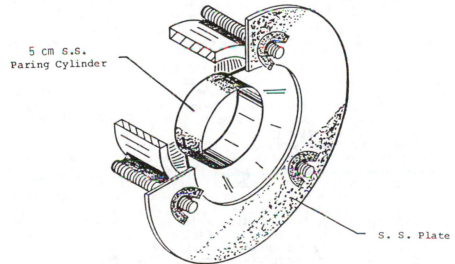

5 cm S.S.
Paring Cylinder

S. S. Plate

Figure 6. Core Paring Tool.

of 60 x 90 x 120 cm shown in Figure 7. The glove box can be prepared for sampling in approximately 30 minutes by filling it with the desired number of presterilized sample containers and sterile stainless steel core paring devices and then purging it with nitrogen gas to reduce internal oxygen below detectable limits.

During preparation for field sampling, a sufficient number of quart and pint glass sample containers are sterilized in the laboratory. Sterilization is done by washing the containers and sealable lids and then autoclaving at a temperature of 120°C at 1 atmosphere pressure for 60 minutes. As the containers and lids are removed from the autoclave using sterile equipment, they are placed in a laboratory environmental chamber or glove box. When filled to capacity, the chamber is sealed and the interior air is flushed from the box by purging with pressurized nitrogen gas for 30 minutes at a rate of 2500 L/hr at a pressure slightly in excess of atmospheric. This procedure displaces gases inside the sample containers and fills them with nitrogen as the chamber fills. After 30 minutes of purging, the containers and lids are wrapped in sterile foil inside the chamber while under a positive pressure of nitrogen atmosphere. The lids are then placed on the containers and screwed down hand tight. The chamber is then opened, and the containers are removed and packed for transport to the field.

In the field, the glove box is loaded with a sufficient number of presterilized sample containers and sterile stainless steel core paring devices to collect a minimum of 300 cm of cored sediment (three separate 100 cm samples). Prior to placement inside the glove box, at least three paring devices are rinsed in a 95 percent ethanol bath, placed in a stainless steel pan, and ignited to fire-burn dry the excess ethanol. They are then carefully wrapped in sterile foil and placed inside the glove box. The glove box is then closed and purged with pressurized nitrogen gas as previously described for laboratory work, reducing oxygen levels below detectable limits in about 30 minutes. A positive pressure of nitrogen flowing through the box is maintained during all sampling activities.

Quality assurance tests of the field glove box were conducted by measuring a series of 1000 microliter samples of vented gas with a Varian Model 90-P gas chromatograph equipped with a thermal conductivity detector. These tests verified that the air-oxygen level inside the box after 30 minutes purging is less than 0.02 percent on a volume per volume basis (10).

After the extruder mounted sampler is inserted into the glove box through the iris diaphragm, the plastic wrapped cutting shoe and core catcher basket are removed. A presterilized foil-wrapped paring tool and holding bracket are unwrapped and screwed onto the sample tube. About 10 cm of sample is then extruded through the 5 cm diameter paring tool and then carefully broken away, exposing an aseptic face. Samples are then routinely collected in the sterile sample containers, sealed, and numbered inside the glove box. Paring the core is necessary to remove possible contamination of the disturbed core wall.

When the stored sample containers have been filled and sealed inside the glove box (normally after three 100 cm sampling events), the box must be opened, samples removed, the box thoroughly cleaned, reloaded with sample containers, and prepared for repurging. Normally if the samples are to be analyzed for volatile organic compounds, or microbiological parameters, they are removed from the box through the

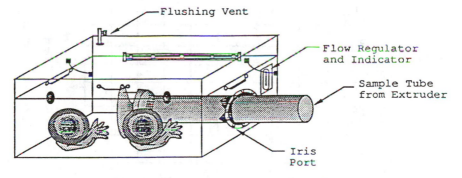

Figure 7.   Field Sampling Glove Box.

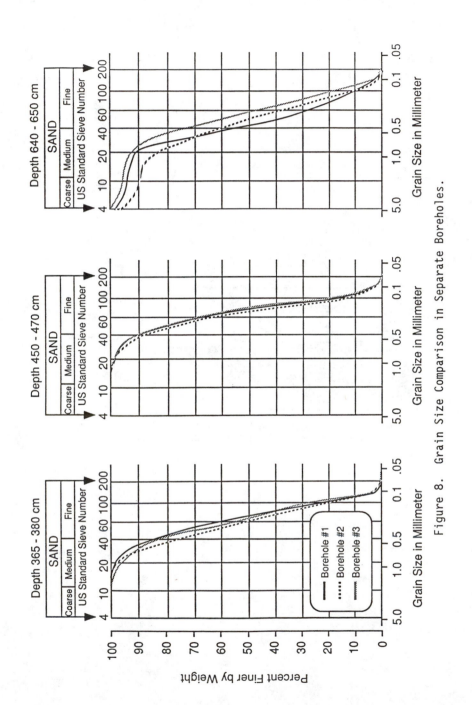

Figure 8. Grain Size Comparison in Separate Boreholes.

iris diaphragm after each sampling event, labeled, and packed in ice for transport to the laboratory, thus minimizing time of exposure to ambient temperatures and pressures.

Additional innovative sampling activities can be performed inside the glove box for detailed site characterization or research activities. Often small duplicate subsamples are desired for quality assurance and very precise analysis of petroleum hydro-carbons. Small 25 ml sterile disposable syringes approximately 1 cm in diameter can be inserted directly into the soil core exposed through the paring ring. A syringe can be pushed into the core through the paring ring while pulling a vacuum to hold the sample during retrieval. The subsample can then be placed in 40 ml sterile VOA bottles containing 5 ml of < pH 2 water to preserve the sample. Pairs of miniature cores of each sample interval are normally taken for quality assurance documentation. This method fixes core material for later analysis without refrigeration.

The glove box is also equipped with special sample container lids with sealed vents plumbed through the top of the box. This allows a quick analysis of the head space over volatile hydrocarbon samples using a portable combustible hydrocarbon analyzer meter. Careful sampling and portable analysis can detect the interface of vapors within a 4 cm core section using these techniques.

Quality Control of Piston Sampling

There has been speculation concerning the reliability of core recovery and capability of collecting duplicate soil samples from precise stratigraphic horizons. Core recovery monitoring was conducted at a site in Traverse City, Michigan, during a 1988 site characterization. Samples were collected both above and below the water table in a homogeneous noncohesive sand which consistently heaved below the water table. During the study 77 cores, 100 cm long, were collected, with greater than 97% core recovery.

Evaluation of the repeatability of cores from identical intervals in adjacent boreholes was performed by drilling three boreholes in a triangular pattern about one meter apart. Each borehole was sampled at three identical depth intervals to determine the repeatability of in situ coring with the specially designed piston sampler. Cohesionless soils were collected, both above and below the water table. Grain size distribution analyses of three core intervals in the separate boreholes were compared in Figure 8. The shallowest depth was about 60-80 cm above the water table, while the intermediate and deepest intervals were below the water table at depths of 60-120 cm and 300-365 cm, respectively. The data presented in Figure 8 indicates that the sampler has an excellent capability of collecting cores of replicate physical properties in heaving noncohesive soils, supporting the preface that integrity with respect to in situ soil sampling can be routinely performed with a combination of clam-shell hollow-stem auger drilling and piston core sampling.

Literature Cited

1. Perry, C.A.; Hart, R. Grnd. Wat. Mon. Rev. 1985, 5, 70-3.
2. McRay, K.B. Grnd. Wat. Mon. Rev. 1986, 6, 37-8.
3. Hackett, G. Grnd. Wat. Mon. Rev. 1987, 7, 51-62.
4. Keely, J.F.; Boateng, K. Grnd. Wat. 1987, 25, 300-13.
5. Hackett, G. Grnd. Wat. Mon. Rev. 1988, 8, 60-8.
6. Keely, J.F.; Boateng, K. Grnd. Wat. 1987, 4, 427-39.
7. Leach, L.E.; Beck, F.P.; Wilson, J.T.; Kampbell, D.H. Proc. 2nd Nat. Outdoor Action Conf. 1988, 31-51.
8. Zapico, M.M.; Vales, S.; Cherry, J. Grnd. Wat. Mon. Rev. 1987, 7, 74-82.
9. Wilson, J.T.; Leach, L. In Situ Reclamation of Spills From Underground Storage Tanks: New Approaches for Site Characterization, Project Design and Evaluation Performance, U.S. Environmental Protection Agency, National Technical Information Service: Springfield, VA, PB89-219976, 1989.
10. Vandegrift, S.A.; Kampbell, D.H. J. Chromatographic Sci. 26, 566-69.

RECEIVED September 17, 1990

Chapter 21

Techniques for Collecting Soil Samples in Field Research Studies

Frank A. Norris, Russell L. Jones, S. Dwight Kirkland, and
Terry E. Marquardt

Rhone-Poulenc Ag Company, P.O. Box 12014, Research Triangle
Park, NC 27709

A variety of techniques have been developed for collecting soil
samples in unsaturated zone studies on the movement and degradation
of agricultural chemicals.   The appropriateness of a sampling
technique will depend on the desired depth of the soil core, soil
properties, and the location of the agricultural chemical residues in the
soil profile.   Determining the amount of soil residues immediately
after a chemical has been sprayed on the soil surface is often difficult.
Data from a study comparing several techniques for measuring the
amount of an agricultural chemical sprayed onto the soil surface
indicate that using filter paper disks or collecting soil samples using 77
mm diameter tubes provided the best measurement of the amount
applied.

Soil sampling is used in environmental fate studies to measure the amount of an
agricultural chemical in the soil as a function of depth at a specific time. During the
past decade, the nature of unsaturated zone environmental fate studies has expanded
from sampling surface soils in simple dissipation studies to complete sampling of the
soil profile down to the water table in groundwater research studies. The numbers of
such studies, as well as the number of soil samples collected during a study, has been
increasing rapidly.   Therefore, much attention has been devoted to comparing
existing soil sampling methods and making modifications to improve the
performance of these methods.

The choice of sampling techniques is dependent on the location of residues in the
soil profile, the nature of the soil, and the properties of the agricultural chemical
being studied. The objective of this paper is to discuss the advantages and problems
of these sampling techniques and is an update of the information in a previous
summary (1). The information in this paper is intended to apply only to relatively
non-volatile compounds which are stable during the time required for sample

0097–6156/91/0465–0349$06.00/0
© 1991 American Chemical Society

collection. This paper is not a discussion of the design of studies which use soil sampling; this topic is discussed in a companion paper (Jones and Norris in this work).

Sampling Procedures

A variety of sampling techniques have been used to collect soil cores in unsaturated zone studies of agricultural chemical residues. Basically the techniques can be divided into two types. One approach is to collect soil in intact cores or core segments. The other is to remove soil from a hole as it is deepened.

Collection of soils in intact cores or core segments is usually performed by pushing sampling tubes or split spoon samplers into the soil. Depending on the diameter of the sampling tube and the sampling depth, these tubes are inserted manually, using hydraulic equipment, or sometimes hydraulically inside a hollow stem auger.

The most common of the sampling devices which extract soil as the hole is deepened is the bucket auger. Another procedure has been to place a tube in the soil and then remove the soil in this tube by use of an implement such as a spoon. Another technique (2) uses a golf cup cutter.

Sampling Tubes. The collection of soil samples in intact segments is probably the most widely used sampling technique. In general, sample tubes used to collect soil samples in environmental fate studies should be greater than about 50 mm to insure a representative sample. This will provide more than the amount of soil usually required for an analysis. Sample tubes of this diameter may be inserted manually to a depth of about 0.5 m, or with rather simple equipment down to a depth of 1 to 2 m. Below this depth drilling equipment (hollow stem auger and split spoon sampler) is usually required. The cost and elapsed time required to collect samples using drilling equipment usually makes collection of more than a few cores impractical. Regardless of the equipment used to insert the sampling tube, each tube should be carefully washed prior to each use.

Sampling tubes are usually used to collect an entire soil core which is then subdivided into segments of the desired depth increments. Two problems can arise as a result. First, soil near the top of the core is often spread along the sides of the tube as the tube is pressed downward. This has been confirmed by the observation of pesticide granules along the side of the tube, and also in recent dye studies (V. Clay, Mobay, personal communication, 1989). The movement of surface soil along the sides of the sampling tube has been shown to result in contamination of deeper increments, especially when residues at the soil surface are relatively high (Rhone-Poulenc, unpublished data). One approach to eliminating this contamination when different depth increments are being collected in an single intact sample core is to remove the outside of the entire core, perhaps after freezing. Another procedure is to collect the first depth increment, enlarge the hole diameter, and insert a lining to prevent surface soil from falling into the hole prior to the collection of the remaining depth increments. The other problem associated with sampling tubes is that the sample core length in the tube may be either longer or shorter than the actual depth of the hole as a result of compaction or expansion, making it difficult to divide the soil core properly into the desired depth increments. This problem is most likely to arise when soil texture changes or rock is encountered in the soil profile. Because of possible of cross-contamination between depth increments from a single intact core and compaction or expansion of the core, the authors' preference is to re-insert the tube into the hole for each depth increment with the upper 50 mm of soil from each tube (except for the uppermost sample from each hole) discarded to reduce any residues introduced into the hole during the raising and lowering of the auger.

Bucket Augers. In field research studies involving the collection of soil samples at depths greater than about 1.2 m, bucket augers have been the most widely used collection technique. This technique has been used by the authors to manually collect samples as deep as 7.8 m in a variety of soils. Normally the authors use a conventional bucket auger, but in some coarse soils a sand bucket auger is needed to keep the sample from falling out of the bucket as it is raised out of the hole. A bucket auger technique should not be used to collect samples just after an agricultural chemical has been sprayed onto the surface of the soil, since the shape of the auger tends to push the upper 10 mm of soil towards the outside of the hole. The major concern with the use of a bucket auger has been the introduction of soil into the bottom of the hole during the raising and the lowering of the auger. This problem can be minimized by using only trained personnel to collect samples, discarding the upper 50 mm of each bucket (except the uppermost bucket from each core), and by using a clean auger for each sample. The authors clean augers using a detergent with ammonia and a toilet brush, followed by a clean water rinse. An advantage of the bucket auger is the ability to collect a larger diameter core (often an 83 mm diameter auger is used) than obtained with most sample tube procedures, perhaps resulting in a more representative sample.

Other Excavation Techniques. These procedures, such as removal of soil inside a tube placed in the soil with an implement such as a spoon or the golf cutter approach (2), are usually restricted to no deeper than the upper 0.5 m. They may be followed by other sampling techniques such as tube samplers or bucket augers to obtain deeper samples.

Sampling Immediately After Application

Collection of soil cores immediately after application of an agricultural chemical requires special consideration. The purpose of the initial post-treatment sampling is to demonstrate that an application has been made and confirm the application rate. However, because of the variability of soil samples, an application rate based on the amount of material applied is usually more accurate than a rate determined by soil analyses. If the agricultural chemical is incorporated into the soil, then the same technique used to collect soil cores during the rest of the study is usually most appropriate. However, if the chemical is sprayed or broadcast onto the soil surface, then a different sampling technique is usually required.

In the authors experiences, the collection of soil samples just after a spray or broadcast application has often resulted in considerably less residues than would be predicted from the amount of chemical actually applied. To determine which sampling techniques are most appropriate, a study was performed in which 10 different techniques (Table I) were used to assess the amount of chemical applied. These techniques included one conventional soil core sampler, four bucket auger procedures, two sample tube techniques, two buried dish procedures, and one filter paper measurement procedure. Sand was used in some of these procedures to cover the soil surface (either before or after inserting the sampling device) because a previous study indicated that it would improve recovery. Twisting the auger while placing it onto the soil was tested also to see if this would minimize the outward movement of the upper 10 mm of soil due to the shape of the auger.

The test was carried out on a Norfolk loamy sand soil at the Rhone-Poulenc research facility near Clayton, North Carolina. An application of aldoxycarb [2-methyl-2-(methylsulfonyl)propionaldehyde 0-(methylcarbamoyl)oxime], chosen because aldoxycarb is easily analyzed, was made at a nominal rate of 3.36 kg ha^{-1} to a narrow strip of land approximately 50 m in length using a single pass of a tractor-mounted sprayer with four spray nozzles mounted 0.5 m apart. The treatment area was divided into four subplots each 10 m long, with a 5 m buffer on each end. In each subplot there were 40 sampling locations: 10 transects one meter apart, each with four spray positions. Two spray positions were located directly underneath the middle two spray nozzles and the other two spray positions were located halfway between spray nozzles. For each of the sampling techniques, one sample was collected from each of the four spray positions in each of the four subplots for a total of 16 samples per sampling technique. For each spray position, the order of the sampling techniques within a subplot was selected randomly. A subsequent analysis of the data indicated no significant variations in residue concentrations as a result of plot location.

The individual analyses and averages are shown in Table II. Statistical analyses by a variety of tests show that the data is divided into two distinct groups. The first group with lower residues consisted of the standard core sampler and the four bucket auger methods. The second group consisted of the two buried dishes, the two sampling tubes, and the filter paper. Twisting the auger as it was placed in the soil may have slightly increased the recovery but the recovery was still not acceptable. Pouring sand over the soil did not seem to significantly increase recovery in any of the techniques. Therefore, the use of sampling tubes, buried dishes, or filter paper were the most acceptable procedures for determining the amount applied to soil. Although this study indicates that filter paper was a good method, other more recent experiments conducted by the authors with other agricultural chemicals have been less satisfactory, perhaps due to the chemical not being totally adsorbed, inability to extract the chemical from the paper, or degradation of the chemical on the paper.

Table I. Techniques Used in Immediate Post-Application Sampling Study

1. Conventional soil core sampler pressed to a depth of 80 mm.

2. Bucket auger (83 mm diameter) placed on the ground before twisting to a depth of 80 mm.

3. Bucket auger, same as 2, except auger was twisted as it was placed on the ground.

4. Bucket auger, same as 3, except approximately 25 mm of soil was placed on top of soil in auger before withdrawing the auger from the hole.

5. Bucket auger, same as 2, except approximately 25 mm of sand was placed on the sample site before augering.

6. Plastic dish (80 mm in diameter, 80 mm high) buried flush with soil surface filled with soil removed from hole.

7. Plastic dish, same as 6, except dish covered with a layer of sand prior to removal from ground.

8. Copper tubing (77 mm inside diameter, 80 mm long) pressed into soil immediately after application.

9. Copper tubing, same as 8, except approximately 25 mm of sand was placed on sample site before pressing the tube into the soil.

10. Filter paper (Whatman Grade No. 1; 90 mm diameter) placed on ground surface prior to application.

Table II. Results of Immediate Post-Application Sampling Study

Method No.:	1	2	3	4	5	6	7	8	9	10
Description:	Soil Core Sampler	Bucket Auger, no twist	Bucket Auger, twist	Bucket Auger, sand added	Bucket Auger, into sand	Buried Dish	Buried Dish, sand added	Sample Tube	Sample Tube, into sand	Filter Paper
Sample Diameter (mm)	25.4	89	89	89	89	89	80	77	77	90
Aldoxycarb Residues: Individual Analyses (mg)										
core 1	0.140	0.99	1.24	1.60	1.30	2.04	1.75	1.31	1.32	2.17
core 2	0.099	0.94	1.59	1.05	1.52	1.46	1.23	1.68	1.60	1.99
core 3	0.095	1.07	1.13	1.14	1.42	1.88	1.36	1.49	1.62	1.98
core 4	0.100	1.14	1.19	1.28	1.14	1.50	1.45	1.71	1.14	2.17
core 5	0.081	0.75	0.81	1.01	1.15	1.37	1.31	1.65	1.14	1.97
core 6	0.160	0.80	0.80	1.21	0.96	1.81	1.62	1.92	1.04	1.78
core 7	0.080	0.83	0.90	1.32	0.85	1.80	1.48	1.41	1.53	1.66
core 8	0.130	0.78	0.91	1.00	0.84	0.50	1.47	1.80	1.67	1.51
core 9	0.110	1.19	0.91	1.32	1.13	1.80	1.22	1.70	1.31	2.03
core 10	0.100	0.86	1.30	1.40	1.42	1.76	1.31	1.31	1.02	2.08
core 11	0.084	1.25	1.22	1.08	0.93	1.28	1.53	2.18	1.81	1.77
core 12	0.130	0.98	1.64	1.23	1.25	1.47	1.56	1.67	1.38	1.69
core 13	0.098	1.17	0.97	1.53	1.22	0.83	1.73	1.09	1.39	2.14
core 14	0.086	0.78	1.37	1.41	1.20	0.85	1.48	1.79	2.20	1.44
core 15	0.093	0.85	0.74	1.44	1.40	2.10	0.95	1.19	1.28	1.87
core 16	0.073	1.13	0.90	0.95	0.84	1.22	1.03	1.38	1.22	2.07
coefficent of variation (%)	24	18	25	16	19	31	16	18	22	12
Average of Analyses (mg)	0.10	0.97	1.10	1.25	1.16	1.48	1.41	1.58	1.42	1.90
(kg/ha)	2.05	1.56	1.77	2.01	1.87	2.94	2.80	3.39	3.04	2.98

Therefore, suitability studies should be performed prior to application with the specific chemical under study.

Recommendations

The authors' assessment of the various sampling procedures is summarized in Table III. A variety of sampling techniques will produce acceptable results; therefore the choice of technique may depend on the sampler's past experience, available equipment, and the expected depth of the chemical in the soil. The authors' preferences are: For measuring the amount of an agricultural chemical sprayed on the soil surface, the use of filter paper in addition to 77 mm diameter sampling tubes is recommended. At other times, when an agricultural chemical is concentrated in the upper 10 mm of soil, the use of the 77 mm diameter sampling tubes has produced satisfactory results. When deeper cores are required, but there are still residues concentrated at the soil surface (for example, just after an application but residues from previous applications are present throughout the soil profile), the sampling tubes can be used to collect a sample from the upper 0.15 m of soil followed by a bucket auger procedure to collect samples from deeper depths. After a soil incorporated application or at any sampling interval where residues are not concentrated in the upper 10 mm of soil, the authors prefer the bucket auger technique described elsewhere (*3*). One exception is in stoney soil where some type of split spoon sampler (*4*) is appropriate. For soil cores deeper than about 7.8 m, the use of hollow stem augering combined with split spoon sampling (*5*) is recommended.

Several important guidelines should always be followed to reduce potential contamination during the collection of soil samples regardless of which sampling technique is used. Samples should always be collected by conscientious, trained personnel. Careful attention must be given to proper labeling of samples. Because of the large number of soil samples usually collected at a sampling interval, it is quite easy to confuse the identity of some samples. Therefore, care must be taken to make sure that each sample is placed into the proper prelabeled container. Also concern about cleanliness must be paramount in the minds of everyone. All sampling equipment should be thoroughly washed between samples (the wash water should not be discarded in the test plot). All sampling equipment and sample containers should be kept off the ground and (as much as possible) kept away from dust which might contain agricultural chemical residues. Hands should not be placed inside the sample bags during collection or processing. Sample containers (before and after addition of samples) should not be transported in vehicles used to transport agricultural chemicals. Nor should containers or collected samples be stored in the same areas used to store agricultural chemicals, analytical standards, or application equipment.

Table III. Summary of Sampling Techniques

Technique	Suitable for Sampling Surface Residues?	Maximum Sampling Depth (m)	Comments
Filter Paper	yes	-	can be used only to measure the amount sprayed on the soil surface
Buried Dishes	yes	-	can be used only to measure the amount applied to the soil surface.
Core Sampler	no	0.5	the small diameter appears to result in poor recovery of surface residues
Golf Cup Cutter	yes	0.1	
Excavation of Soil Inside a Ring	yes	0.5	
Bucket Auger	no	7.8	
Sampling Tubes			
Manual Placement	yes	0.5	
Hydraulic Placement	yes	2	
Hollow Stem Auger	no	>10	

Literature Cited

1. Jones, R. L. In *Environmental Fate of Pesticides*; Hutson, D. H.; Roberts, T. R, Eds.; Progress in Pesticides in Biochemistry and Toxicology, vol. 7; John Wiley and Sons: Chichester, U.K., 1990; pp. 27-46.
2. Day, E. W.; Decker, O. D.; Griggs, R. D.; West, S. D. *Proc. Sixth International Congress of Pesticide Chemistry (IUPAC)*, 1986, abstract 6B-16.
3. Jones, R. L.; Hornsby, A. G.; Rao, P. S. C. *Pestic. Sci.* 1988, *23*, 307-325.
4. Jones, R. L.; Rourke, R. V.; Hansen, J. L. *Environ. Toxicol. Chem.* 1986, *5*, 167-173.
5. Pacenka, S.; Porter, K. S.; Jones, R. L., Zecharias, Y. B.; Hughes, H. B. F. *J. Contam. Hydrol.* 1987, *2*, 73-91.

RECEIVED September 5, 1990

AGROCHEMICAL SURFACE LOSS TECHNIQUES

Chapter 22

Soil-Pan Method for Studying Pesticide Dissipation on Soil

B. D. Hill, D. J. Inaba, and G. B. Schaalje

Agriculture Canada, Lethbridge, Alberta T1J 4B1, Canada

To predict the amount of pesticide that could leach through the soil and contaminate groundwater requires information about the residue levels at the soil surface over time. A soil-pan method has been developed to estimate surface residues and their dissipation rates. An indoor spray chamber is used to apply the pesticide to soil contained in metal flats, the treated flats are moved outdoors and set into a field, and the soil is sampled over the season by taking four cores per flat. Using this method, it was determined that the emulsifiable concentrate formulation of deltamethrin dissipated faster than the Flowable formulation. When the soil-pan method was compared with a field-plot method, the dissipation of lambda-cyhalothrin was faster in the soil pans. Monitoring the soil temperature and moisture indicated that both were slightly higher in the soil pans than in the adjacent field plots. At present, the soil-pan method is best suited for the direct comparison of different treatments.

The amount of pesticide that potentially can leach through the soil is a function of the amount of residues at the surface. Therefore, when determining the potential for groundwater contamination, a first consideration is to estimate the amount of surface residues. Surface residues vary with the amount of pesticide initially deposited and the rate of surface dissipation.

The methods available for studying the surface dissipation of pesticides on soil range from large field plots with aerial application (1), small field plots with ground-rig application (1), and outdoor microplots with pipet application (1-5) to indoor and laboratory incubations (5-7). Ideally, these methods should meet the following criteria:

1. an even and accurate pesticide application,
2. representative and precise sampling with a manageable number of samples,

0097–6156/91/0465–0358$06.00/0
Published 1991 American Chemical Society

3. convenient sample handling and sample preparation for analysis,
4. results should be representative of actual field situations.

With field studies, it is difficult to achieve criteria 1-3. Multi-nozzle spray boom applications are usually uneven and extensive sampling with subsampling is required. With the indoor and laboratory incubations, there are always concerns about whether criterion 4 has been met. The outdoor microplot method, originally developed by Smith (2,3) for soil-incorporated herbicides, uses an accurate pipet application, whole-plot (20 x 20 cm) sampling to reduce error, and is conducted in an actual field situation. However, we found that for surface-applied pesticides, the high water volume from pipet application washed the pesticide into the soil and altered its dissipation compared with nozzle application (1).

In an attempt to rectify the pipet application problem and yet retain the advantages of the microplot method, we developed a soil-pan method for studying surface residues. This method features a single nozzle spray application, field dissipation, a reduced sample size with no subsampling, and minimum sample handling. This paper describes the soil-pan method and compares dissipation results from it with results from field plots for the pyrethroid insecticides deltamethrin [(S)-α-cyano-3-phenoxybenzyl (1R,3R)-cis-2,2-dimethyl-3-(2,2-dibromovinyl)cyclopropanecarboxylate] and lambda-cyhalothrin (formerly PP321), a 50:50 mixture of [(S)-α-cyano-3-phenoxybenzyl (1R,3R)-cis-2,2-dimethyl-3-(2-chloro-3,3,3-trifluoroprop-1-enyl)cyclopropanecarboxylate] and [(R)-α-cyano-3-phenoxybenzyl (1S,3S)-cis-2,2-dimethyl-3-(2-chloro-3,3,3-trifluoroprop-1-enyl)cyclopropanecarboxylate]. Deltamethrin is widely used in western Canada to control grasshoppers, cutworms, flea beetles and alfalfa weevil; lambda-cyhalothrin is currently under development to control the same insects.

Materials and Methods

Soil Type. The Lethbridge sandy clay loam (Typic Haploboroll, fine loamy, mixed, mesic) contained 24.2% clay, 20.5% silt and 55.3% sand with CEC=20.1 meq 100 g^{-1}, 2.2% organic matter and pH=7.9.

Soil Pan Method. The soil pans (50 x 35 x 9 cm metal flats with drainage holes in the bottom) were prepared as follows: Soil from the 0-9 cm layer of a fallow field was transferred into the pans and loose-packed by tapping the bottom and sides of the pan. The soil was levelled off at the top of the pans and the pans temporarily set into the fallow field. The soil pans were then equilibrated for 7-10 days under field conditions. If no significant rainfall occurred, irrigation (2-3 cm) was used to further settle the soil in the pans and to re-crust the surface. Pesticide treatments were applied by transferring the pans (without disturbing the soil) to an indoor spray chamber equipped with a single, travelling nozzle. The spray chamber was calibrated and optimized so that the spray pattern, as indicated by water-sensitive paper (8), was even across the 35-cm pan width. After spraying, the soil pans were immediately returned outdoors and set into the fallow until level with the field surface. To sample, four strata were visually identified across the length of the pan and one core sample (0-2.5 cm x 2.38 cm i.d.) was taken at random within each stratum. The holes left by sampling were not filled in, but were marked with a small stake so that they could be avoided on subsequent samplings. The soil pans were left in the field until the end of the experiment, at which time the pans and the remaining treated soil were easily removed.

1986 Soil-Pan Experiment. An initial, direct comparison of two treatments was conducted to establish that the soil-pan method was viable. The dissipation of two deltamethrin formulations, an emulsifiable concentrate (EC) and a Flowable (FL), was determined on Lethbridge soil. The treatments were applied at 10 g ha^{-1} using 125 L ha^{-1} volume with five replicate pans per treatment (EC, FL, and unsprayed control). The pans were set into the field in a closely grouped, completely randomized design. Samples were taken at intervals over 0-16 weeks.

1988 Soil-Pan versus Field-Plot Experiment. A pan versus plot experiment was conducted to determine whether the soil-pan method gave representative results compared with a larger field-plot experiment. Prior to the pesticide applications, the soil in the pans and the field plots was prepared (cultivated, raked, watered and equilibrated) as identically as possible. The field-plot experiment consisted of four replicates, each 2 x 4 m. Lambda-cyhalothrin (FL formulation) was applied at 15 g ha^{-1} in 125 L ha^{-1} volume using a bicycle sprayer with four nozzles on a 2-m boom. The plots were sampled by bulking two cores per site from eight sites per replicate (sites chosen at random within a stratified design). Pans were sprayed and sampled as before (four replicates) and set into the field-plot experimental site.

1989 Soil-Pan versus Field-Plot Experiment. The 1988 pan versus plot experiment was repeated using an EC formulation of lambda-cyhalothrin instead of the FL formulation. As a precaution against possible wind and water erosion, and loss of surface residue from the pans, the soil level was left 2.5 cm below the lip of the pans.

Sample Handling and Residue Analysis Method. All soil samples were stored at -40°C until analysis. Samples from the soil-pan method were thawed and the whole four-core sample was analyzed directly. The composite samples (16 cores) from the field-plot experiments were thawed, air-dried overnight, ground, mixed and 50 g subsamples taken. The subsamples were re-frozen and re-thawed before analysis.

The residue analysis method has been described previously in detail (5). Briefly, samples were extracted by shaking with acetone/hexane, liquid-liquid partitioned into hexane, cleaned up on acid alumina microcolumns, and quantified by EC-GLC using a DB-1 capillary column.

Results and Discussion

1986 Soil-Pan Experiment. The dissipation of deltamethrin was biphasic (Figure 1) with mean DT50=3.2 weeks and mean DT90=23 weeks (DT50 and DT90 are the times required for 50% and 90% of the initial residue to disappear). The EC formulation degraded significantly ($P<0.01$) faster than the FL formulation. The mean CV between replicates on a given sample date was an acceptable 12.4%. (It was decided to reduce the number of replicates to four in future pan studies.) The soil-pan method proved viable and effective for this direct comparison of two treatments. This experiment did not determine directly whether the results of the soil-pan method were representative of field dissipation; however, the dissipation rate and variability compared well (Table I) with results from our previously reported studies (*1,5*).

Table I. Deltamethrin Dissipation on Lethbridge Soil

Method, Plot Size	Application Method	DT50, weeks	Mean CV (%) Between Reps
Microplot[a]	pipet	6.4	10.0
Field-plot[a]	ground-rig	3.7	6.3
Soil-pan	spray chamber	3.2	12.4
Large-scale[a]	aerial	2.0	14.5

[a]Previously reported studies (*1,5*).

1988 Soil-Pan versus Field-Plot Experiment. Results of the soil-pan method were not representative of the field-plot dissipation. The lambda-cyhalothrin dissipated significantly ($P<0.01$) faster (Figure 2) in the pans (DT50=0.8 weeks, DT90=9.3 weeks) than in the field plots (DT50=2.2 weeks, DT90=25 weeks). However, experimental error was similar even though four times as many cores were taken from the field plots. The mean CV between replicates was 17.2% for the pans compared with 15.4% for the plots.

The following postulate was formed to explain the faster dissipation in the pans. Because most of the difference in dissipation rates occurred between 0 and 2 weeks (Figure 2), when surface processes (volatilization, photolysis, physical loss) have their greatest effect, it seemed reasonable to assume that some surface process was exaggerated in the pans. There were no apparent differences in the condition of the soil surface (crusted with some cracking) or soil moisture between the pans and the plots. Soil temperatures (at 0.5-cm depth) inside the pans were only slightly higher than those in the field plots (Figure 3). A more likely explanation for surface losses of residues in the pans was erosion. There was a significant rainfall event (19 mm in less than 1 h on day 7) which could have washed some of the surface residues out of the pans. (Leaching should not have been a factor because in separate laboratory trials, lambda-cyhalothrin did not leach past the top 2.5 cm.) Also, because a FL formulation was used (where the dried residues can remain on the soil surface adsorbed to suspension

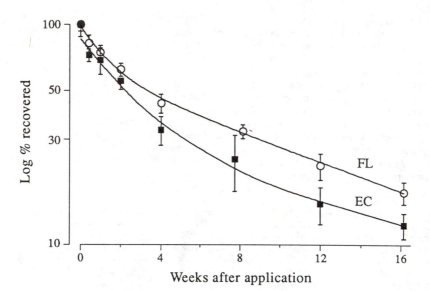

Figure 1. Dissipation of deltamethrin (EC and FL formulations) on Lethbridge soil using the soil-pan method (1986). Each value is a mean of five replicates ± SD.

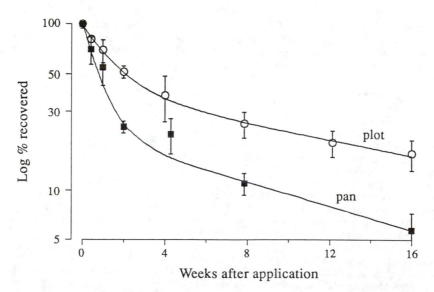

Figure 2. Dissipation of lambda-cyhalothrin on Lethbridge soil using the soil-pan method compared with the field-plot method (1988). Each value is a mean of four replicates ± SD.

and charge agents from the formulation), the lambda-cyhalothrin residues may have been susceptible to wind erosion and this effect could have been exaggerated in the pans.

1989 Soil-Pan versus Field-Plot Experiment. Once again, the soil-pan method results were not representative of the field-plot results. The lambda-cyhalothrin dissipated significantly ($P<0.01$) faster (Figure 4) in the pans (DT50=1.7 weeks, DT90=8.9 weeks) than in the field plots (DT50=2.9 weeks, DT90=14 weeks). Experimental error was again similar in spite of taking only four cores from the pans. The mean CV between replicates was 14.5% for the pans compared with 14.7% for the plots.

The results of this 1989 experiment had one major difference compared with the 1988 results. In this experiment, the initial dissipation rates (0-1 weeks) between the pans and the plots were similar, but the later dissipation rates (2-16 weeks) were different. For a pyrethroid like lambda-cyhalothrin, the 2-16 weeks dissipation is mainly the result of microbial degradation (9-11). Differences in the rate of microbial degradation may have been related to subtle temperature and moisture differences (Figure 5) between the soil in the pans and the soil in the field. The soil in the pans did take longer to dry out after a rainfall. This problem may be overcome by using wooden pans with screened bottoms in future experiments.

Assessment of the Soil-Pan Method. Theoretically, the soil-pan method has several advantages over the use of larger field plots:

1. the spray chamber gives a more accurate and even spray application,
2. only four cores are required for a representative and precise sample,
3. the whole four-core sample (~50 g) is analyzed; subsampling is avoided,
4. only a small field area is required and the location is flexible,
5. different soil types can be brought in and tested at the same location,
6. pans are easily hand-weeded over the season,
7. treated soil is easily removed from the site at the end of the experiment.

The most salient feature of the soil-pan method is that it uses a spray nozzle application to simulate the surface distribution of residues from a field spray operation, yet adequate sampling consists of taking only four cores. The sampling procedure (four cores per $0.2 \, m^2$) may at first seem inadequate, but it is actually more intensive than the guidelines (12 cores per $36 \, m^2$) suggested by Taylor et al. (12,13) to achieve a CV <20%. The soil-pan method should allow the determination of surface residues accurately and quickly and thus facilitate predictions of the potential for groundwater contamination.

The soil-pan method does have certain limitations:

1. For pesticides that readily leach, the depth of the pans would have to be increased and/or lysimeters installed beneath the pans.
2. Unless the area of the pans is increased, the number

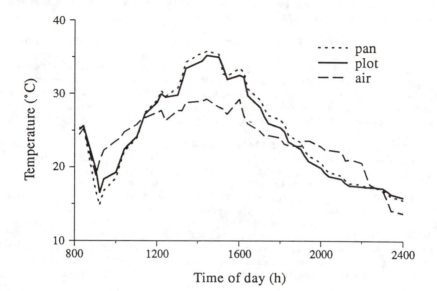

Figure 3. Soil temperatures at 0.5-cm depth in pans versus field plots (1988). Air temperature at 1.5-m height is also indicated.

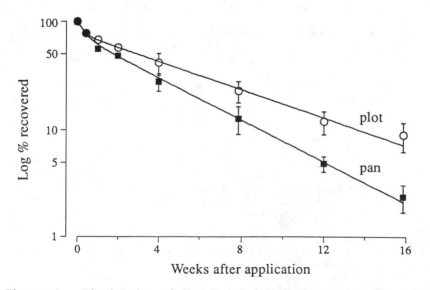

Figure 4. Dissipation of lambda-cyhalothrin on Lethbridge soil using the soil-pan method compared with the field-plot method (1989). Each value is a mean of four replicates ± SD.

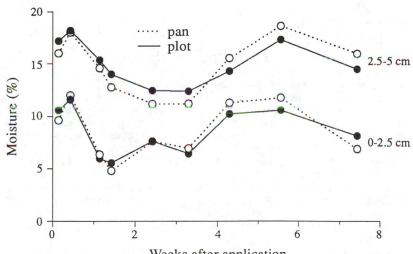

Figure 5. Soil moisture at two depths in pans versus field plots (1989).

of samplings is limited to about seven dates (i.e., 28 cores at four cores per date).

3. There may be slight temperature and moisture differences between the soil inside the pan and the rest of the field.

4. Dissipation in the pans may not always be representative of field dissipation.

The most serious limitation of the soil-pan method for studying the dissipation of surface residues is the question of whether the results are representative of field dissipation. It is surprising that the minor differences in temperature (Figure 3) and moisture (Figure 5) between the pans and the plots caused dissipation to be faster in the pans. A change in the design of the pans is being considered to improve the agreement between soil pans and field plots. At its present stage of development, the soil-pan method is best suited for the direct comparison of different treatments.

Literature Cited

1. Hill, B. D.; Schaalje, G. B. *J. Agric. Food Chem.* **1985**, *33*, 1001-1006.

2. Smith, A. E. *Weed Sci.* **1971**, *19*, 536-537.

3. Smith, A. E. *J. Agric. Food Chem.* **1972**, *20*, 829-831.

4. Hill, B. D. *J. Agric. Food Chem.* **1981**, *29*, 107-110.

5. Hill, B. D. *J. Environ. Sci. Health* **1983**, *B18*, 691-703.

6. Laskowski, D. A.; Swann, R. L.; McCall, P. J.; Dishburger, H. J.; Bidlack, H. D. *Test Protocols for Environmental Fate and Movement of Toxicants;* Proceedings of a Symposium, 94th Ann. Meeting; AOAC: Arlington, VA, 1981; pp 122-149.

7. Laskowski, D. A.; Swann, R. L.; McCall, P. J.; Bidlack, H. D. *Res. Rev.* **1983**, *85*, 139-147.

8. Hill, B. D.; Inaba, D. J. *J. Econ. Entomol.* **1989**, *82*, 974-980.

9. Kaufman, D. D.; Haynes, S. C.; Jordan, E. G.; Kayser, A. J. *ACS Symp. Ser.* **1977**, *42*, 147-161.

10. Williams, I. H.; Brown, M. J. *J. Agric. Food Chem.* **1979**, *27*, 130-132.

11. Chapman, R. A.; Tu, C. M.; Harris, C. R.; Cole, C. *Bull. Environ. Contam. Toxicol.* **1981**, *26*, 513-519.

12. Taylor, A. W.; Caro, J. H.; Freeman, H. P.; Turner, B. C. *ACS Symp. Ser.* **1985**, *284*, 25-35.

13. Taylor, A. W.; Freeman, H. P.; Edwards, W. M. *J. Agric. Food Chem.* **1971**, *19*, 832-836.

RECEIVED October 9, 1990

Chapter 23

Rainfall Simulation for Evaluating Agrochemical Surface Loss

Robert L. Hill[1], Christoph M. Gross[2], and J. Scott Angle[1]

[1]Department of Agronomy, University of Maryland, College
Park, MD 20742
[2]Soil Conservation Service, U.S. Department of Agriculture,
Annapolis, MD 21401

Agrochemical surface loss can be a serious point source of
groundwater contamination through direct channeling of
runoff to groundwater. Agrochemical surface loss
measurements may also be used to estimate potential
agrochemical reserves available for leaching to groundwater.
Surface losses are a critical component of groundwater
contamination models. Rainfall simulators are effective
research tools to evaluate potential agrochemical surface loss.
The choice of a simulator is determined by simulator
properties, research objectives, and the agrochemical loss to
be evaluated. Whether nutrient and/or pesticide loss will occur
in the aqueous or sediment-associated fraction of runoff is
largely dependent on the inherent agrochemical properties
and the timing of the rainfall event after agrochemical
application. The general techniques presented could be
incorporated into research studies to evaluate potential
contamination effects of agrochemical surface loss.

Agrochemicals carried via surface runoff can be a serious and often
overlooked source of groundwater contamination. Abandoned wells,
improperly-protected wellheads, and geological formations characterized by
karst topography can provide direct channels for surface runoff losses to
contaminate groundwater. The contamination resulting from these
agrochemical losses can be serious since this direct channeling of runoff
bypasses the natural filtering and degradation which occurs as
agrochemicals move through the soil (1). These direct channels may also
link surface aquifers to deeper aquifers and, thus, increase the significance
of any potential contamination. While the direct channeling of agrochemicals
from the soil surface to groundwater is recognized as a serious source of

0097–6156/91/0465–0367$06.00/0
© 1991 American Chemical Society

groundwater contamination, the magnitude of losses from channeled surface sources to groundwater has not been adequately documented.

Agrochemical surface loss measurements may also be used to estimate potential agrochemical reserves available for leaching to groundwater. Modeling procedures often use a mass balance-type approach to account for agrochemicals in the soil system and the potential for leaching of agrochemicals to groundwater. Since the dilution effects of groundwater on agrochemical concentrations within groundwater are not well-understood, an accounting of the fate of the modeled agrochemical after introduction to the soil is often a critical component of model integrity. Surface loss of agrochemicals may be an important factor in the accounting process.

Because runoff and the contaminants it carries are critically important to soil erosion processes and water supply quality, scientists have been attempting to evaluate the effects of rainfall on surface losses for many years. Unfortunately, soil and water loss evaluations based on natural rainfall occurrence and events usually take 10 to 25 years (2,3). There have been attempts to simulate rainfall since the 1930's such that experimental control could be maintained over the intensity, duration, and occurrence of rainfall events. It was realized early that the application of rainfall in a controlled manner could greatly shorten the time necessary to evaluate the effects of different management and cultural practices on surface losses.

Early attempts at simulating rainfall were concerned with rainfall intensity and duration. It was not until the early 1940's with the work of Laws (4), and Laws and Parsons (5), that the relationships between raindrop fall velocity, drop size, distance of fall, rainfall intensity and drop size distribution were characterized. Rainfall simulators designed during the late 1940's - early 1950's attempted to more nearly approximate an appropriate drop size for natural rainfall. Wischeimer in 1958 and 1959 (6,7) documented the relationship between the kinetic energy of rainfall and surface soil loss. Previous simulators utilizing an individual drop formation technology did not approximate the necessary kinetic energies. The design of rainfall simulators using pressurized nozzles allowed rainfall to be produced with characteristics very similar to natural rainfall (2).

There are several reviews which address the historical development of rainfall simulators (8-10). There are also many different types and designs of simulators. A listing of approximately 65 simulators used throughout the world is given in the Proceedings of the Rainfall Simulator Workshop (11).

The primary purpose of this chapter is to list some of the advantages and limitations of using simulated rainfall, to give a brief overview of simulator properties and general types of simulators which can be selected for various purposes, and to discuss factors which should be considered when evaluating agrochemical runoff losses. The general techniques presented could be incorporated into research studies to evaluate the potential contamination effects of agrochemicals lost in runoff which have been directly channeled into groundwater or used to obtain agrochemical surface loss estimations for fine-tuning groundwater contamination models.

Advantages and Limitations of Using Simulated Rainfall

An important advantage of using simulated rainfall is the control which the scientist has over experimental conditions. Timing of rainfall application, rainfall intensity, and storm duration can all be controlled. The simulated rainfall will be uniformly distributed over the experimental area and will possess a constant range of drop sizes. The reproducibility of simulated rainfall offers advantages in experimental design over natural rainfall, which may vary in intensity several times during a rainfall event and which may be spatially variable over relatively short distances.

Simulated rainfall allows a large number of treatments or management practices to be evaluated for the same rainfall conditions in a relatively short time. Comparisons may be made under the same treatment or management practice for different rainfall intensities and durations. Intricate measurements may be made in a controlled environment which would be nearly impossible during natural rainfall events. Studies may be designed for the laboratory or field, depending upon research goals.

Personnel can be assembled in a timely fashion for the experimental field work portions of a study. Although several technical support people may be needed during the simulated rainfall event, the number of man hours is greatly reduced as compared to studies dependent on natural rainfall. For most major rainfall simulation field studies, a minimum crew of 3 to 4 people is usually required. Although plot preparation time is generally less for simulated rainfall studies, the actual time spent raining is small compared to the time required for preparation before and after the simulated storm.

Rainfall simulators can be expensive to design, construct, and calibrate, but the cost per unit of data collected is usually low when compared to the unit data cost of long-term experiments dependent on natural rainfall (*12*). Some researchers have concluded that the amount of research data that can be generated by a rainfall simulation program is usually limited by the amount of subsequent analysis that can be done (*13*). Meyer (*14*) suggests that 2 to 4 months of field research work produces enough samples and data to keep researchers busy the remaining 8 to 10 months of the year. Caution is suggested to ensure that large amounts of time and money not be placed in simulator design, construction, and calibration if that is not a primary research goal (*13*). Initial time and costs for a simulator can be reduced by reviewing the properties and capabilities of existing rainfall simulators and adopting a previously-tested design. Meyer (*14*) stated that the cost and time to construct a simulator and the personnel needed to operate an effective simulated rainfall research program are the greatest obstacles facing potential users.

The results obtained from a rainfall simulation program are primarily qualitative (2), but offer valid relative comparisons which may help provide insight for regulatory decisions. Natural rainfall experiments still need to be conducted and at some point related to the results obtained from simulated rainfall studies to obtain the maximum benefit from both types of data. Young and Burwell (3) found that runoff from simulated rainfall plots was about the same as from natural rainfall plots if similar intensity-duration patterns were used. Sediment loss was less for the simulated rainfall storms, but were in relative proportion to the erosion index (kinetic energy times maximum 30-minute intensity) of the simulated rainfall versus the natural rainfall. Barnett and Dooley (15) found in a comparison of 19 natural storms and 35 simulated storms that if sediment losses were based on erosion index values of the respective storms, no real differences existed between the natural and simulated data when the data was used for prediction purposes. More studies of this type need to be conducted particularly for comparisons of agrochemical losses.

The nature of using simulated rainfall is generally limited to relatively small land areas because of the large amounts of water required as the study area is increased. Water availability may, in itself, limit the scale of a rainfall simulation project and may help make decisions as to the type of simulator to use. Regardless of the scale of simulation, processes evaluated for small land areas may not be the same compared to field scale because of differences in operational scale. The inadequate conversion of data from rainfall simulation studies to large land areas because of simulator differences is regarded as a weak link in model testing (13). It should be noted that these limitations become insignificant when compared to complete dependence on natural rainfall events.

Properties of Rainfall Simulators

The choice of a rainfall simulator design largely depends on the research goals for which it will be used. Simulators have been used in the laboratory and field to gather basic information on runoff, erosion, infiltration, soil crusting behavior, and agrochemical loss and effectiveness. The criteria for choosing a given simulator will depend on desired rainfall characteristics, the plot size and process under study, and simulator portability and cost (16). Several researchers (2,9,11,14,16-20) have described these criteria in detail and include:

1.   Drop size distribution and impact velocities near those of natural rainfall,
2.   Uniform rainfall intensity with random drop size distribution over the entire application area,
3.   Reproducible storms of desired rainfall intensities and durations,
4.   Continuous application over the entire plot area,
5.   Sufficient areal coverage to meet study requirements,
6.   Near vertical raindrop impact,

7.   Satisfactory operation under moderate winds and high temperatures for field units,
8.   Portability for field units,
9.   Reasonable cost.

It is unlikely that any one simulator would satisfy all research criteria. The nature of the research study determines the importance and range of the design criteria. Simulators have been used on areas ranging from a few square centimeters in laboratory studies to more than a hectare. Cost and technical limitations will generally prevent any one simulator from meeting all the design criteria.

The question of which rainfall characteristics should be used for the comparison of simulated rainfall to natural rainfall is still a matter of debate. The importance and interaction of rainfall characteristics such as drop size distribution, impact velocity, and kinetic energy is not well understood. Meyer (*14,18*), in a comparison of several rainfall energy parameters for natural and simulated rainfall, suggested that both the drop size distribution and drop fall velocity of natural rainfall should be closely simulated and that any disproportionate change in either property is not advised.

Types of Rainfall Simulators

Rainfall simulators can be categorized in terms of the mechanism which is used to produce raindrops: individual droppers or nozzles. This classification system has been used in the world inventory of rainfall simulators, previously mentioned, where a listing of the various types of simulators includes the method of raindrop formation, a drop size description, intensity range, plot size, and literature reference (11). Simulators using individual droppers have been widely used on small plot (<1.5 m^2) studies while nozzle-type simulators have been used on both large and small plot studies.

Drop-type Simulators.   Drop-type simulators use yarn or capillary tubing of various materials to form raindrops which fall from a tip at near zero velocity. Recent tubing materials used have included brass, copper, polyethylene, Teflon, glass, stainless steel, and hypodermic needles. Raindrop size is controlled by tubing size, tubing material characteristics, and the velocity of flow through the tube. Droppers generally produce only one size or a very limited range of sizes of raindrops. There have been recent attempts to construct simulators with interchangeable tips so that different size drops might be formed (*21*) or that use air pressure to change drop size in the formation process (*22*). Drop-type simulators are also usually limited to a single or a very narrow range of rainfall intensities. Intensity is usually controlled by varying the distance between the droppers or changing the pressure head on the drop forming feed mechanism.

An additional limitation with this type of simulator is that the impact velocities of raindrops is typically less than similar-sized drops of natural rainfall since the raindrop leaves the dropper at near-zero velocity. The fall height is generally insufficient for raindrops to attain terminal velocity (eg. a 4-mm drop would require ca. 8 meters to attain 95% of its terminal velocity). If the simulators were of sufficient height to attain terminal velocity, the falling raindrops would be easily affected by wind currents, particularly for field research conditions. Because of the intricacies involved in the method of drop formation, these simulators are usually designed for laboratory or small plot studies of less than a few square meters. They can be useful where a precise control of drop size is important.

Nozzle-type Simulators. Nozzle-type simulators use pressurized nozzles to produce rainfall. Bubenzer (*11*) stated there were at least 17 different nozzles being used on various simulators, but only 5 nozzles were widely used (*10*). The nozzles form a range of drop sizes which can approximate the distributions observed for natural rainstorms. The water is forced from the nozzle at positive pressure which reduces the fall distances necessary to obtain near terminal velocity. The increased pressure, while facilitating higher impact velocities, results in reduced drop size (*23*). To overcome this problem, large orifice nozzles are used to produce the desired raindrop size distributions and energy levels. These high output nozzles generally result in intensity rates which are unrealistically high. Specific examples of methods used by various simulators to intermittently apply rainfall and reduce the intensity will be cited later in the text. The use of intermittently-applied rainfall has not been found to adversely affect the results of rainfall simulation studies as researchers have observed close similarities between results from natural rainfall and intermittent simulated rainfall (*3,15*). Although the following discussion of nozzle-type simulators is not comprehensive, it includes commonly-used simulators and tries to present simulators which differ in nozzle-type and/or mechanism of rainfall application.

    Type F Simulator. The Type F rainfall simulator was developed by the Soil Conservation Service during the late 1930's and represented the first attempt to develop a standard device for the study of rainfall effects on soil behavior (*9*). The Type F nozzle produces a raindrop size distribution similar to high intensity rainfall (*2*). The median drop size is 3.4 mm, which is larger than the median size observed in naturally occurring storms (*10*). The water jet from this nozzle is different from most modern simulators in that the spray is directed upwards at a 7 degree angle from vertical and allowed to fall from an average height of ca. 2.4 m. This height is not sufficient to reach terminal velocity for a large proportion of the drops. Meyer (*18*) stated the kinetic energy per unit of rainfall from the Type F nozzle is 56% of natural rainfall at equivalent rainfall intensities. Since the water is sprayed upwards, it is subject to wind drift. This simulator runs continuously and does not require intermittent application. It was designed for plots which are 1.8 meters wide with lengths in multiples of 3.7 m.

Rainulator. The rainulator developed by Meyer and McCune (2) represented the first rainfall simulator which applied rainfall with characteristics which closely resembled natural rainfall. This simulator uses the Spraying Systems Co. Veejet 80100 nozzle, which produces a flat, fan-type spray pattern that is sprayed downward over the plot area. The Veejet 80100 nozzle produces a raindrop size distribution equivalent to a 2 to 13 mm hr^{-1} rainfall event although only 78% of the kinetic energy per unit volume of rain is produced. Reduction of rainfall intensity is achieved by reciprocating movement of the nozzle back and forth across the plot area. The rainulator produces rainfall intensities of 64 or 127 mm hr^{-1}. The plot width is limited to 4.3 m or less, but because of the modular design of the unit, the only restriction on plot length is the number of modular units available and the water supply (24). The first unit covers an effective length of 4.6 m with additional units covering 6.1 m. The rainulator gives satisfactory performance in winds less than 24 km hr^{-1}.

Rotating-boom Simulator. The trailer-mounted rotating-boom simulator uses the same Veejet 80100 nozzle as the rainulator (25). The nozzles are mounted 2.7 m above the ground in the downward position on ten booms which support a total of 30 nozzles. The nozzles are positioned on radii of 1.5, 3.1, 4.6, 6.1, and 7.6 m with 2, 4, 6, 8, and 10 nozzles on each radius, respectively. Rainfall is intermittently applied by rotating the vertical stem to which the booms are attached at 3.5 to 4.0 rpm. Rainfall intensities of 64 or 127 mm hr^{-1} per hour are possible by operating either 15 or 30 nozzles. Water requirements are 246 and 492 L min^{-1} for intensities of 64 and 127 mm hr^{-1}, respectively. The simulator accommodates paired rectangular plot areas up to 10.7 m in length with an overall width of 12.2 m or less for the pair of rectangular plots including an appropriate-sized alleyway for the 2.4-m trailer width. The rotating-boom simulator can be moved in the field by merely disconnecting the water coupling. Local highway moves may be accomplished by removing the booms and cables (ca. 30 minutes) and reassembling at the new site (< 2 hours). Long distance moves on a truck or trailer can be accomplished in relatively brief time periods. The rotating-boom simulator was selected for use in the United States Department of Agriculture - Agricultural Research Service - Soil Conservation Service - Forestry Service and United States Department of the Interior - Bureau of Land Management Water Erosion Prediction Project(WEPP) because of the similarity of rainfall produced to natural rainfall and the portability of the unit for use throughout the United States.

Small-plot Simulators. Most notable among the nozzle-type simulators which have been developed for use in the laboratory or on small plot areas are the Purdue sprinkling infiltrometer (17), the rotating disk simulator (23), and the interrill rainfall simulator (20).

Purdue Infiltrometer. The Purdue sprinkling infiltrometer uses a single full-cone spray nozzle operated from a height of 2.7 m to rain on an area slightly larger than 1 m². There is a choice of a Spraying Engineering Co. model 5B, 5D, or 7LA nozzle that can be used to produce intensities of 64, 83, and 114 mm hr⁻¹, respectively. The design includes a runoff recirculation device for picking up and accumulating runoff. This simulator has been used extensively for soil infiltration measurements. Designs for modification of both the simulator (26) and the runoff collection device (27) have been published.

Rotating Disk Simulator. The rotating disk simulator uses a slotted rotating disk to reduce nozzle application rates to realistic levels. The disk rotates below either a Spraying Systems Co. Fulljet 1-HH-12 or 1.5-H-30 nozzle which is constantly spraying. Simulated rainfall may only be sprayed to the soil surface through the slot in the disk. By varying the aperture angle of the slot and the angular velocity of the disk, different rainfall intensities may be obtained. The disk is shaped to a shallow cone with side slopes of 1:12 and is 40 cm in diameter. Simulated rainfall which is not sprayed through the slot to the soil surface is recycled. One of the main objectives of the rotating disk simulator was to achieve a relatively low intensity rainfall composed of large drops and high impact velocities. Kinetic energy per unit of rainfall was stated to be 100% of natural rainfall at 50 mm hr⁻¹ intensity (23).

Interrill Simulator. The interrill rainfall simulator uses rapidly oscillating nozzles to produce a range of rainfall intensities. Either Veejet 80100 or Veejet 80150 nozzles may be used. The Veejet 80150 nozzle approximates the distribution for 26 to 51 mm hr⁻¹ rainfall. Moore et al. (16) reported that the Veejet 80150 nozzle produced rainfall with kinetic energies within 10% of natural rainfall for rainfall intensities over the approximate range of 23 to 120 mm hr⁻¹. Two nozzles are mounted side by side 3 m above the soil surface on a horizontal shaft which oscillates in a 90 degree arc back and forth across the plot area. A clutch-brake is used to pause the nozzles after each pass across the plot area. The nozzles are continuously spraying and water is recycled during each pause by using catchment pans. By controlling the duration of each pause a range of rainfall intensities is possible. Intensity may also be adjusted by using either one or both nozzles. Meyer and Harmon (28) adapted this design for larger plot areas by using additional nozzles. Somewhat similar is the Kentucky rainfall simulator (16) which also uses the Veejet 80150 nozzle and a similar oscillatory nozzle motion for rainfall application. An advantage of the Kentucky simulator is in its modular design for use over variably sized land areas.

The choice of a rainfall simulator depends upon research objectives and the natural process which the researcher desires to simulate. The types of simulators discussed offer certain advantages and limitations for different types of research. Unfortunately, budgetary constraints of the research project often dictates the choice of a rainfall simulator with cost usually being proportional to the rainfall area.

Nature of Agrochemical Surface Losses

Nutrients and pesticides are both dissolved in the aqueous fraction of runoff as well as attached to soil particles being transported by runoff. Soluble and sediment-bound losses are not equal in magnitude and the distribution between the two fractions depends upon the characteristics of the agrochemical, soil, and the rainfall event which is responsible for the runoff event.

Agrochemicals that are highly soluble in water are rapidly leached into the soil where they are not subject to surface loss. Only when a significant runoff event occurs soon after application of the contaminant will surface losses of soluble materials be excessive. Wauchope (29) stressed the importance of single storms on pesticide losses by denoting critical or catastrophic status to some types of rainfall events. A critical rainfall event occurs within 2 weeks of pesticide application, involves at least one cm of rain, and results in a runoff volume which is 50% or more of the precipitation. A catastrophic rainfall event has surface losses of 2% or more of the applied pesticide. The occurrence of a severe rainfall event can result in increased agrochemical losses of any surface-applied compound. The loss of pesticides formulated as wettable powders could be typically 3 times the long-term anticipated loss of 2 to 5% if a critical event was the first rainfall occurrence (29). Most rainfall simulation studies usually involve the simulation of critical or catastrophic events and present a worst case scenario. Wauchope (29) stated that "single-event losses of 5% or more occur almost exclusively in small-plot simulation studies".

The quantity of a soluble chemical or compound lost via surface routes usually accounts for only a small percentage of that constituent which is present in the soil system. This concept may be demonstrated with nitrates, a common agricultural fertilizer. Following application to the soil, nitrates leach rapidly into the soil and out of the surface layer which is subject to erosion. Under typical cropping systems, up to 99% of all nitrates lost from these systems are lost via leaching and subsurface flow (30,31). Bauder and Schneider (32) and Timmons (33) have also shown that the loss of nitrates via leaching is directly related to the quantity of water percolating through the soil profile. In a simulated rainfall laboratory study using 3 Ohio soils, Hoyt et al. (34) reported that nitrate accounted for approximately 50% of mineral N in runoff and slightly less than 100% in leachate. The ratio of N loss in runoff to leachate was small except for one soil which had relatively smaller amounts of water leaching through the soil and, therefore, higher rates of runoff. Runoff from the less permeable soil accounted for an average 24% of the mineral N losses with 72% of the mineral N runoff losses being in the nitrate form.

Nutrient losses in runoff are primarily controlled by the solubility of the material in water and the affinity of the dissolved material for soil colloids. Angle et al. (35) has shown that surface losses of soluble N (primarily as nitrates) are only a small fraction of the total amount of nitrogen lost from small watersheds cropped to corn (Zea mays L.). Most of the N was lost as organic N which was tightly adsorbed to soil particles. Similar results were shown for surface losses of phosphorus. In the same study, losses of soluble, dissolved P were only about 10% of the total amount of P lost from the watersheds. The majority of the P was lost in a sediment-bound form. Numerous other studies under natural (36-38) and simulated rainfall (39-44) conditions have shown that most P is lost from agricultural land when attached to sediment in runoff. Baker and Laflen (45) reported that for a rainfall simulation study the average ratio of available P in runoff sediment to P ion in runoff solution was approximately 2300.

While the long-term consequences of nutrient losses are most important as related to sediment-bound nutrients, losses of soluble nutrients may be extremely important on a limited and short-term scale. Most sediment-attached nutrients, upon discharge into receiving bodies of water, require dissolution and release before adverse environmental consequences can occur (47,48). Soluble nutrients, on the other hand, are immediately available in receiving water and thus are important in the eutrophication of water (49,50). Therefore, while much smaller losses of soluble nutrients can be expected in runoff, the environmental consequences of the loss of soluble nutrients may be extremely significant. This may be particularly true if these losses are being directly channeled into groundwater reserves.

Solubility of pesticides is also important in determining whether the primary route of loss will be via leaching or runoff. Pesticides characterized by low water solubilities will remain on or near the soil surface and are thus subject to potential surface loss. The interaction of the pesticide with soil colloids is equally important in determining the route of loss. Adsorption coefficient (k_d) values have been established for nearly all pesticides and are used to predict the potential mobility within the soil profile. To simulate soil-surface-applied pesticide loss in the sediment phase of runoff, Wauchope (51) applied copper and zinc solutions with estimated K_d's of 186 and 20, respectively. Differences in K_d did not affect simulated rainfall runoff losses; essentially identical surface loss loads of copper and zinc were highly correlated to eroded sediment. Overall losses of copper and zinc ranged from 3 to 7%, which was in good agreement with losses expected from strongly soil-bound pesticides.

Most pesticides lost via the soil surface are lost in the aqueous phase of surface flow. While many pesticides are tightly bound to erodible soil colloids, runoff water volume is several orders-of-magnitude greater than runoff sediment volume. Therefore, while pesticides concentrations may be lower in the solution phase of runoff, the total loss may be greater. Catastrophic pesticide losses are generally a result of unusually large runoff volumes, not unusually high pesticide concentrations (29). For example,

Leonard et al. (*52*) have shown that nearly all triazine herbicides are lost in the aqueous phase of runoff as opposed to the sediment-bound phase. Glenn and Angle (*53*) examined runoff losses of atrazine [6-chloro-*N*-ethyl-*N'*-(1-methylethyl)-1,3,5-triazine-2,4-diamine] and simazine [6-chloro-*N*,*N'*-diethyl-1,3,5-triazine-2,4-diamine] in runoff when each was applied to corn and reported that approximately 1% of the total amount of herbicide was lost in runoff. Of this amount, nearly all of it was associated with the aqueous phase. Sauer and Daniel (*54*) reported that >90% of the soluble herbicides, atrazine and alachlor [2-chloro-*N*-(2,6-diethyl-phenyl)-*N*-(methoxymethyl)acetamide] were lost in the aqueous fraction during a rainfall simulation study. Aqueous fraction losses accounted for only 40% of the total loss of chlorpyrifos [0,0-diethyl 0-(3,5,6-trichloro-2-pyridinyl) phosphorothioate (9Cl)], a tightly adsorbed and relatively water insoluble insecticide. Dieldrin [1,8,9,10,11,11-hexachloro-4,5-exo-epoxy-2,3-7,6-endo,2,1-7,8-exo-tetracyclo(6.2.1.1^{3,6}.0^{2,7})dodec-9-ene] is another pesticide which is lost primarily in the aqueous fraction of runoff. Caro and Taylor (*55*) reported the total load of dieldrin in the aqueous phase was over 30 times that of dieldrin adsorbed to sediment. Wauchope (*29*) has summarized the distribution of losses for a variety of pesticides, depending on whether they are lost in the aqueous or sediment-bound phase. Pesticides which are relatively water soluble account for the greatest anticipated long-term losses, particularly since these losses can not be reduced using erosion control practices.

The interaction between the timing of the rainfall event and the persistence of the pesticide can also be of importance to pesticide losses. White et al. (*56*) found in a simulated rainfall study that the loss of 2,4-D [(2,4-dichlorophenoxy) acetic acid] in runoff was inversely proportional to the amount of time between application and rainfall occurrence.

Interpretation of Rainfall Simulation Data

Rainfall simulation is a dynamic process and changes are constantly occurring at the soil surface during a rainstorm event. Soil infiltration may increase as the rainfall application rate increases although runoff is occurring (*57,58*). This potential change in infiltration with intensity should be noted and a constant infiltration rate should not be assumed for calculating runoff losses (*59*). The intermittency of simulated rainfall may also affect the amount of rainfall, energy, and, therefore, time before runoff is initiated (*58,60*). As the time period between intermittent applications is increased to 10 seconds or greater, the amount of energy to seal the soil surface and initiate runoff may also be increased. The point of runoff initiation can be very important as herbicide concentrations in water and sediment have been negatively correlated with time to runoff initiation (*61*).

Rainfall intensities and storm durations selected for simulation are frequently based either on an expected return frequency for a given location or the amount of kinetic energy which will be delivered to the soil surface.

The amount of kinetic energy absorbed by the soil surface has a large effect on the soil particle splash-detachment dynamics which occur during a rainfall event. If the rainfall event is based on the amount of kinetic energy applied during the rainfall event and the rainfall simulator produces rainstorms at less than 100% kinetic energy for comparable intensities, then a higher intensity will be necessary to produce the desired kinetic energy. Since most simulators produce less than 100% kinetic energy for comparable rainfall intensities, the amount of runoff and/or sediment loss are obviously not going to be proportional to quantities observed during a natural rainfall event. The experimental design and goals will help dictate which criteria is used for rainfall intensity selection, but care must be exercised in data interpretation. Erosion index, as previously noted, has been found to be a very useful parameter in relating the sediment losses occurring during a simulated rainfall event to natural rainfall.

Care must also be taken in data interpretation because of the variability associated with rainfall simulation studies. There will be slight variations in the simulated rainfall intensities when simulators are used under different weather conditions at different locations and times. Meyer (14) has suggested that soil loss determinations be adjusted for variations in intensity by multiplying measured soil loss times the squared ratio of selected intensity to measured intensity. It has also been reported that when a small plot simulator is used, large variations among replicates can occur because relatively small plot areas are sampled which limits soil movement to interrill erosion (62). The process of soil loss can be very important if sediment-associated agrochemical losses are being evaluated.

Losses from rainfall simulation studies reflect edge-of-plot concentrations and may be affected by anomalies occurring within the plot area or the border effects of the plot. The smaller the plot area the greater the effect of any anomaly or border. Losses may also reflect the time of sampling as agrochemical runoff concentrations may vary an order-of-magnitude or more during a single runoff event (29). Caution should be taken in data extrapolation to large field size areas as processes on different scales may not be the same. Evaluation of soil erosion using small plot simulators is a good example of how processes differ over different size land areas since small plot simulation only considers the interrill component of erosion. While interrill erosion is a very dynamic process taking into account raindrop impact-soil splash relationships, the scouring action of water occurring in rill erosion is not adequately represented and may not be totally represented even when large area simulators are used. The process of surface agrochemical loss from large land areas is even less well-understood.

The uniformity of simulated rainfall, while being a desired characteristic for statistical evaluation, in itself presents problems in data interpretation when comparing simulated rainfall to natural rainfall. Natural rainfall varies over time and distance usually several times during the same rainstorm event.

Direct comparison of simulated rainfall events to natural rainfall events is questionable because of the many differences and problems associated with rainfall simulation. Other factors unique to the soil type, physical properties of the soil and its surface at the time of evaluation, and the capabilities of the selected rainfall simulator may also influence experimental results and should be considered during data interpretation. An alternative approach, when considering agrochemical losses for comparison studies of different management treatments, is to evaluate the agrochemical loss as a relative potential loss for statistical comparisons. Conclusions based on this type of approach offer little information of the magnitude of losses which might be occurring, but can offer a basis for decisions on the relative effectiveness of different management practices.

Summary

Rainfall simulation can be a valuable research methodology to apply reproducibly uniform intensities or energy levels of rainfall under controlled experimental conditions. It can be used to study intricate physical processes occurring at the soil surface or for comparison testing of a broad range of management or other practices in a relatively brief time frame. The design, construction, and calibration of a rainfall simulation can be expensive and time consuming. Rainfall simulation results may not be directly extrapolated to larger land areas, but must be evaluated within the constraints of the rainfall simulation study.

Rainfall simulation is a viable research avenue which may be used to estimate potential agrochemical surface loss. Agrochemical surface loss measurements may be important in estimating the contribution of abandoned wells, improperly-protected wellheads, or karst topography on the point source contamination of groundwater. An accounting of agrochemical surface loss may also be important in modeling efforts to predict potential groundwater contamination from an introduced agrochemical.

Obviously, rainfall simulation is not the answer to all environmental research problems involving rainfall and resultant processes. Each scientist must evaluate the goals and objectives of their research project to see if rainfall simulation will offer a viable research alternative to provide the answers which are needed. Rainfall simulation has many limitations, but there are also many advantages particularly when a total dependence on natural rainfall is the alternate choice.

Literature Cited

1   *The Importance of Sealing an Abandoned Well*, Alliance for a Clean Rural Environment (ACRE), 1990, Fact Sheet Number 6.
2   Meyer, L. D.; McCune, D. L. *Agric. Engr.*, 1958, *39* (10), 644-648.

3   Young, R. A.; Burwell, R. E. *Soil Sci. Soc. Am. Proc.*, 1972, *36* (5), 827-830.
4   Laws, J. O. *Trans. Am. Geophys. Un.*, 1941, *22*, 709-721.
5   Laws, J. O.; Parsons, D.A. *Trans. Am. Geophys. Un.*, 1943, *24*, 452-460.
6   Wischmeier, W. H.; Smith, D. D. *Trans. Am. Geophys. Un.*, 1958, *39*, 284-291.
7   Wischmeir, W. H. *Soil Sci. Soc. Am. Proc.*, 1959, *23*, 245-249.
8   Mutchler, C. K.; Hermsmeir, L. F. *Trans. Am. Soc. Agric. Engr.*, 1965, *8*, 67-68.
9   Hall, M. J. *Water Resour. Res.*, *6* ((4), 1104-1114.
10   Bubenzer, G. D., *An overview of rainfall simulators*, Paper No. 80-2033, Am. Soc. Agric. Engr.: St. Joseph, MO, 1980; 11pp.
11   Bubenzer, G. D. *Proceedings of the Rainfall Simulator Workshop*, USDA: Washington, D.C., 1979; ARM-W-10, 120-130.
12   Neff, E. L., *Proceedings of the Rainfall Simulator Workshop*. USDA: Washington, D.C., 1979; ARM-W-10, 3-7.
13   Mech, S. J. *Trans. Am. Soc. Agric. Engr.*, 1965, *8* (1), 66-75.
14   Meyer, L. D. In *Soil Erosion Research Methods*, R. Lal (ed.), Soil and Water Cons. Soc.: Ankeny, IA, 1988; 75-95.
15   Barnett, A. P.; Dooley, A. E. *Trans. Am. Soc. Agric. Engr.*, 1972, *15* (6), 1112-1114.
16   Moore, I. D.; Hirsch, M. C.; Barfield, B. J. *Trans. Am. Soc. Agric. Engr.*, 1983, *26*, 1085-1089.
17   Bertrand, A. R.; Parr, J. F. *Design and operation of the Purdue sprinkling infiltrometer*, Research Bulletin No. 723, Purdue University: West Lafayette, IN, 1961; 16 pp.
18   Meyer, L. D. *Trans. Am. Soc. Agric. Engr.*, 1965, *8*, 63-65.
19   Meyer, L. D. *Proceedings of the Rainfall Simulator Workshop*, USDA: Washington, D.C., 1979; ARM-W-10, 35-44.
20   Meyer, L. D.; Harmon, W. C. *Trans. Am. Soc. Agric. Engr.*, 1979, *22*, 100-103.
21   Römkens, M. J. M.; Glenn, L. F.; Nelson, D. W. *Soil Sci. Soc. Amer. Proc.*, 1975, *39* (1), 158-160.
22   Brackensick, D. O.; Rauls, W. J.; Hamm, W. R. *Trans. Am. Soc. Agric. Engr.*, 1979, *22* (2), 320-325, 333.
23   Morin, J.; Goldberg, D.; Seginer, I. *Trans. Am. Soc. Agric. Engr.*, 1967, *10*, 74-77, 79.
24   Meyer, L. D. *Soil Sci. Soc. Amer. Proc.*, 1960, *24* (4), 319-322.
25   Swanson, N. P. *Trans. Am. Soc. Agric. Engr.*, 1965, *8* (1), 71-72.
26   Dixon, R. M.; Peterson, A. E. *Construction and operation of a modified spray infiltrometer and a flood infiltrometer*, Research Rpt. *15*, Wisconsin Agric. Exp. Sta.: Madison, WI, 1964; 31pp.
27   Dixon, R. M.; Peterson, A. E. *Soil Sci. Soc. Amer. Proc.*, 1968, *32*, 123-125.

28   Meyer, L. D.; Harmon, W. C. *Trans. Am. Soc. Agric. Engr.*, 1985, *28*, 448-453.

29   Wauchope, R. D. *J. Environ. Qual.*, 1978, *7*, 459-472.

30   Alberts, E. E.; Spomer, R. A. *J. Soil Water Conserv.*, 1985, *40*, 153-157.

31   Hubbard, R. K.; Sheridan, J. M. *J. Environ. Qual.*, 1983, *12*, 201-205.

32   Bauder, J. W.; Schneider, R. P. *Soil Sci. Soc. Amer. J.*, 1979, *43*, 348-352.

33   Timmons, D. R. *J. Environ. Qual.*, 1984, *13*, 305-309.

34   Hoyt, G. D.; McLean, E. O.; Reddy, G. Y.; Logan, T. J. *J. Environ. Qual.*, 1977, *6*, 285-290.

35   Angle, J. S.; McClung, G.; McIntosh, M. S.; Thomas, P. M.; Wolf, D. C. *J. Environ. Qual.*, 1984, *13*, 431-435.

36   Reddy, G. Y.; McLean, E. O.; Hoyt, G. D.; Logan, T. J. *J. Environ. Qual.*, 1978, *7*, 30-34.

37   Sharpley, A. N. *J. Environ. Qual.*, 1980, *9*, 521-526.

38   Sharpley, A. N.; Menzel, R. G.; Smith, S. J.; Rhoades, E. D.; Olness, A. E. *J. Environ. Qual.*, 1981, *10*, 211-215.

39   Andraski, B. J.; Mueller, D. H.; Daniel, T. J. *Soil Sci. Soc. Amer. J.*, 1985, *49*, 1523-1527.

40   Munn, D. A.; McLean, E. O.; Ramirez, A.; Logan, T. J. *Soil Sci. Soc. Amer. Proc.*, 1973, *37*, 428-431.

41   Sharpley, A. N. *Soil Sci. Soc. Amer. J.*, 1985, *49*, 1527-1534.

42   Mueller, D. H.; Wendt, R. C.; Daniel, T. C. *Soil Sci. Soc. Amer. J.*, 1984, *48*, 901-905.

43   Barisas, S. G.; Baker, J. L.; Johnson, H. P.; Laflen, J. M. *Trans. Am. Soc. Agric. Engr.*, 1978, *21*, 893-897.

44   McIsaac, G. F.; Mitchell, J. K.; Hirsch, M. C. *Nutrients in runoff and eroded sediment from tillage systems in Illinois*, Paper No. 87-2066, Am. Soc. Agric. Engr.: St. Joseph, MO, 1987; 19pp.

45   Baker, J. L.; Laflen, J. M. *Trans. Am. Soc. Agric. Engr.*, 1983, *26*, 1122-1127.

46   Ahuja, L. R.; Lehman, O. R.; Sharpley, A. N. *Soil Sci. Soc. Amer. J.*, 1983, *47*, 746-748.

47   Fumumai, H.; Ohgak, S. *J. Environ. Qual.*, 1988, *17*, 205-212.

48   Fumumai, H.; Ohgak, S. *Water Sci. Technol.*, 1988, *14*, 215-226.

49   Sharpley, A. N.; Menzel, R. G. *Adv. Agron.*, *41*, 297-324.

50   Sharpley, A. N.; Smith, S. J. *J. Environ. Qual.*, 1989, *18*, 313-316.

51   Wauchope, R. D. *J. Environ. Qual.*, 1987, *16*, 206-211.

52   Leonard, R. A.; Langdale, G. W.; Fleming, W. A. *J. Environ. Qual.*, 1979, *8*, 223-229.

53   Glenn, D. S.; Angle, J. S. *Agric. Ecosyst. Environ.*, 1987, *18*, 273-280.

54   Saver, T. J.; Daniel, T. C. *Soil Sci. Soc. Amer. J.*, 1987, *51*, 410-415.

55   Caro, J. H.; Taylor, A. W. *J. Agric. Food Chem.*, 1971, *19*, 379-384.

56   White, A. W.; Asmussen, L. E.; Hauser, E. W.; Turnbull, J. W. *J. Environ. Qual.*, 1976, *5*, 487-490.

57   Moldenhauer, W. C.; Burrows, W. C.; Swartzendruber, D. *Int. Congr. Soil Sci. Trans.* 7th, 1960, *1*, 426-432.

58   Sloneker, L. L.; Moldenhauer, W. C. *Soil Sci. Soc. Amer. Proc.*, *38*, 157-158.

59   Young, R. A. *Proceedings of the Rainfall Simulator Workshop.* USDA: Washington, D.C., 1979; ARM-W-10, 108-112.

60   Sloneker, L. L.; Olson, T. C.; Moldenhauer, W. C. *Soil Sci. Soc. Amer. Proc.*, 1974, *38*, 985-987.

61   Baker, J. L.; Laflen, J. M.; Hartwig, R. O. *Trans. Am. Soc. Agric. Engr.*, 1982, *25*, 340-343.

62   Meuller, D. H.; Wendt, R. C.; Daniel, T. C. *Soil Sci. Soc. Amer. J.*, 1984, *48*, 896-900.

RECEIVED September 17, 1990

INDEXES

Author Index

Affiliation Index

Subject Index

Production: Kurt Schaub
Indexing: Deborah Steiner
Acquisition: A. Maureen Rouhi
Cover design: Tina Mion

Printed and bound by Maple Press, York, PA

Paper meets minimum requirements of American National Standard
for Information Sciences—Permanence of Paper for Printed Library
Materials, ANSI Z39.48–1984   ∞

Other ACS Books

Chemical Structure Software for Personal Computers
Edited by Daniel E. Meyer, Wendy A. Warr, and Richard A. Love
ACS Professional Reference Book; 107 pp;
clothbound, ISBN 0–8412–1538–3; paperback, ISBN 0–8412–1539–1

Personal Computers for Scientists: A Byte at a Time
By Glenn I. Ouchi
276 pp; clothbound, ISBN 0–8412–1000–4; paperback, ISBN 0–8412–1001–2

Biotechnology and Materials Science: Chemistry for the Future
Edited by Mary L. Good
160 pp; clothbound, ISBN 0–8412–1472–7; paperback, ISBN 0–8412–1473–5

Polymeric Materials: Chemistry for the Future
By Joseph Alper and Gordon L. Nelson
110 pp; clothbound, ISBN 0–8412–1622–3; paperback, ISBN 0–8412–1613–4

The Language of Biotechnology: A Dictionary of Terms
By John M. Walker and Michael Cox
ACS Professional Reference Book; 256 pp;
clothbound, ISBN 0–8412–1489–1; paperback, ISBN 0–8412–1490–5

Cancer: The Outlaw Cell, Second Edition
Edited by Richard E. LaFond
274 pp; clothbound, ISBN 0–8412–1419–0; paperback, ISBN 0–8412–1420–4

Practical Statistics for the Physical Sciences
By Larry L. Havlicek
ACS Professional Reference Book; 198 pp; clothbound; ISBN 0–8412–1453–0

The Basics of Technical Communicating
By B. Edward Cain
ACS Professional Reference Book; 198 pp;
clothbound, ISBN 0–8412–1451–4; paperback, ISBN 0–8412–1452–2

The ACS Style Guide: A Manual for Authors and Editors
Edited by Janet S. Dodd
264 pp; clothbound, ISBN 0–8412–0917–0; paperback, ISBN 0–8412–0943–X

Chemistry and Crime: From Sherlock Holmes to Today's Courtroom
Edited by Samuel M. Gerber
135 pp; clothbound, ISBN 0–8412–0784–4; paperback, ISBN 0–8412–0785–2

For further information and a free catalog of ACS books, contact:
American Chemical Society
Distribution Office, Department 225
1155 16th Street, NW, Washington, DC 20036
Telephone 800–227–5558